ELECTRONIC AMPLIFIER CIRCUITS
Theory and Design

Contents

1

Basic Amplifier Definitions

1-1 Amplifier Gain. An amplifier is designed for the purpose of increasing the level of voltage, current, or power. The amount of this increase is known as the *gain* of the amplifier. The method of specifying gain quantitatively is important and should be keyed to the intended application of the amplifier. For example, in an oscilloscope amplifier *voltage* gain is important because both the input and output are at a high impedance level. On the other hand a telephone repeater amplifier has a limited available input power and must supply power to a specified line impedance. Consequently a gain expressed as a ratio of available input power to output power is appropriate. Four commonly used ways of specifying gain are illustrated in Fig. 1-1. Although the first three examples are given in terms of power, the gain may also be given in terms of voltage or current; e.g., insertion voltage gain $= V_2/V_1'$. The "actual" power gain of an amplifier ($= P_2/P_1$) involves certain ambiguities and is not in common use.

1-2 Frequency Response. Speaking first in terms of steady-state sine-wave signals, the gain of an amplifier is never absolutely constant with respect to frequency; thus a specified value for the magnitude of gain is meaningless unless it is known at what frequency that gain was measured. In general, "gain" with no qualification means the *mid-frequency* gain of an amplifier. In Fig. 1-2 a typical curve is shown for gain as a function of frequency (usually called the *frequency response*) of a *lowpass* amplifier. Such an amplifier is commonly called an *audio* or *video* amplifier depending upon the application; the video amplifier requires a higher upper-frequency limit. Strictly speaking, a lowpass amplifier should amplify from zero frequency (d-c) to some upper limit; however, to avoid the problems of d-c amplification most practical lowpass amplifiers are a-c-coupled and have the dashed response curve. The frequencies at which the power gain drops

1

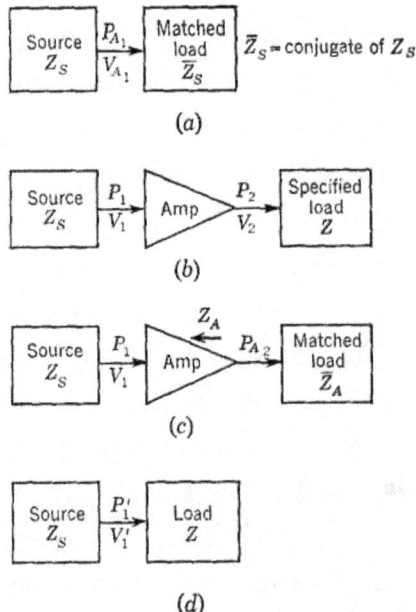

Fig. 1-1 Gain definitions. (a) Available power from source $\triangleq P_{A,1}$. (b) Transducer power gain $\triangleq P_2/P_{A,1}$. (c) Available power gain $\triangleq P_{A,2}/P_{A,1}$. (d) Insertion power gain $\triangleq P_2/P_1'$. Two additional definitions of gain are illustrated implicitly in (b), the so-called *voltage gain* V_2/V_1 and the *actual power gain* P_2/P_1. The symbol $\triangleq$ means "defined as."

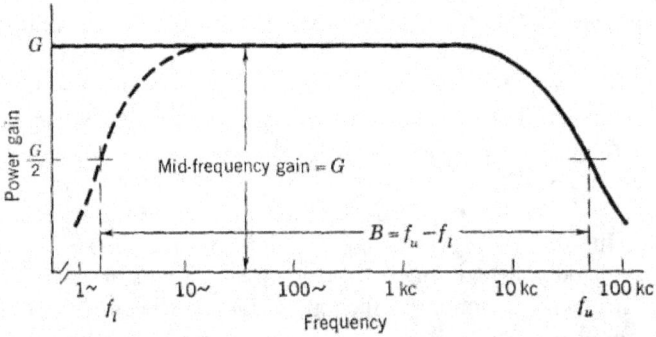

Fig. 1-2 Typical frequency response of lowpass amplifier.

to one-half the midband value are known as the *band-edge* (or *cutoff*) frequencies—indicated in Fig. 1-2 as f_l and f_u. The *bandwidth* of the amplifier is the difference between these two frequencies: $B \stackrel{\Delta}{=} f_u - f_l$. In a typical lowpass amplifier the frequency f_l is so low that $B \cong f_u$. In cases where G is measured as a voltage gain, the upper and lower band-edge frequencies are taken to be where the voltage gain drops to $1/\sqrt{2}$ of the midband value.[1]

For many purposes gain is conveniently plotted vs. frequency with both scales logarithmic. The magnitude of gain is then usually expressed in decibels,

$$db = 10 \log G_{\text{power}} \qquad (1\text{-}1)$$

Often voltage or current gain is given in decibels, too, although this is, strictly speaking, not in conformity with the historical definition of the

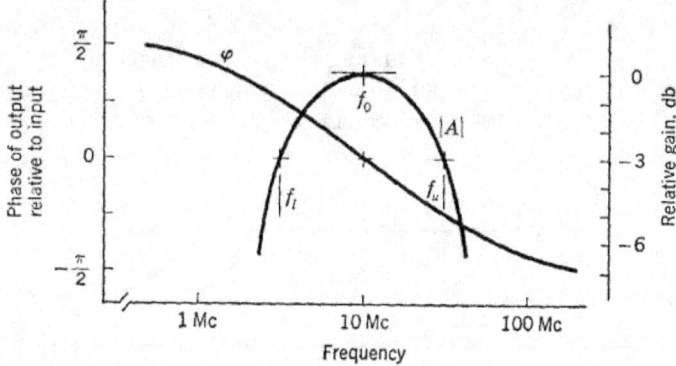

Fig. 1-3 Typical phase and frequency response of a single-stage bandpass amplifier.

decibel unless the impedance levels across which the voltages are measured are equal. If V_1 and V_2 are measured across R_1 and R_2 at the input and output of the amplifier, respectively, the decibel gain is correctly given as

$$db = 20 \log \frac{V_2}{V_1} + 10 \log \frac{R_1}{R_2} \qquad (1\text{-}2)$$

In common practice and throughout this book, the second term is omitted; the result is thus proportional to a voltage ratio only and is not necessarily the exact power ratio.[2] Note that the gain at the band edges is 3 db less than at midband; hence, these are often called the -3-db frequencies.

[1] Other definitions of bandwidth will be given later for special applications; these will be assigned different designations to prevent confusion.

[2] For many purposes gain in decibels is given relative to band center. In this case, the second term in Eq. (1-2) is not needed.

A *bandpass* (or *filter*) amplifier has a frequency response which is similar to that shown dashed in Fig. 1-2. The main difference lies in the fact that both the low- and high-frequency band edges are carefully controlled in order to amplify a specified band of frequencies, rather than merely making the low-frequency band edge as low or the high-frequency edge as high as possible. We shall find that a practical difference is that the effects at f_u and f_l can only be treated separately in the lowpass amplifier.

The phase shift of the output voltage of an amplifier relative to the input voltage is also a function of frequency. In Fig. 1-3 is shown such a function (known as the *phase response*), together with the amplitude response for a typical bandpass amplifier.

1-3 Step Response. If the steady-state amplitude and phase responses of an amplifier are known, the output from the amplifier for any input waveform, such as a transient, can presumably be found. However, it is frequently more convenient to deal directly with the output of the amplifier in response to a standard input or driving waveform without taking the intermediate step of finding the sine-wave characteristics. An input waveform commonly used is a voltage which changes instantaneously from one value to another—a *step* voltage (Fig. 1-4a). Such an input wave-

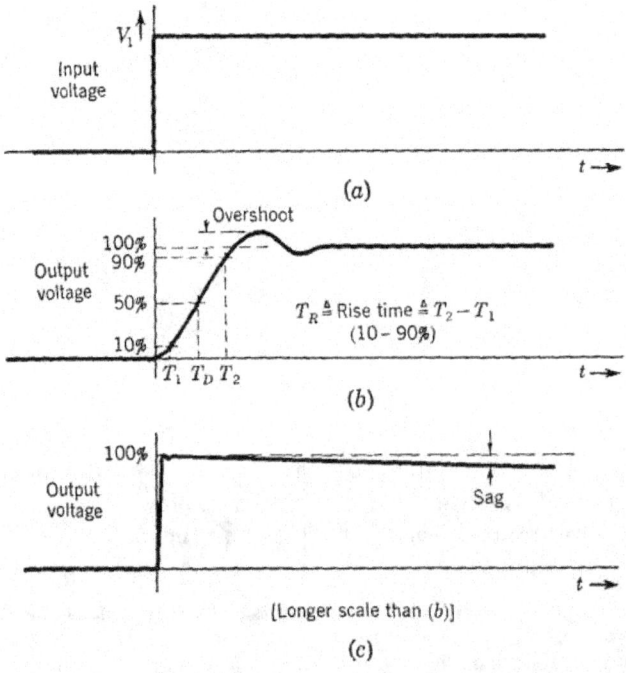

Fig. 1-4 Lowpass amplifier response to a step voltage input.

form contains components at all frequencies, and since the amplifier cannot uniformly amplify all frequencies, the output is not a faithful reproduction of the input. The rising portion of the output (Fig. 1-4b) does not jump to its final value instantaneously but has a finite *rise time*. Since it is difficult to define exactly where the rise begins and ends, the rise time is usually specified as the time for the output to go from 10 to 90 per cent of the final value of the output. The output may rise beyond the 100 per cent level and oscillate about it. The amount by which the output exceeds the 100 per cent value is known as *overshoot* and is usually expressed in per cent also.

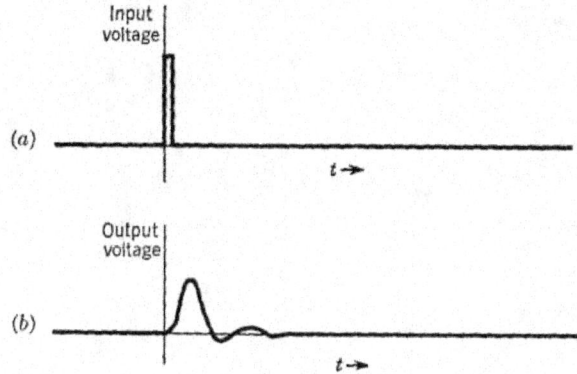

Fig. 1-5 A typical amplifier impulse response.

The output of an amplifier usually does not rise immediately after the application of the input step; thus, the amplifier acts to delay the signal passing through it. The *delay time* T_D is arbitrarily defined as the time for the output to rise to 50 per cent of its final value measured from the time at which the step occurs.

If the time scale is lengthened out to permit inspection of a longer time interval of the output, the output of a typical amplifier is seen to deviate from the 100 per cent level. Indeed, if the amplifier is not capable of amplifying zero frequency (a d-c amplifier), the ultimate value of the output must be zero. The failure of the amplifier output to remain at the 100 per cent level is known as *sag*. This is usually expressed as per cent in a given time. Note that the presence of sag makes the exact 100 per cent value somewhat uncertain. In a practical amplifier the 100 per cent point may be taken most accurately as the output amplitude immediately after the oscillations causing overshoot have disappeared.[1]

[1] If the amplifier has no overshoot, the 100 per cent value is the maximum value of the output. In certain amplifiers with f_l and f_u close together the 100 per cent value is very uncertain. However, the step response is not ordinarily used to characterize such an amplifier.

1-4 Impulse Response. Another method of describing the transient behavior of an amplifier is to use an *impulse* to drive the amplifier; the resulting output is the impulse response. For measurement purposes the impulse, which ideally has zero width (duration) but finite area, can be satisfactorily approximated by a very narrow pulse, as shown in Fig. 1-5*a*. The impulse response of the amplifier is the time derivative of the step response, as depicted for a typical amplifier in Fig. 1-5*b*. Rise time and delay time can be defined mathematically by using the impulse response (see Sec. 4-9); however, it is difficult to use the impulse response for the measurement of these quantities.

2

Equivalent Circuits for
Linear Active Devices

The amplifiers discussed in this book all employ some form of active two-port device to provide gain in the amplifier. Historically the first such active device used in electronics was the vacuum-tube triode. The triode is, of course, still in common use, but for the type of amplifier with which we shall be primarily concerned a later development, the pentode, is almost universally used because of its relative freedom from plate-to-grid feedback. For many applications a relative newcomer, the transistor, is replacing the vacuum-tube types because of the greater inherent reliability, lower power consumption, and smaller size. However, the complete replacement of the tube by the transistor does not seem likely, for the latter has shortcomings at high temperatures and high radiation intensities and in the production of high power at high frequencies. Also there are specific circuit applications where the tube seems superior to the transistor, e.g., as a high input-impedance device, or where low noise operation is necessary with high source impedances. Consequently, throughout the following discussion of equivalent circuits for both kinds of devices, it is well to keep in mind that the equivalent circuits convey implications as to what each device is best suited for; the reader should compare the representations of the different devices with each other.[1]

2-1 Representation of a Linear Active Device with the Two-port Circuit Parameters. The active device we wish to represent may be indicated as shown in Fig. 2-1, where the device has two input terminals, or an input *port* as it is often called, and two output terminals, or an output

---

[1] For an interesting discussion, see D. G. Fink, Transistor vs. Vacuum Tubes, *Proc. IRE*, April, 1956, p. 479.

port. Usually two of the four terminals are common, i.e., connected together inside the box. The functional relationship between the voltages

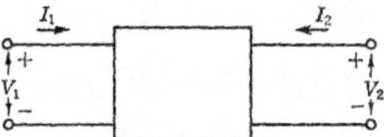

Fig. 2-1 The terminal conventions for a general two-port network.

and currents may be expressed in many different ways, but the following illustrate three of the common and useful ways:

The z, or *impedance*, parameters

$$V_1 = z_{11}I_1 + z_{12}I_2 \qquad\qquad (2\text{-}1)$$

$$V_2 = z_{21}I_1 + z_{22}I_2 \qquad\qquad (2\text{-}2)$$

The y, or *admittance*, parameters

$$I_1 = y_{11}V_1 + y_{12}V_2 \qquad\qquad (2\text{-}3)$$

$$I_2 = y_{21}V_1 + y_{22}V_2 \qquad\qquad (2\text{-}4)$$

The h, or *hybrid* (series-parallel), parameters

$$V_1 = h_{11}I_1 + h_{12}V_2 \qquad\qquad (2\text{-}5)$$

$$I_2 = h_{21}I_1 + h_{22}V_2 \qquad\qquad (2\text{-}6)$$

Note that only four quantities are needed to specify one pair of voltages or currents if the other pair is known. In some cases fewer than the four quantities will suffice. Each of the parameter sets leads directly to an equivalent circuit, as shown in Fig. 2-2. Note that the hybrid parameters have mixed dimensions: $h_{11} =$ ohms, $h_{22} =$ mhos, while h_{12} and h_{21} are dimensionless voltage and current ratios, respectively.

The z's are also known as the open-circuit impedance parameters because open-circuiting the 2-2 terminals makes $I_2 = 0$; then

$$z_{11} = \left.\frac{V_1}{I_1}\right\} \qquad\qquad (2\text{-}7)$$
$$\qquad\qquad\qquad I_2 = 0$$
$$z_{21} = \left.\frac{V_2}{I_1}\right\} \qquad\qquad (2\text{-}8)$$

Likewise opening the 1-1 terminals (making $I_1 = 0$) gives

$$z_{12} = \left.\frac{V_1}{I_2}\right\} \qquad\qquad (2\text{-}9)$$
$$\qquad\qquad\qquad I_1 = 0$$
$$z_{22} = \left.\frac{V_2}{I_2}\right\} \qquad\qquad (2\text{-}10)$$

The y's are a set of short-circuit parameters, for a similar set of relations may be written when V_2 and V_1 are, respectively, set to zero instead of I_2 and I_1.

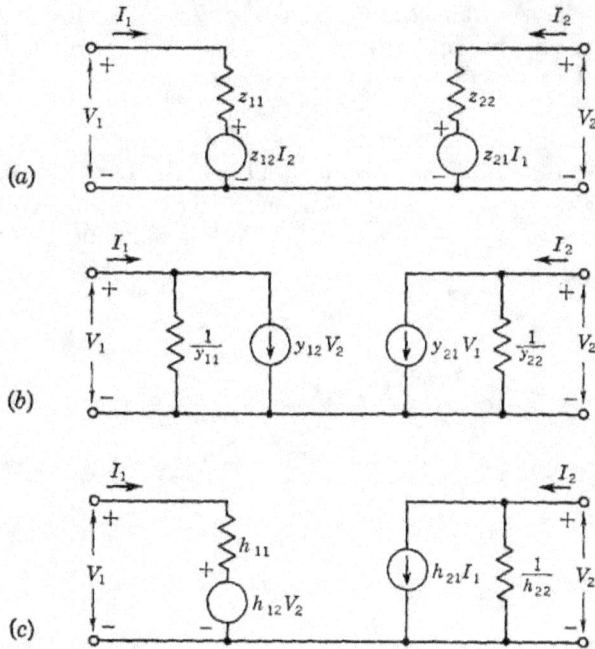

(a)

(b)

(c)

Fig. 2-2 Equivalent circuits for the two-port parameters. (a) z parameters. (b) y parameters. (c) h parameters.

The h's, however, may be found by short-circuiting the 2-2 terminals and opening the 1-1 terminals,

$$h_{11} = \left.\frac{V_1}{I_1}\right\}$$
(2-11)

$$V_2 = 0$$

$$h_{21} = \left.\frac{I_2}{I_1}\right\}$$
(2-12)

$$h_{12} = \left.\frac{V_1}{V_2}\right\}$$
(2-13)

$$I_1 = 0$$

$$h_{22} = \left.\frac{I_2}{V_2}\right\}$$
(2-14)

This parameter set will be found to be useful in transistor measurements, for the necessary open- and short-circuit terminations can be well approximated over wide frequency ranges.

The use of one set of parameters is no more or less accurate than the use of another. However, the precision of physical measurement of one set or another varies considerably. Also, one set may lead to a simpler representation than another set, thus leading to simpler circuit calculations. Hence, in the following equivalent-circuit representations of an active device, no attempt will be made to express the representation in all possible forms; rather, the forms most easily understood from a physical point of view and those easiest to use will be discussed.

2-2 Representation of Vacuum Tubes at Low Frequencies.[1] From the analysis point of view it does not matter whether the tube shown in Fig. 2-3 is a triode, tetrode, or pentode, for only the two-ports are

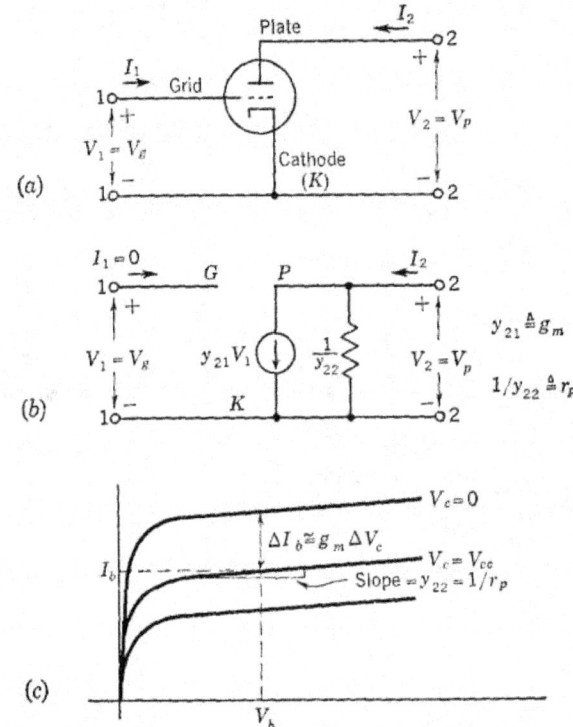

Fig. 2-3 (a) Generalized negative-grid vacuum tube at low frequencies. (b) Its representation by y parameters. (c) The relation to the graphical characteristics of the tube.

important; all other electrodes are at ground potential for a-c voltages. Hence any of the circuit parameters defined in (2-1) through (2-6) could be used to describe the small-signal properties of the tube. In Fig. 2-3b the

[1] Frequencies low enough for tube capacitances to be neglected.

y parameters are used to represent the tube. Since $I_1 = 0$ in a negatively biased tube at low frequencies, y_{11} and y_{12} have disappeared, y_{21} is seen to be the more familiar quantity g_m, and y_{22} is the reciprocal of the plate resistance, r_p. The quantities g_m and r_p are constant with respect to the voltages and currents only if these make small excursions about an operating point; i.e., these parameters provide only a small-signal representation because of the inherent nonlinearities of the tube. Both g_m and r_p are functions of the operating point of the tube, specified as V_b, V_c, and I_b. (In the case of a multigrid tube, the d-c grid potentials are specified as V_{c1}, V_{c2}, etc., for grids 1 and 2, respectively.)

The Thévenin equivalent of the circuit of Fig. 2-3b shown in Fig. 2-4b is equally valid. The circuit of Fig. 2-4a is usually associated with pentodes, while that of Fig. 2-4b is associated with triodes, for the magnitudes of g_m, r_p, and μ lead to useful approximations with these associations. For small

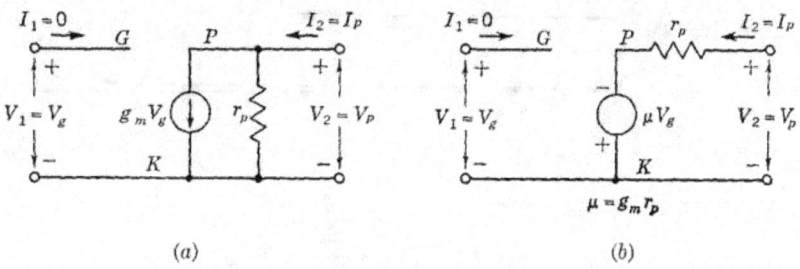

(a)   (b)

Fig. 2-4 Alternative equivalent circuits for pentode or triode.

triodes the range is 1 to 100 kilohms for r_p and 5 to 100 for μ. In pentodes typical values are 1,000 to 20,000 μmhos for g_m and 100 kilohms to several megohms for r_p.

Note that, although the equivalent circuits are shown as two-port networks, this is not necessary. Any one of the three terminals may be considered common, but if necessary, all three terminals may be above ground. The only precaution is observing that V_g is always measured on the grid with respect to the cathode.[1]

2-3 Representation of a Transistor at Low Frequencies. The choice of an equivalent circuit for the junction transistor at low frequencies is complicated by the large number of such circuits which are both possible and useful. We shall start, then, with a very simple circuit which can be related to the terminal properties of the device and which can be elaborated to include all the major low-frequency effects. The resulting more complicated circuit can be related to the physics of the device and will be useful

---

[1] For example, if the grid is the common terminal, then $V_g = -V_1$; for the plate as the common terminal, $V_g = V_1 - V_2$.

later when high-frequency effects will be included.   The circuit also can be easily related to the general circuit parameters (z's) so that a tie between the theory of the device and the network parameters is illustrated.

If the common-base characteristics of an *NPN* transistor [1] (Fig. 2-5) are inspected, the collector current is seen to be almost independent of collector voltage (on the assumption that operation is restricted to the linear region,

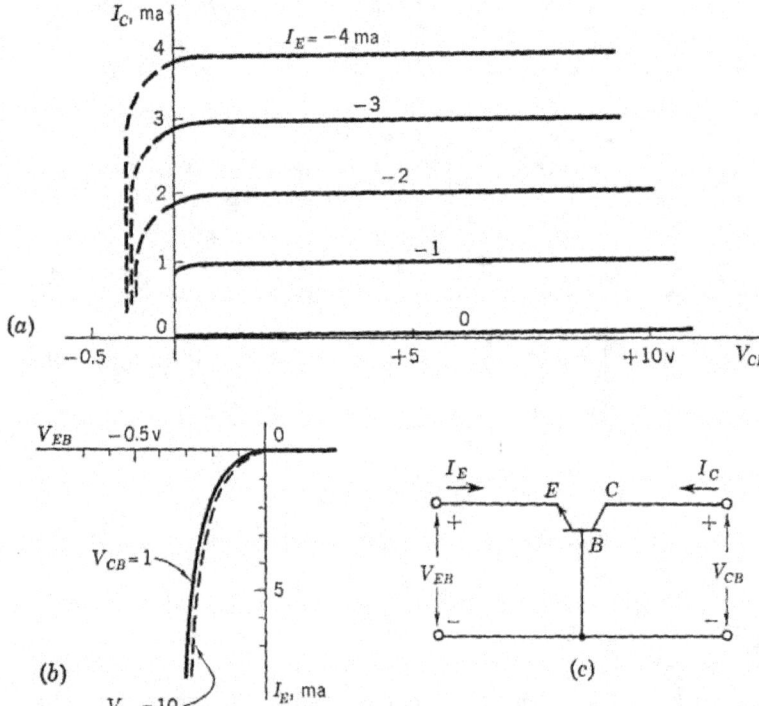

Fig. 2-5 Common-base *NPN*-transistor terminal characteristics.

$V_{CB} > 0$) [2] and proportional to emitter current.   A change of ΔI_E gives a collector current change of $\alpha' \Delta I_E$, where α' represents the fraction of emitter current that reaches the collector.   The α' parameter commonly

[1] The following discussion will be in terms of *NPN* transistors; however, the discussion is applicable to *PNP* transistors by interchanging the words "electron" and "hole" and reversing the direction of all voltages and currents.

[2] The following notation regarding currents and voltages will be used: Capital letters and subscripts indicate static or d-c values—I_E, V_{BE}, etc.; capital letters and lower-case subscripts are a-c, root-mean-square (rms) values for time-varying signals—I_b, V_{be}, etc.; lower-case letters and subscripts represent instantaneous values of the varying components—i_b, v_{be}, etc.   See IRE Standards on Letter Symbols for Semiconductor Devices, 1956, *Proc. IRE*, vol. 44, pp. 934–937, July, 1956.

ranges between 0.9 and 1.0 and is proportional to (1) the emitter efficiency (the fraction of emitter current that is carried by electrons into the base region), (2) the transport factor (the fraction of electrons emitted into the base region that reach the collector), and (3) the collector multiplication factor (the ratio of the electrons entering the collector region to those reaching the collector). The first two factors are less than 1, whereas the collector multiplication factor may be larger than unity, particularly at high collector-to-base voltages.

In the emitter-base circuit relatively high currents flow with small applied voltages; hence the emitter must represent a rather small (and nonlinear because of the curvature of the emitter characteristic) resistance. At a given operating point (I_E, V_{EB}) the emitter characteristic may be represented by a linear resistance r_{in} which has a value

$$r_{in} = \frac{\partial V_{EB}}{\partial I_E} \tag{2-15}$$

The quantities r_{in} and α' lead to the simplest representation of the transistor shown in Fig. 2-6. Although this transistor equivalent is the most rudimentary, it is nonetheless useful for rough calculations of many circuits. However, certain imperfections in the circuit are noticeable from a closer inspection of the transistor terminal characteristics: these are that the collector current curves are not absolutely independent of the collector voltage—the collector current increases very slightly with increasing collector voltage—and the input voltage V_{EB} decreases very

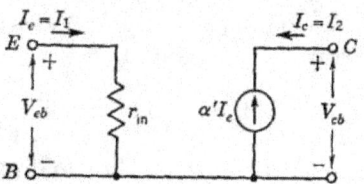

Fig. 2-6 The simplest transistor representation.

slightly with increasing V_{CB} and constant I_E. The change in I_C with changing V_{CB} can be taken into account with a shunt conductance across the current generator of value $g = \partial I_C / \partial V_{CB}$. The change in V_{EB} with changing V_{CB} can be accounted for by adding a voltage generator which is controlled by V_{CB} in series with r_{in}. The value of this generator is $\mu = \partial V_{EB} / \partial V_{CB}$. The addition of g and μ to the equivalent circuit of Fig. 2-6 gives the circuit of Fig. 2-7, which is identical to the h-parameter equivalent circuit shown in Fig. 2-2c. Consequently the previously defined parameters become, in the h-parameter notation,

$$h_{11b} = r_{in} \tag{2-16}$$

$$h_{21b} = \alpha' \tag{2-17}$$

$$h_{12b} = \mu \tag{2-18}$$

$$h_{22b} = g \tag{2-19}$$

The third subscript b indicates the common terminal of the transistor, in this case the base. All four of these parameters can be found from the graphical characteristics, but only with limited accuracy. Alternatively, small-signal a-c measurements can be used. (See Sec. 15-3.)

This equivalent circuit is very similar in form to an equivalent circuit which may be derived from the physical theory of the transistor [1] and which

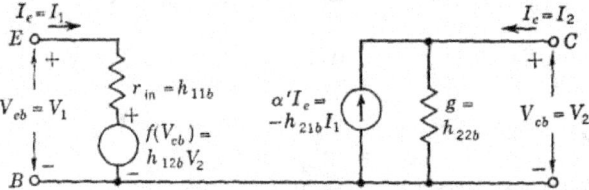

Fig. 2-7 A modification of the simple equivalent circuit (cf. Figs. 2-2c and 2-6).

is shown in Fig. 2-8. The portion which is drawn in solid lines represents the *intrinsic* transistor, i.e., an ideal transistor which is assumed to have zero ohmic resistance, particularly in the base region. Because of the actual base resistance r_b' the complete transistor may have terminal characteristics (i.e., parameters) which differ markedly from the intrinsic transistor. The identification of each parameter in Fig. 2-8 is:

r_e' The incremental resistance of the emitter-base diode. This is approximately equal to kT/qI_E, where k is Boltzmann's constant, T is the junction temperature in degrees Kelvin, and q is the electronic charge.

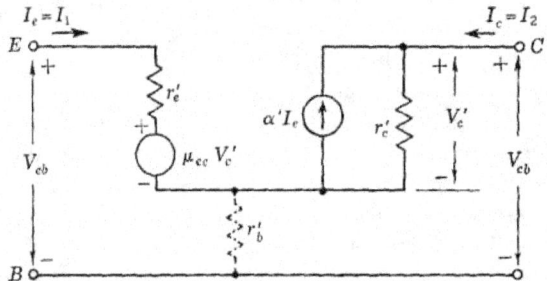

Fig. 2-8 A model of the junction transistor based upon the theory of the device.

At room temperature $r_e' \cong 25/I_E$, where r_e' is in ohms and I_E is in milliamperes.

α' The fraction of the incremental emitter current which arrives at the collector in the intrinsic transistor. (Note that this is not quite $\Delta I_C/\Delta I_E$ for the whole transistor because of the resistances r_b' and r_c'.)

[1] J. M. Early, Effects of Space-charge Layer Widening in Junction Transistors, *Proc. IRE*, vol. 40, p. 1401, November, 1952.

r_c' The incremental collector resistance. This arises because increasing
the collector voltage increases the width of the depletion region sur-
rounding the collector junction. In turn this causes the effective
width of the base region to decrease. The change in base width
caused by changes in V_C' is called base-width modulation. The decreased base width caused by increased V_C' reduces the current lost in
the base, thereby slightly increasing the collector current. This is a
small effect, so that the resistance r_c' is typically several megohms.

μ_{ec} The $\mu_{ec} V_C'$ generator represents the reverse transfer through the
transistor, i.e., the effect that output voltage has on the input. This
generator is also a consequence of base-width modulation: the decreased base width caused by increased V_C' in this case very slightly
affects the emitter diode, giving a reduced value of V_{EB} for the constant I_E. This is also a small but not always negligible effect. Typical
values of μ_{ec} are 10^{-3} to 10^{-4}.

In each case an equation can be written which expresses the circuit
constant in terms of the physical properties of the device. Hence one
important function of this circuit is that it gives a link between device
theory and circuit theory which is not provided by the general circuit
parameters.

In addition to the preceding intrinsic transistor parameters each region
of the transistor possesses an ohmic resistance to current flow. In the usual
transistor only the base resistance r_b' is of importance; this is included as a
dotted resistor in Fig. 2-8 to emphasize that it is not a part of the intrinsic
transistor. The resistance r_b' is sometimes called the base spreading resistance or the extrinsic base resistance. Since r_b' is really a distributed

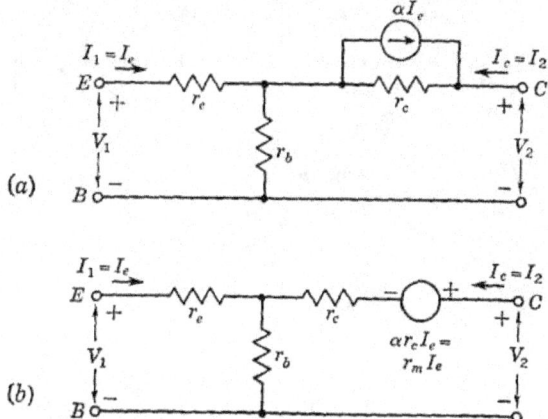

Fig. 2-9 T equivalent circuits with I_e as the control parameter.

resistance which is only approximated by a single, lumped value, it does not exactly represent the effects of the actual base resistance. However, the approximation is good in most cases, with the possible exception of grown-junction transistors at high frequencies.

2-4 The T Equivalent Circuits. The circuit shown in Fig. 2-8, while perfectly valid at low frequencies, has five quantities which define the four current and voltage variables. Since, in general, only four constants are necessary, the extra constant represents an unnecessary complication. A very commonly used equivalent circuit is the T equivalent circuit shown in Fig. 2-9a. Several equivalent forms of this circuit may be made, one of which is shown in Fig. 2-9b. The values for this circuit are related to those of the circuit of Fig. 2-8 by the equations [1]

$$r_b = r_b' + \mu_{ec} r_c' \tag{2-20}$$

$$r_c = r_c'(1 - \mu_{ec}) \cong r_c' \tag{2-21}$$

$$r_e = r_e' - \mu_{ec}(1 - \alpha')r_c' \tag{2-22}$$

$$\alpha = \frac{\alpha' - \mu_{ec}}{1 - \mu_{ec}} \cong \alpha' \tag{2-22a}$$

[1] The transformation from one equivalent circuit to another may be accomplished in many ways; however, some ways are much simpler than others. To illustrate a simple procedure for this case, consider applying unit current ($I_2 = 1$) to the output terminals of Figs. 2-9a and 2-8. For the circuits to be equivalent, the resulting voltages V_1 and V_2 must be equal,

$$V_2 = 1(r_b + r_c) = (r_b' + r_c')1$$

$$V_1 = 1(r_b) = (r_b' + \mu_{ec} r_c')1 \tag{2-20}$$

(Note that $V_c' = 1r_c'$.) Therefore

$$r_c = r_c'(1 - \mu_{ec}) \cong r_c' \tag{2-21}$$

(Usually $\mu_{ec} \ll 1$.) Now apply unit current to the input terminals ($I_1 = 1$), and equate the resulting V_1 and V_2 in each case,

$$V_1 = 1(r_e + r_b) = (r_e' + r_b' + \mu_{ec}\alpha' r_c')1$$

(Note that $V_c' = \alpha' r_c' I_1$.) Substituting for r_b, we get

$$r_e = r_e' - \mu_{ec}(1 - \alpha')r_c' \tag{2-22}$$

$$V_2 = 1(r_b + \alpha r_c) = (r_b' + \alpha' r_c')1$$

Substituting for r_b and r_c, we obtain

$$\alpha = \frac{\alpha' - \mu_{ec}}{1 - \mu_{ec}} \cong \alpha' \tag{2-22a}$$

Other ways of accomplishing the desired transformation can be easily shown to be considerably more complicated. In this case the z parameters have been equated; each quantity times a 1 is actually a z parameter; for example, $r_b + r_c = z_{22b}$.

Note that $\alpha \cong \alpha'$ and $r'_c \cong r_c$ but that $r_b \neq r'_b$ and $r_e \neq r'_e$. Actually r_e is about half r'_e; that is, $r_e \cong kT/2qI_E \cong 13/I_E$ (I_E in milliamperes). The resistance r_b is considerably larger than r'_b.

The parameters of the T equivalent are functions of the operating point of the transistor to varying degrees. Typical variations of the values of r_e, r_b, r_c, and α with emitter current and collector voltage are shown in Figs. 2-10 and 2-11. Careful note should be taken of typical values for these

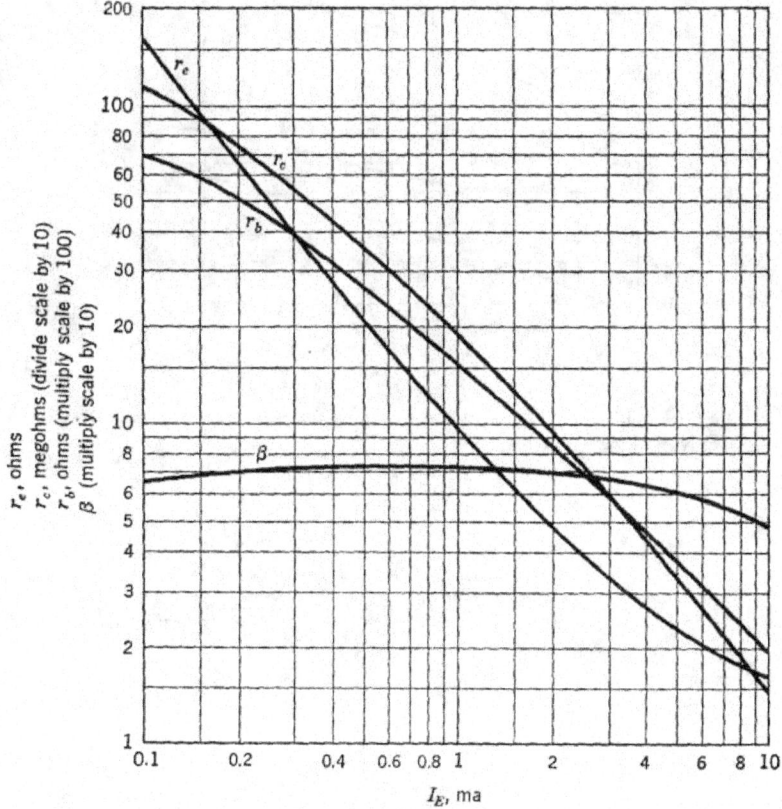

Fig. 2-10 Parameters of a typical transistor as a function of emitter current (transistor type 2N123 with $V_{CB} = -5$ volts).

four parameters, for in later work approximations will often be made which are valid only if the parameters have certain values. For example, if a small load resistance is connected from collector to base, the resistance r_c may often be neglected. In this case the qualification for the word "small" is that R_L, the load, must be much smaller than the several megohms of r_c.

Although all the discussion to this point has shown the transistor connected in the common-base configuration, this does not imply that the

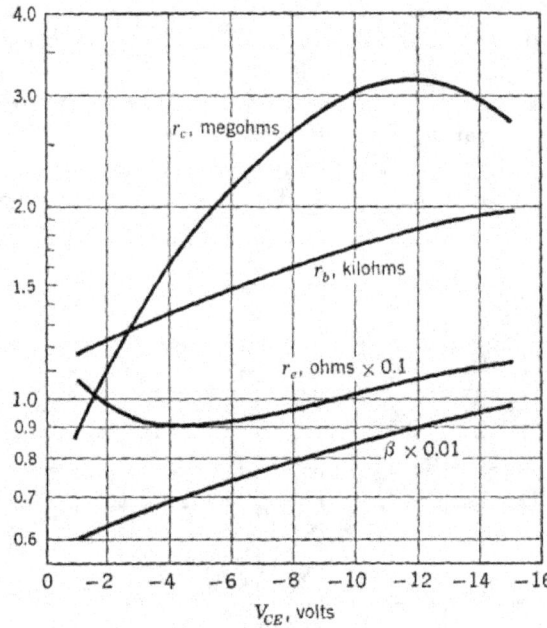

Fig. 2-11 Parameters of a typical transistor as a function of collector voltage (transistor type 2N123 with $I_E = 1$ ma).

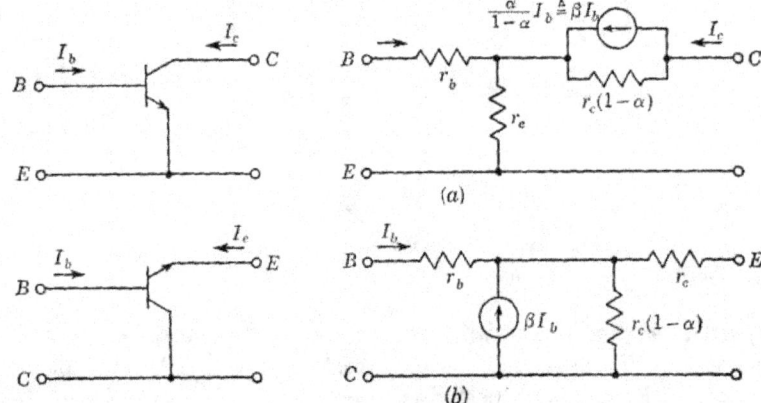

Fig. 2-12 Equivalent circuits with base current as the control current. (a) Drawn in the common-emitter configuration. (b) Drawn in the common-base configuration.

transistor must be used in that way or, indeed, that the equivalent circuits are valid only in that configuration. The T equivalent, or example, is equally valid with any terminal grounded or even with none grounded; however, for the common-emitter and common-collector configurations (Fig. 2-12a and b), it is often more convenient to have the controlled current generator a function of the input, or base, current rather than of the emitter current. The circuits in this figure can be easily derived by noting that $I_e + I_b + I_c = 0$. One convenient property of these T equivalent circuits is that once the parameters are known for one configuration they are known for all, whereas with the two-port parameters more complicated

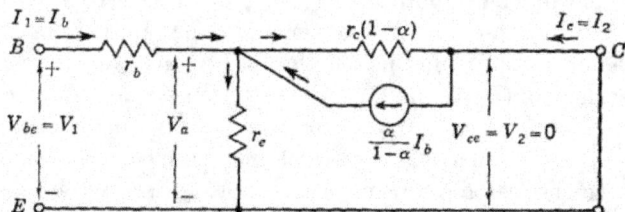

Fig. 2-13 Circuit for finding h_{11e} and h_{21e} in terms of the T parameters.

transformations are necessary to go from, say, common-base to common-collector parameters. The one new parameter which does arise in Fig. 2-12 is[1]

$$\frac{\alpha}{1-\alpha} \overset{\Delta}{=} \beta \overset{\Delta}{=} h_{21e} \overset{\Delta}{=} h_{fe} \tag{2-23}$$

This quantity arises so often that it is given a special name, β.

One difficulty with the T equivalent circuit is that the important parameters cannot be measured directly because the internal node is inaccessible. On the other hand the hybrid parameters can be easily measured, as shown in Chap. 15, Amplifier Measurements. Therefore, the equivalences between the h's and the T equivalent circuit are useful.[2,3]

[1] Two systems of h-parameter notation are in common use. These are: $h_{11} \equiv h_i$, $h_{12} \equiv h_r$, $h_{21} \equiv h_f$, and $h_{22} \equiv h_o$. The subscripts stand for *input, reverse, forward,* and *output,* respectively. The second subscript is then used to denote the common terminal; hence, h_{fb} is the parameter h_{21b}. The equivalences are also shown in Table 2-1.

[2] The relations needed to go from quantities measured in terms of h parameters to the T equivalents are in Chap. 15 on Amplifier Measurements, since it is in this connection that the relations are usually needed.

[3] A simple method to find h_{11e} and h_{21e} is to short-circuit the output terminals of the T as shown in Fig. 2-13 and apply an input current, I_b.

$$V_{be} = I_b h_{11e} = I_b r_b + V_a \qquad \text{since } V_{ce} = 0 \tag{2-24}$$

$$V_a = \left(I_b + \frac{\alpha I_b}{1-\alpha}\right)\frac{r_e r_c(1-\alpha)}{r_e + r_c(1-\alpha)}$$

The h parameters for all the configurations of the transistor are tabulated in Table 2-1. (The second subscript gives the common terminal; that is, h_{11b} is the input impedance with the base grounded.) In addition, typical values of the parameters for a small transistor operating at $I_E = 0.5$ ma are given. One important point to note in Table 2-1 is the approximations which are made to give the approximate equations. As shown in Eqs. (2-29) and (2-30), $r_e \ll r_c(1 - \alpha)$ is the condition which must be fulfilled for the approximate equations to be valid; this condition is *usually* well fulfilled, since r_e is typically about three orders of magnitude smaller than $r_c(1 - \alpha)$.

2-5 Representation of the Vacuum Tube at High Frequencies. Two somewhat different approaches may be taken to represent a vacuum tube for circuit purposes at high frequencies. The first is to take the low-frequency equivalent circuit (such as that of Fig. 2-4) and simply add the lumped parameters (L, R, and C) to the external terminals to account for the effects of leads, interelectrode capacitances, etc., and then to make a first-order approximation with an additional lumped element(s) for the effects due to the electron stream in the tube. A second approach is to think of the tube simply as a two-port network and measure the two-port network parameters—a convenient set at high frequencies being the short-circuit admittance parameters (y_{11}, etc.). The latter approach has the advantage of giving results easily applied to a single-frequency network solution but has the disadvantage that measurements must usually be made over the range of frequencies of interest, thereby necessitating many measurements. Also, the latter method gives little insight into the physical

$$V_a = \frac{I_b}{1 - \alpha} \frac{r_e r_c (1 - \alpha)}{r_e + r_c(1 - \alpha)} \tag{2-25}$$

$$h_{11e} = r_b + \frac{r_e r_c}{r_e + r_c(1 - \alpha)}$$

$$\cong r_b + \frac{r_e}{1 - \alpha} \quad \text{for } r_e \ll r_c(1 - \alpha) \tag{2-26}$$

$$h_{21e} = \frac{I_c}{I_b} = \frac{\alpha}{1 - \alpha} - \frac{V_a}{r_c(1 - \alpha)}$$

$$= \frac{\alpha}{1 - \alpha} - \frac{r_e r_c}{[r_e + r_c(1 - \alpha)][r_c(1 - \alpha)]} \tag{2-27}$$

$$h_{21e} \cong \frac{\alpha}{1 - \alpha} \quad r_e \ll r_c(1 - \alpha) \tag{2-28}$$

h_{22e} and h_{12e} are found by opening the 1-1 terminals and applying a *voltage* to the 2-2 terminals. Then

$$h_{22e} = \frac{I_2}{V_2} = \frac{1}{r_c(1 - \alpha) + r_e} \cong \frac{1}{r_c(1 - \alpha)} \quad \text{since } I_b = 0 \tag{2-29}$$

$$h_{12e} = \frac{V_1}{V_2} = \frac{r_e}{r_e + r_c(1 - \alpha)} \cong \frac{r_e}{r_c(1 - \alpha)} \tag{2-30}$$

Table 2-1 *

	Exact equations	Approximate equations	Typical values
$\left.\begin{array}{l}h_{11b}\\ h_{ib}\end{array}\right\}$	$= r_e + \dfrac{r_b r_c(1-\alpha)}{r_b + r_c}$	$\cong r_e + r_b(1-\alpha)$	35 ohms
$\left.\begin{array}{l}h_{12b}\\ h_{rb}\end{array}\right\}$	$= \dfrac{r_b}{r_b + r_c}$	$\cong \dfrac{r_b}{r_c}$	2×10^{-4}
$\left.\begin{array}{l}h_{21b}\\ h_{fb}\end{array}\right\}$	$= -\alpha - \dfrac{r_b(1-\alpha)}{r_b + r_c}$	$\cong -\alpha$	-0.98
$\left.\begin{array}{l}h_{22b}\\ h_{ob}\end{array}\right\}$	$= \dfrac{1}{r_c + r_b}$	$\cong \dfrac{1}{r_c}$	$0.4\ \mu mho$
$\left.\begin{array}{l}h_{11e}\\ h_{ie}\end{array}\right\}$	$= r_b + \dfrac{r_e r_c}{r_e + r_c(1-\alpha)}$	$\cong r_b + \dfrac{r_e}{1-\alpha}$	1,750 ohms
$\left.\begin{array}{l}h_{12e}\\ h_{re}\end{array}\right\}$	$= \dfrac{r_e}{r_c(1-\alpha) + r_e}$	$\cong \dfrac{r_e}{r_c(1-\alpha)}$	5×10^{-4}
$\left.\begin{array}{l}h_{21e}\\ h_{fe}\end{array}\right\}$	$= \dfrac{\alpha}{1-\alpha} - \dfrac{r_e r_c}{[r_e + r_c(1-\alpha)][r_c(1-\alpha)]}$	$\cong \dfrac{\alpha}{1-\alpha} = \beta$	49
$\left.\begin{array}{l}h_{22e}\\ h_{oe}\end{array}\right\}$	$= \dfrac{1}{r_c(1-\alpha) + r_e}$	$\cong \dfrac{1}{r_c(1-\alpha)}$	$20\ \mu mhos$
$\left.\begin{array}{l}h_{11c}\\ h_{ic}\end{array}\right\}$	$= r_b + \dfrac{r_e r_c}{r_e + r_c(1-\alpha)}$	$\cong r_b + \dfrac{r_e}{1-\alpha}$	1,750 ohms
$\left.\begin{array}{l}h_{12c}\\ h_{rc}\end{array}\right\}$	$= \dfrac{(1-\alpha)r_c}{r_e + (1-\alpha)r_c}$	$\cong 1$	1
$\left.\begin{array}{l}h_{21c}\\ h_{fc}\end{array}\right\}$	$= \dfrac{-r_c}{r_e + r_c(1-\alpha)}$	$\cong \dfrac{-1}{1-\alpha} \cong -\beta$	-50
$\left.\begin{array}{l}h_{22c}\\ h_{oc}\end{array}\right\}$	$= \dfrac{1}{r_e + r_c(1-\alpha)}$	$\cong \dfrac{1}{r_c(1-\alpha)}$	$20\ \mu mhos$

* Values are computed for $r_c = 2.5$ megohms, $r_b = 500$ ohms, $r_e = 25$ ohms, and $\alpha = 0.98$.

behavior of the tube unless the admittances are carefully interpreted. The method is nevertheless valuable at uhf and microwave frequencies, where the concepts of leads, voltages, and currents tend to become ambiguous.[1]

For the areas covered herein the first approach seems most fruitful, and it will be developed here. Figure 2-14a shows that a lumped [2] approxima-

[1] L. C. Peterson, Equivalent Circuits of Linear Active Four-terminal Networks, *Bell System Tech. J.*, vol. 27, no. 4, p. 593, October, 1948.

[2] By lumped is meant the representation of a distributed circuit by a finite number of inductors, capacitors, etc.; e.g., a lead wire into a tube is really a transmission line and therefore cannot be exactly represented by a single inductor. The approximation is excellent, however, as long as the lead is very short compared with a wavelength at the highest frequency considered.

tion to the tube, exclusive of effects due to the electron stream, would
have an inductance in series with each electrode caused by the inductance
of the lead wires and a capacitor representing the capacitance between that
electrode and each of the other electrodes. Thus, in the triode tube there
are three inductors and three capacitors; these may be superimposed upon
the low-frequency equivalent circuit with the result shown in Fig. 2-14b.
The resulting circuit is more complicated than necessary except at very
high frequencies. At medium frequencies, say to 20 or 30 Mc, the inductances have relatively little effect in small receiving tubes, and the important
elements are the interelectrode capacitances; therefore the circuit can be
simplified to contain only the three capacitors, together with μ and r_p.*

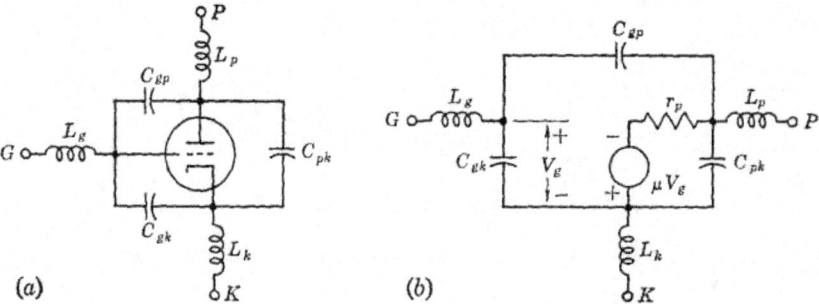

Fig. 2-14 Triode h-f equivalent circuits.

One important effect caused by the small lead inductances, however, is that
a circuit which operates normally at low frequencies may oscillate at very
high frequencies where the lead inductances begin to have appreciable reactance. A simple example of the causes of such an effect is given in Prob.
2-7.

The situation in the case of the pentode is similar, but complicated by
the addition of the two extra electrodes. Figure 2-15 shows the major
reactances of interest. Some of those associated with the suppressor grid
are not shown, for they have little effect. Although all shown may become
of interest at very high frequencies, the circuit is usually simplified as
shown in Fig. 2-15b for medium-frequency use. Here the inductances are
ignored, and the screen is assumed to be at a-c ground potential. Hence
$C_{\text{in}} = C_{g_1 k} + C_{g_1 g_2}$ (since $C_{g_1 p} \ll C_{\text{in}}$) and $C_{\text{out}} = C_{p g_3} + C_{p g_2}$. The feedback capacitor $C_{g_1 p}$ is included in the equivalent circuit, but it may some-

---

* Note that the inductances shown, for example, L_K, if in the grounded lead particularly, should include the inductance of the external lead from the tube to the ground
plane. This fact is important both in calculations and in physical layout, since a long
ground lead or a bypass capacitor with series inductance may greatly increase the total
effective lead inductance.

times be neglected if the voltage gain of the stage is low.   In the pentode the first inductance likely to become important as frequency increases is the cathode lead inductance.   The effect of this inductance may be represented by means of a frequency-dependent conductance, which is discussed in connection with transit-time loading.   The capacitances in either the triode or the pentode are affected by the space charge in the tube; hence the

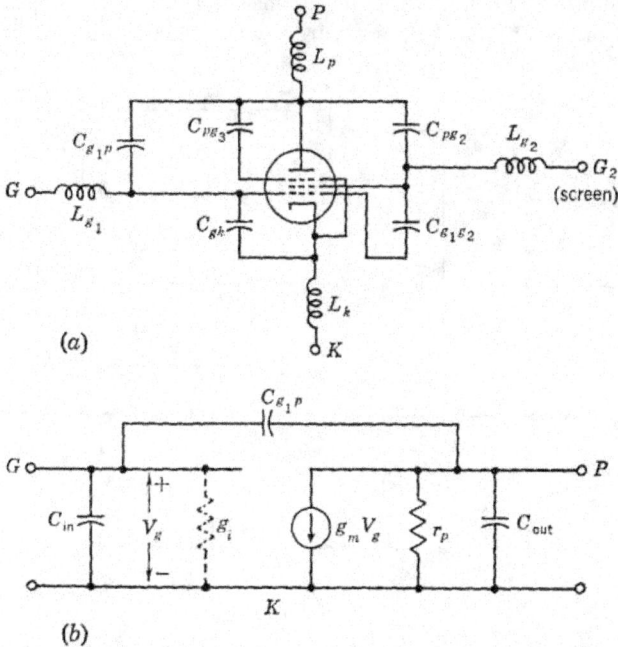

Fig. 2-15 High-frequency equivalent circuits for the pentode.

capacitances are dependent upon the d-c operating point of the tube.   The capacitance most affected by change in operating point is C_{gk} (or C_{in} in the case of the pentode).   The other capacitances are practically independent of the operating point of the tube.   Since the input capacitance forms an important part of the grid tuned circuit in a high-frequency amplifier, the variation of input capacitance with operating point can cause serious detuning if the operating point is varied to provide gain control.[1]   A reduction of this effect may be obtained by adding a small unbypassed cathode resistor; however, this resistor also reduces the gain since the effective transconductance becomes

$$g'_m = \frac{g_m}{1 + g_m R_k} \qquad (2\text{-}31)$$

[1] For data see F. Langford-Smith, "Radiotron Designer's Handbook," 4th ed., sec. 23.5, Radio Corporation of America, New York, 1952.

At low frequencies the input admittance of a vacuum tube is essentially a purely capacitive susceptance; however, at higher frequencies a conductance component becomes important. This component arises from three major sources. First, there is some dielectric loss associated with the insulators in the base and interior of the tube. This conductance varies directly (approximately) with frequency. Second, there is a conductance due to the finite transit time of electrons past the control grid. This component of input conductance varies with the square of frequency.

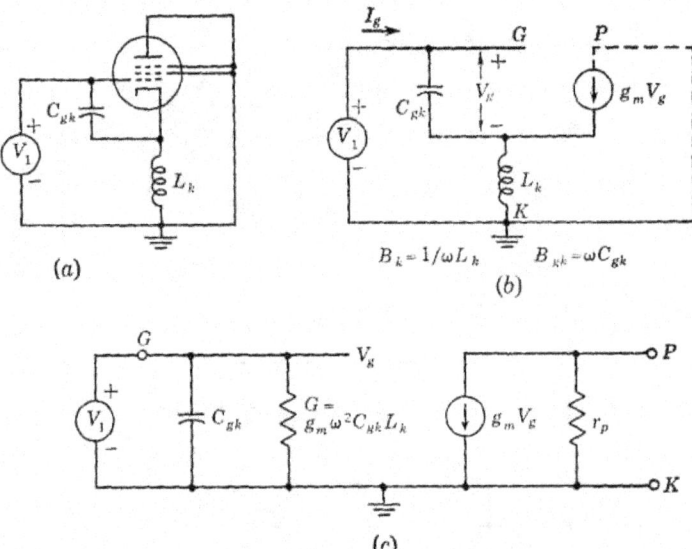

Fig. 2-16 Equivalent circuit for the calculation of input conductance due to cathode lead inductance.

Third, there is a component which is caused by the cathode lead inductance. The value of this component may be calculated with the aid of the circuit shown in Fig. 2-16.

The equation for the input loop (Fig. 2-16b) is

$$V_1 = \frac{I_g}{jB_{gk}} + \frac{I_g + g_m V_g}{-jB_k} \quad \text{and} \quad V_g = \frac{I_g}{jB_{gk}} \tag{2-32}$$

$$\frac{I_g}{V_1} = Y_{11} = \frac{B_{gk}B_k}{-jB_k + jB_{gk} + g_m} \tag{2-33}$$

Usually $B_k \gg B_{gk}$, so that Eq. (2-33) may be simplified.

$$Y_{11} \cong B_{gk}\frac{B_k}{-jB_k + g_m} \cong jB_{gk}\left(1 - j\frac{g_m}{B_k} + \cdots\right) \tag{2-34}$$

$$Y_{11} \cong j\omega C_{gk} + g_m\omega^2 C_{gk}L_k + \cdots \tag{2-35}$$

The higher-order terms may usually be ignored because $g_m/B_k \ll 1$. The last equation shows that the input admittance due to C_{gk} and L_k may be represented by the parallel combination of C_{gk} and a conductance of value $g_m\omega^2 C_{gk}L_k$ connected between grid and ground, as shown in Fig. 2-16c.

Because the conductances due both to transit time and to L_k vary as frequency squared, it is difficult to measure experimentally the two effects separately. Consequently, the two are usually lumped together, and the total input conductance is expressed in the empirical relation

$$g_i = k_c f + k_h f^2 \tag{2-36}$$

where g_i = total input conductance

$k_c f$ = "cold" input conductance (heater power off)

$k_h f^2$ = "hot" input conductance (due to transit time and L_k)

Values for k_h for some modern triodes are given in Chap. 13.[1] The complete pentode equivalent circuit for high frequencies is then that of Fig. 2-15, where $C_{\text{in}} = C_{g_1 g_2} + C_{g_1 k}$, as before, and g_i is given by Eq. (2-36). The preceding has assumed that the screen lead inductance is negligible. Above 100 Mc this may not be true, especially if care is not taken in bypassing the screen to ground through a capacitor. The effect of the inductance is to cause a *negative* input conductance to appear from the control grid to ground. Uncontrolled, this effect may lead to oscillation, but the effect is sometimes used to reduce excessive grid loading at very high frequencies.[2]

2-6 High-frequency Equivalent Circuits for Transistors. The simple "low-frequency" equivalent circuits for transistors thus far presented are valid up to frequencies of only a few kilocycles for audio transistors to frequencies of perhaps a megacycle or more for transistors useful in the vhf range. The basic limitation to transistor action at high frequencies is the relatively low velocity of the current carriers, which leads to transit-time effects at relatively low frequencies compared with vacuum tubes, including a large part of the useful frequency spectrum of the transistor. As has been seen in the case of the vacuum tube, an equivalent circuit which even approximately takes into account the effects of transit time is relatively complicated. The same is true for the transistor; for this reason we shall not attempt to make an "exact" equivalent circuit, if indeed this is possible, but rather we shall attempt to find several useful, relatively simple equivalent circuits which reasonably well represent the operation of the transistor *in the type of operation under consideration.* The latter point is very important because, for example, a representation that is perfectly adequate for a video amplifier calculation may be very poor for a 456-kc intermediate-frequency amplifier.

---

[1] For values for older tubes, see *ibid.*, p. 929.

[2] Problem 2-7 illustrates the calculation of this effect in a triode tube. The effect in the pentode is the same if the pentode screen is treated as the triode plate.

A simple picture of the problem caused by the transit time in a transistor may be derived by considering the processes taking place in the base of a transistor when a very narrow pulse (ideally, an impulse) of current is supplied to the emitter. At the initial instant the electron density (in an NPN unit) in the base at the edge of the emitter junction is increased. Ideally this increased density would travel from emitter to collector with unchanged form, but at a finite velocity. The time for the pulse to travel from emitter to collector is the transit time in question. If the pulse arrived with the same shape with which it began, there would be no frequency problems to worry about—the transistor would have acted as an ideal delay line of infinite bandwidth. However, the mechanism of carrier propagation in the base is, at least in part, a diffusion process. This is analogous to the process of heat flow in a metal. If one momentarily touches a hot object to the end of a metal rod, the hot spot will slowly move down the rod, but while the hot spot is of zero length in the beginning, the spot begins to broaden out as it travels down the rod. The same is true of the distribution of electrons in the base: initially the distribution is very narrow, corresponding to the narrow input-pulse length; as the electrons diffuse across the base, their random velocities cause the distribution to broaden—the longer the transit time, the more the broadening. In Fig. 2-17 is shown the result of the process; the input pulse is square and narrow, but the output pulse has nonzero rise time and a rounded shape. The rounding of the output current pulse shows that some of the high frequencies of the original pulse have been lost; i.e., the current gain of a transistor falls off at high frequencies. It may be seen from this qualitative discussion that the high-frequency response could be improved by anything which would reduce the transit time, since this would decrease the spreading of the pulse. Two things may be done to accomplish transit-time reduction. One is to narrow the width of the base—the current gain times bandwidth is inversely proportional to the square of the base width. Second, the velocity of the carriers in the base can be increased by arranging the impurities in the base region in such a way that an electric field is set up to increase the velocity of the minority carriers (electrons) through the base, as in a drift or graded-base transistor.

An expression for the current gain α' as a function of frequency may be derived from the theory of the transistor; however, the expression involves transcendental functions and is thus not convenient for representation with a lumped network. Instead a simplified form can be used,

$$\alpha'(j\omega) = \frac{\alpha_0' \epsilon^{-j\omega T}}{1 + j\omega/\omega_\alpha} \qquad (2\text{-}37)$$

where α_0' is the low-frequency value of I_c/I_e. The function $\epsilon^{-j\omega T}$ has unity magnitude and thus does not affect the magnitude of α', which is $\alpha_0'/\sqrt{2}$ at

$\omega = \omega_\alpha = 2\pi f_\alpha$. The frequency f_α is usually called the alpha-cutoff frequency. The magnitude and phase of α' are plotted as a function of frequency in Fig. 2-18. In this figure the logarithm of $|\alpha'|$ is plotted with a logarithmic frequency scale.

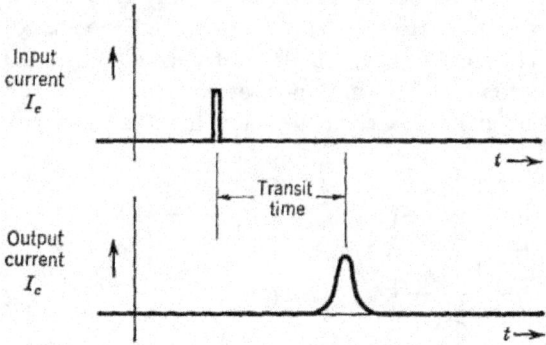

Fig. 2-17 The effect of transit time and diffusion on the short-circuit current output of a transistor.

The function $\epsilon^{-j\omega T}$ represents a phase shift proportional to frequency or, in the time domain, a pure time delay T. The relative size of T depends on the type of transistor. For a simple junction transistor $T \cong 0.2/\omega_\alpha$ and is often neglected; in a drift transistor T is much more important and may be $1/\omega_\alpha$ or more. Note that $T = 1/\omega_\alpha$ gives a phase shift in the numerator

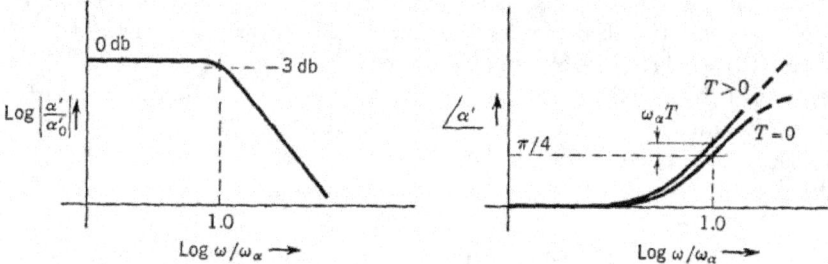

Fig. 2-18 α' versus frequency.

of Eq. (2-37) of 1 radian at $\omega = \omega_\alpha$, or a total phase shift in α' of $1 + \pi/4$ radians. The phase of α' is plotted in Fig. 2-18 for two values of T.

The frequency dependence of alpha given in Eq. (2-37) may be incorporated into the equivalent circuit of Fig. 2-8 resulting in a frequency-dependent current generator $\alpha' I_e$. The reverse transfer generator $\mu_{ec} V'_c$ has a similar frequency dependence, since it also depends upon effects transmitted through the base region; however, the effect of this generator is nearly always negligible at high frequencies, as will be shown.

The flow of emitter current results in a distribution of carriers in the base region. This distribution cannot be altered instantaneously because of the previously mentioned transit-time effects. Hence a sudden change in emitter current gives rise to a gradual change in the charge distribution in the base. The initial change in emitter current must first change the charge distribution at the emitter-base junction. This effect may be approximated by placing a capacitor of value $1/\omega_a r_e'$ in parallel with r_e'

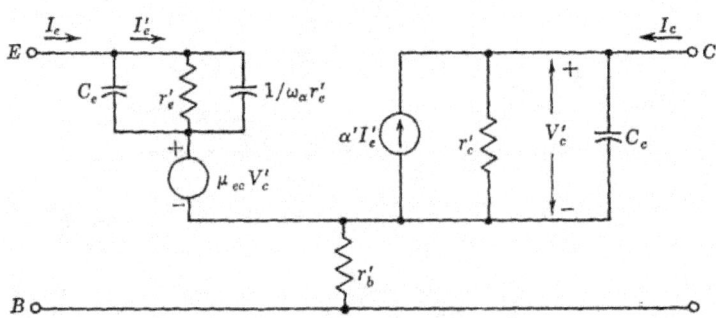

Fig. 2-19   The T equivalent circuit for high frequencies.

(Figs. 2-8 and 2-19). This capacitor is generally called the emitter diffusion capacitance.

One other effect of major importance in transistor frequency response arises from the space-charge region existing at the reverse-biased collector-base junction. In this region the junction field removes the carriers, leaving behind the fixed charges of impurity atoms fixed in the crystal lattice. The amount of fixed charges exposed on each side of the junction must satisfy Poisson's equation for the amount of potential across the junction. Thus an increasing potential widens the space-charge region to expose more fixed charges and gives rise to a current of electrons from the n region and holes from the p region. This current is a displacement current similar to the current to a capacitor when an increasing potential is applied. Consequently the effect of the space-charge region is approximated in the equivalent circuit by means of the capacitor C_c (Fig. 2-19).

The collector capacitance depends principally upon the area of the junction and on the density of impurities present on either side of the junction. Because of the space-charge layer-widening effect, the outer boundaries of the space-charge region move farther apart as the reverse bias is increased. Since incremental increases and decreases of charge occur at the outer boundaries, this means that the incremental capacitance C_c *decreases* with an increase in reverse bias. Effectively, the "plates" of the capacitor are moved farther apart. Collector capacitance of typical small transistors may range from 10 to 30 pf for alloy-type units; most drift transistors usually have considerably smaller capacities, 2 to 10 pf being representative. Special types of construction can result in $C_c < 1$ pf; in this range, attention must be paid to mounting structure and lead capacitance as well, which can be appreciably larger. In the remainder of the discussion, however, it will be assumed that any parasitic lead capacities, etc., are negligible or else lumped with C_c.

A similar space-charge layer capacitance exists at the emitter junction, shown in Fig. 2-19 as C_e. The principal effect of this capacitance is to divert to the base lead some of the emitter current that would otherwise have flowed through the base to the collector by means of the minority carriers. A portion of I_e flows through C_e, leaving a fraction I_e' which controls the collector current generator $\alpha' I_e'$. The effect of C_e is thus to lower the effective alpha-cutoff frequency as measured at the terminals of the transistor. The amount of the decrease is small in many transistors because the impedance of C_e is large compared with the low impedance of r_e' in parallel with $1/\omega_a r_e'$; however, in very-high-frequency units where the capacitance $1/\omega_a r_e'$ is small and comparable to C_e, an appreciable fraction of I_e is diverted to the base, and the effect of C_e cannot be ignored. Note that the a-c impedance of r_e' in parallel with $1/\omega_a r_e'$ is inversely proportional to the direct current I_E; consequently there is some advantage to operating at a reasonably high d-c emitter current to make I_e'/I_e as large as possible.

If the frequency ω_a appearing in Eq. (2-37) and in the expression for emitter diffusion capacitance is understood to be the effective alpha-cutoff frequency, the effects of both C_e and transit time across the base being taken into account, then C_e need not be considered separately; that is, C_e is lumped into the capacitance $1/\omega_a r_e'$, as in Fig. 2-20, where ω_a is essentially the alpha-cutoff frequency measured for the complete device.

The resulting T equivalent circuit shown in Fig. 2-20 is more detailed than is really needed for most high-frequency applications. Calculation of

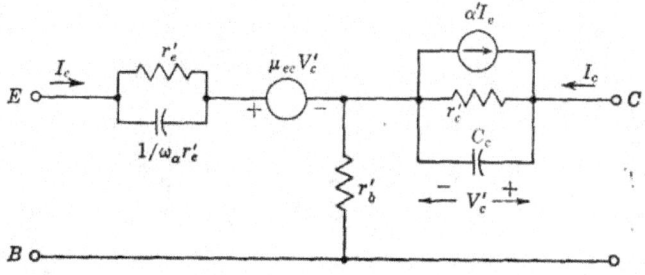

Fig. 2-20 The T equivalent with C_e lumped with $1/\omega_a r_e'$. Here ω_a is the measured cutoff frequency and includes the effect of C_e.

the h parameters for this circuit is one way to show what approximations may be made and the restrictions on their validity. In Fig. 2-21 h_{11b} and h_{21b} may be found by assuming a known value of I_e. Define

$$Z_c \triangleq \frac{r_c'}{1 + j\omega r_c' C_c} \qquad Z_e \triangleq \frac{r_e'}{1 + j\omega/\omega_\alpha} \tag{2-38}$$

Then

$$V_c' = -(1 - \alpha')I_e \frac{r_b' Z_c}{r_b' + Z_c}$$

$$\cong -I_e(1 - \alpha')r_b' \qquad \text{since } Z_c \gg r_b', \text{ usually} \tag{2-39}$$

$$V_1 = Z_e I_e + \mu_{ec} V_c' - V_c'$$

$$\cong Z_e I_e - V_c' \qquad \text{since } \mu_{ec} \ll 1 \tag{2-40}$$

$$h_{11b} = \frac{V_1}{I_1} = Z_e + r_b'(1 - \alpha') \tag{2-41}$$

$$h_{21b} = \frac{I_2}{I_1} = -\alpha' + \frac{V_c'}{Z_c} = -\alpha' - \frac{(1 - \alpha')r_b'}{r_b' + Z_c}$$

$$\cong -\alpha' \tag{2-42}$$

In using the above expressions it must be remembered that α' is now a function of frequency. If the value of h_{11b} is desired, the value of α' with $\epsilon^{-j\omega T} = 1 - j\omega T + (j\omega T)^2/2 + \cdots \cong 1 - j\omega T$ can be substituted into Eq. (2-41),

$$h_{11b} = Z_e + r_b' \left[1 - \frac{\alpha_0'(1 - j\omega T)}{1 + j\omega/\omega_\alpha}\right]$$

$$= Z_e + r_b'(1 - \alpha_0') \frac{1 + j\omega(1/\omega_\alpha + \alpha_0' T)/(1 - \alpha_0')}{1 + j\omega/\omega_\alpha} \tag{2-43}$$

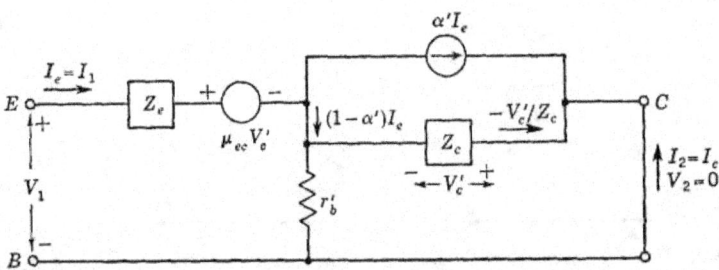

Fig. 2-21 Circuit for finding the high-frequency value of h_{11b} and h_{21b}.

From the last equation h_{11b} can be shown to be the impedance Z_e in series with an RL parallel circuit, as shown in Fig. 2-22. Note that the impedance h_{11b} (and the actual input impedance if the load impedance in the collector is small) is inductive over a considerable frequency range; that is, $|h_{11b}|$ increases with frequency from approximately

$$\omega = \frac{1 - \alpha_0'}{1/\omega_\alpha + \alpha_0' T}$$

to $\omega = \omega_\alpha$.

The quantities h_{22b} and h_{12b} may be obtained by inspection from Fig. 2-20,

$$h_{22b} = \frac{1}{Z_c + r_b'} \cong \frac{1}{Z_c} = \frac{1 + j\omega r_c' C_c}{r_c'}$$

$$\cong j\omega C_c \qquad \text{if } \omega \gg \frac{1}{r_c' C_c} \qquad (2\text{-}44)$$

The calculation of h_{12b} is facilitated if one assumes that $V_2 \cong V_c'$; then

$$h_{12b} = \mu_{ec} + \frac{r_b'}{r_b' + Z_c} \cong \mu_{ec} + \frac{r_b'}{Z_c}$$

$$\cong \left(\mu_{ec} + \frac{r_b'}{r_c'} \right) + j\omega r_b' C_c \qquad (2\text{-}45)$$

$$h_{12b} \cong \frac{r_b'}{r_c'} + j\omega r_b' C_c \qquad (2\text{-}46)$$

$$\cong j\omega r_b' C_c \qquad \text{if } \omega \gg \frac{1}{r_c' C_c} \qquad (2\text{-}47)$$

The approximate form $h_{22b} \cong j\omega C_c$ is useful if one is interested in the high-frequency performance where the capacitive reactance is much less than r_c'. Also, the load impedance appears directly in parallel with h_{22b}, and this must be much smaller than r_c' if the transistor collector circuit is not to be the limitation on the high-frequency response. With low load impedances the effect of r_c' is negligible at *all* frequencies.

The approximate form for h_{12b} shows that it is constant at low frequencies and then increases linearly with frequency. Again with small load impedances the effect of h_{12b} can be ignored at low frequencies; hence only the term $j\omega r_b' C_c$ need be retained in most cases. With these approximations,

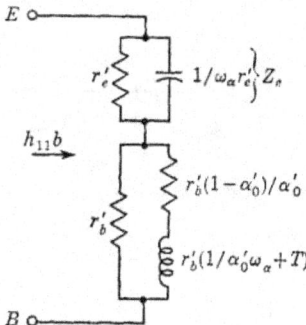

Fig. 2-22 The circuit representation of the short-circuit input impedance h_{11b}.

the equivalent circuit may be modified to the form shown in Fig. 2-23, where the components μ_{ec} and r'_c are omitted.

2-7   The Hybrid-pi High-frequency Equivalent Circuit. Although the equivalent circuit of Fig. 2-23 may be used for the transistor connected in any manner, it is frequently more convenient for the common-emitter

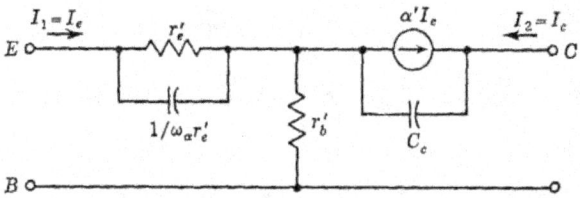

Fig. 2-23 A simplified high-frequency equivalent circuit. The quantity α' is frequency-dependent: $\alpha' = \alpha_0 e^{-j\omega T}/(1 + j\omega/\omega_\alpha)$.

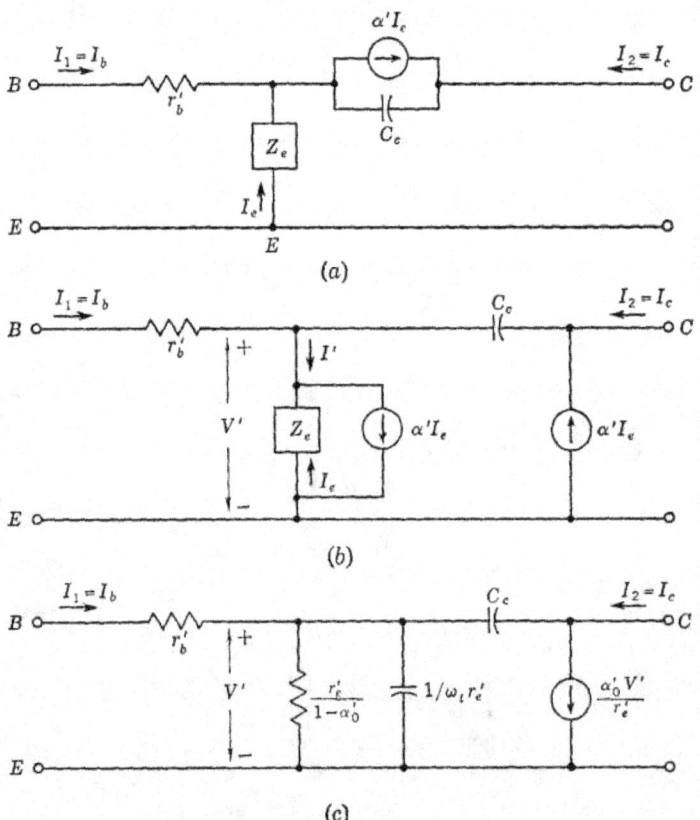

Fig. 2-24 Derivation of the hybrid-pi from the T equivalent circuit.

connection to have a circuit which is controlled by a base quantity rather than by an emitter quantity. (In Fig. 2-23, I_e is the control quantity for the current generator.) An equivalent circuit which has wide application can be easily derived from the simplified T circuit. The latter is shown in Fig. 2-24a, turned around for common-emitter use. The circuit in Fig. 2-24b is identical since the same current leaves the node to the left of C_c and arrives at the node to the right of C_c. The two current generators cancel in the emitter lead so that no current is gained or lost because of their common connection. The combination of Z_e and the current generator can be combined into a new impedance,

$$I' = \alpha' I_e - I_e = -I_e(1 - \alpha') \tag{2-48}$$

$$V' = -I_e Z_e \tag{2-49}$$

$$\frac{V'}{I'} = Z' = \frac{Z_e}{1 - \alpha'} = \frac{r'_e}{1 + j\omega/\omega_\alpha} \frac{1}{1 - \alpha'_0 \epsilon^{-j\omega T}/(1 + j\omega/\omega_\alpha)} \tag{2-50}$$

$$Z' = \frac{r'_e}{1 + j\omega/\omega_\alpha - \alpha'_0 \epsilon^{-j\omega T}} \tag{2-51}$$

An approximate form may be obtained by expanding the exponential in a series and taking only the first two terms. (This assumes that $\omega T < 1$ at the highest frequency of interest.)

$$Z' = \frac{r'_e}{1 - \alpha'_0 + j\omega(1/\omega_\alpha + \alpha'_0 T) - \cdots} \tag{2-52}$$

$$Z' \cong \frac{r'_e}{1 - \alpha'_0 + j\omega/\omega_t} \tag{2-53}$$

where
$$\omega_t \overset{\Delta}{=} \frac{\omega_\alpha}{1 + \alpha'_0 \omega_\alpha T}$$

The impedance Z' given above is that of the parallel combination of a resistor $r'_e/(1 - \alpha'_0)$ and a capacitor $1/\omega_t r'_e$. The current generator at the output of Fig. 2-24b may also be simplified,

$$\alpha' I_e = \frac{-\alpha' V'}{Z_e} = -V' \frac{\alpha'_0 \epsilon^{-j\omega T}}{1 + j\omega/\omega_\alpha} \frac{1 + j\omega/\omega_\alpha}{r'_e} \tag{2-54}$$

$$\alpha' I_e = \frac{(-\alpha'_0 \epsilon^{-j\omega T}) V'}{r'_e} \tag{2-55}$$

The quantity $-\alpha'_0 \epsilon^{-j\omega T}/r'_e$ has a magnitude which is independent of frequency and a phase shift which increases linearly with frequency, i.e., a pure time delay. For many purposes the time delay is not particularly important, and the current generator may be approximated without the exponential function, as shown in Fig. 2-24c.[1]

This circuit, known as the hybrid-pi equivalent circuit, is very useful for circuit work because all the elements in the circuit are reasonably constant with frequency up to frequencies approaching the alpha-cutoff frequency. If a more accurate representation is necessary at low frequencies, which is usually not the case in h-f amplifiers, conductances across C_c and the current generator may be added.[2]

One important new parameter, ω_t, which has considerable practical significance, appears in Eq. (2-53) and in the capacitor of value $1/\omega_t r_e'$ in Fig. 2-24c. To find the physical significance of ω_t, let us investigate the short-circuit current gain h_{21e} of the common-emitter transistor. Consider the collector short-circuited to the emitter in Fig. 2-24c; the voltage V' is then

$$V' = \frac{I_1 r_e'/(1 - \alpha_0)}{1 + [j\omega/(1 - \alpha_0)](1/\omega_t + r_e' C_c)} \cong \frac{I_1 r_e'/(1 - \alpha_0)}{1 + j\omega/\omega_t(1 - \alpha_0)} \qquad (2\text{-}56)$$

The approximation assumes that $C_c \ll 1/\omega_t r_e'$, which is true for almost all high-frequency transistors.

$$I_2 = \frac{\alpha_0}{r_e'} V' + j\omega C_c V' \cong \frac{\alpha_0 V'}{r_e'} \qquad (2\text{-}57)$$

$$\frac{I_2}{I_1} = h_{21e} = \frac{\alpha_0}{1 - \alpha_0} \frac{1}{1 + j\omega/(1 - \alpha_0)\omega_t} = \beta_0 \frac{1}{1 + j\omega/\omega_\beta} \qquad (2\text{-}58)$$

where
$$\beta_0 \overset{\Delta}{=} \frac{\alpha_0}{1 - \alpha_0} \qquad (2\text{-}59)$$

$$\omega_\beta \overset{\Delta}{=} (1 - \alpha_0)\omega_t \qquad (2\text{-}60)$$

The short-circuit current gain is now $\beta_0 = \alpha_0/(1 - \alpha_0)$ at low frequencies, as was also shown in Table 2-1, but the current gain is $\beta_0/\sqrt{2}$ at the frequency $\omega_t(1 - \alpha_0)$. Thus the cutoff frequency ω_β in the common-emitter configuration is $\omega_\beta = \omega_t(1 - \alpha_0) = (1 - \alpha_0)\omega_\alpha/(1 + \alpha_0\omega_\alpha T)$. Note that, compared with the common-base connection, the current gain of the common-emitter connection is increased by $1/(1 - \alpha_0)$, but that the bandwidth is decreased by somewhat more because of the phase-shift factor $\epsilon^{-j\omega T}$. The fact that h_{21e} begins to fall off at a frequency about two orders of

---

[1] An important type of amplifier of which this is not true is the feedback type of amplifier, which is not covered herein.

[2] For an interesting discussion of high-frequency equivalent circuits and many pertinent references, see R. L. Pritchard, Electric-network Representation of Transistors—A Survey, *IRE Trans. on Circuit Theory*, vol. CT-3, p. 5, March, 1956.

magnitude less than ω_α is very important in high-frequency amplifier design, although it is possible to build amplifiers with much wider bandwidths than ω_β.

Because of the importance of ω_t (or $f_t = \omega_t/2\pi$) in the equivalent circuit of Fig. 2-24, it is useful to measure f_t rather than f_α and T. This has the additional advantage of being a considerably easier measurement. Consider the typical curve of $|h_{21e}|$ shown in Fig. 2-25. At frequencies much above f_β, $|h_{21e}|$ falls off at -6 db/octave, and if $\alpha_0' \cong 1$, then $|h_{21e}| = f_t/f$ in this frequency range. The frequency at which $|h_{21e}| = 1$ is f_t. For

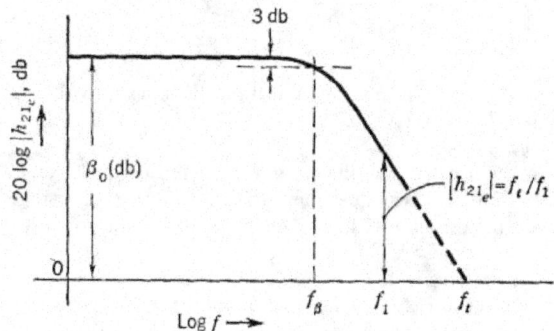

Fig. 2-25 Short-circuit current gain of a common-emitter transistor.

these reasons f_t is often called the gain-bandwidth product of the transistor.

Two important points should be kept in mind in using transistor equivalent circuits. Since the useful circuits are all approximations to some degree, it is unreasonable to expect a single circuit to be the best approximation over the full frequency range of a transistor or for all circuit applications. Also, the accuracy of the representation may vary considerably from one transistor to another, even of the same type. The circuit-application aspect of the equivalent-circuit problem will be further discussed in the sections dealing with amplifier applications. It usually happens that circuits involving very wide-range equivalent circuits, e.g., a fast pulse amplifier, are very noncritical with regard to the accuracy of the transistor representation; hence the model used need be made accurate only at the high-frequency end of the passband.

Another method of using these and other equivalent circuits is to use only the desired form of the equivalent circuit but to obtain the actual element values from measurements taken in the frequency band of interest. This method is useful with transistors which do not exactly have the -6 db/octave slope of $|h_{21e}|$ shown in Fig. 2-25 but which are to be used at frequencies above f_β in, for example, a bandpass amplifier.

PROBLEMS

2-1. A given two-port network has the following h parameters: $h_{11} = 25$ ohms, $h_{12} = 10^{-4}$, $h_{21} = -1$, $h_{22} = 10^{-6}$ mho. What are the values of the z and y parameters for the network?

2-2. Assume that a two-port network characterized by the h parameters is to operate from a source R_g and load R_L. (Assume that the parameters are real.)

a. Derive an equation for the insertion power gain.

b. Derive an equation for the available power gain.

2-3a. Show that the impedance looking into the input of a two-port network terminated in a load impedance Z_L is

$$Z_{in} = h_{11} - \frac{h_{21}h_{12}Z_L}{1 + h_{22}Z_L}$$

b. Derive a similar expression for the admittance looking into the output terminals when the source impedance is Z_g.

2-4. Show that $g_m r_p = \mu$.

2-5. Consider an "iterative" amplifier chain of many identical stages (theoretically an infinite number). A practical question is: Without using interstage transformers, which connection of the transistor (CE, CC, or CB) will give the greatest insertion power gain per stage?

Note first that the effective load Z_L for each stage is equal to the Z_{in} of the succeeding stage and that the generator Z_g is equal to the output $1/Y_{out}$ of the preceding stage (cf. Prob. 2-3). Show that the effective load for each stage in the chain is

$$Z_L = \frac{(\Delta^h - 1) \pm \sqrt{(\Delta^h - 1)^2 - 4h_{11}h_{12}}}{2h_{22}}$$

where

$$\Delta^h = h_{11}h_{22} - h_{12}h_{21}$$

Using the h-parameter values in Table 2-1 for a typical transistor, determine the insertion power gain for each connection (CE, CB, CC). Which gives the maximum?

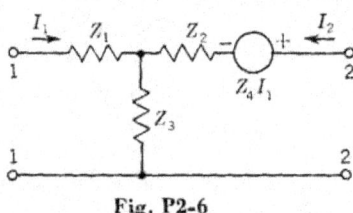

Fig. P2-6

2-6. Show that *any* two-port network expressible in the z parameters may be represented by three impedances and one generator, as shown in Fig. P2-6. Find the values of $Z_1, \ldots, Z_4$ in terms of the z parameters.

2-7. Assume a triode amplifier as shown in Fig. P2-7. (Assume that C_{pk} is part of Z_L.)

a. Find the expression for the input admittance.

b. If Z_L is purely inductive, what is the conductance component of input admittance? What does this tell you about the stability of a triode amplifier at high frequencies? (Note that the inductance could be provided by the plate lead inductance.)

c. If Z_L is purely resistive, show that the total effective input capacity is approximately

$$C_{\text{in}} \cong C_{gk} + C_{gp}(1 + A)$$

where A is the voltage gain of the stage and $C_{gp}(1 + A)$ is often known as the "Miller capacitance."

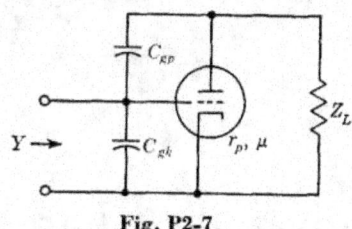

Fig. P2-7

2-8. Using the simplified equivalent circuit of Fig. 2-23, calculate an approximate set of high-frequency h parameters for the common-collector connection of the transistor.

2-9. Derive the value of the current generator and its shunting resistor shown in Fig. 2-12a from the circuit of Fig. 2-9a.

3

The Steady-state Characteristics
of a Lowpass (Video) Amplifier

An elementary but complete amplifier stage is shown in Fig. 3-1 for a pentode stage. All the elements with the exception of the load resistance R_L are "parasitic." That is, they are necessary to set d-c levels, but they

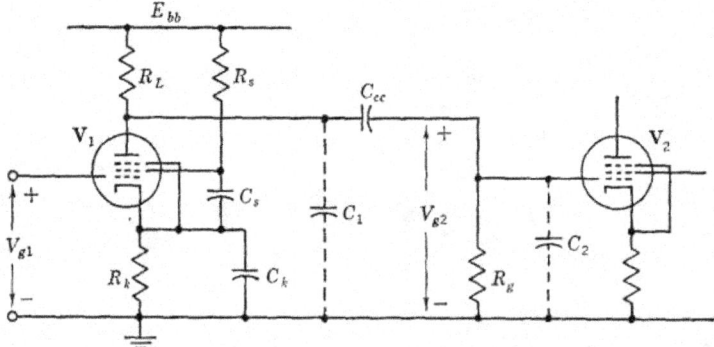

Fig. 3-1 A complete pentode amplifier stage.

impair rather than aid in the production of constant gain or voltage amplification as a function of frequency. The elements C_{cc} and R_g are added to remove the relatively high plate voltage from the grid which is at d-c ground potential. Also, d-c potential changes due to supply variations and tube aging are not amplified along with the signal.[1] The elements R_k

[1] The problem of d-c amplification, which is important in itself, will not be taken up herein, since it is separate and distinct from the problem of obtaining wideband amplification.

and C_k in the cathode circuit and the elements R_s and C_s in the screen circuit are added to bias properly the tube and to maintain these conditions as the tube ages.   All these components affect the amplifier low-frequency response and may affect the high-frequency response if there is appreciable inductance in series with the capacitors.   The capacitances C_1 and C_2,

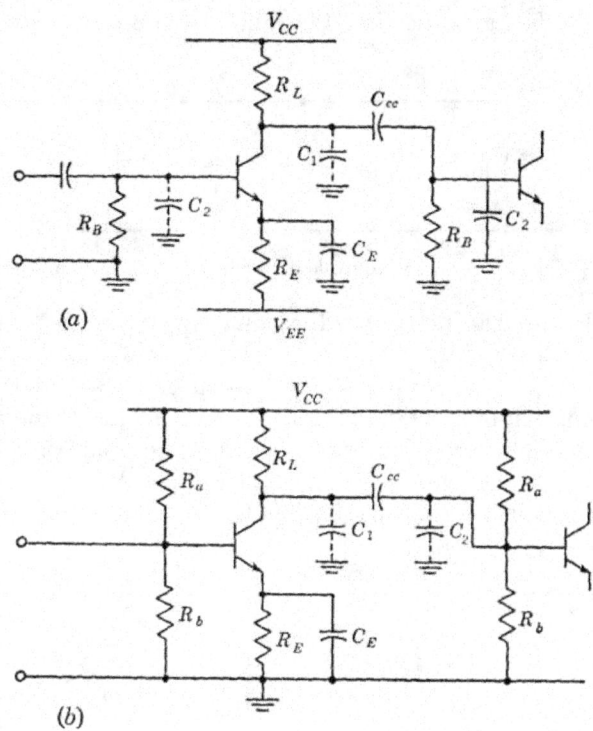

Fig. 3-2 A complete typical transistor stage. Bias is supplied by two d-c supplies in (a) and a single supply in (b).

which include the tube output and input capacitances, respectively, added to the wiring capacitance, reduce the gain of the stage at high frequencies.

In the case of a transistor stage as shown in Fig. 3-2 many of the same elements are present. The stability of the d-c operating point is controlled by the bias resistors R_a, R_b, R_B, and R_E. The latter is bypassed by C_E, which usually imposes the most severe limit on the low-frequency performance of the amplifier. The capacitor C_{cc} performs the same function as in the pentode stage; however, C_{cc} can be profitably omitted in some transistor stages because the collector voltage can be low so that the build-up in d-c voltage from stage to stage is not excessive. Separate means, such as feedback that is effective only at very low frequencies, may be used to reduce the d-c gain and make the amplifier bias stable. Stray capacitances shown as C_1 and C_2 are not usually as important in transistor amplifiers as in vacuum-tube amplifiers because the limitations to high-frequency response come primarily from inside the transistor itself, rather than from the external circuit (see Sec. 2-6).

3-1 Low-frequency Response of Pentode Stage.   Equations may be written which exactly describe the gain vs. frequency of the circuit in

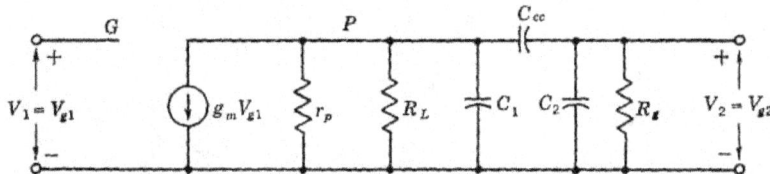

Fig. 3-3 Complete pentode equivalent circuit (omitting bias impedances).

Fig. 3-1 and somewhat less exactly the gain of the circuit in Fig. 3-2; however, the complexity of the equations incorporating all the parasitic effects simultaneously prevents forming many useful general conclusions.   Instead the approach of separating the high- from the low-frequency effects will be used, since this is an extremely accurate approximation in all lowpass wideband amplifiers.[1]   To illustrate, consider Fig. 3-3, which shows the equivalent circuit for a pentode circuit, for the moment omitting con-

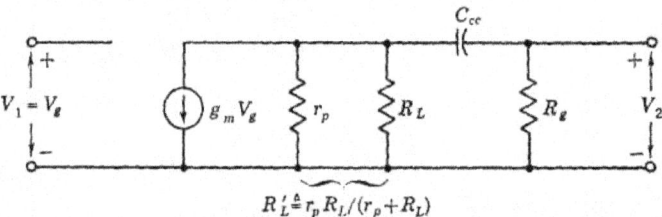

$$R'_L \triangleq r_p R_L/(r_p + R_L)$$

Fig. 3-4 Simplified circuit for determining the low-frequency gain.

sideration of the bias impedances.   Experimentally it is readily observed that over a considerable portion of the passband of the wideband amplifier the gain is almost exactly constant; consequently the frequency-dependent reactances of C_1, C_2, and C_{cc} must have negligible effect in this midband region.   Therefore the reactance of C_1 and C_2 must be much larger than the parallel combination of r_p, R_L, and R_g.   Also, C_{cc} must be acting virtually as a short circuit between plate and grid at the signal frequency.

[1] Analyses of the rare case when this approach is invalid are given by D. C. G. Luck, A Simplified General Method of Resistance-Capacity Coupled Amplifier Design, *Proc. IRE*, vol. 20, p. 1401, August, 1932; and W. F. Curtis, The Limitations of Resistance-coupled Amplification, *Proc. IRE*, vol. 24, p. 1230, September, 1936.

As the frequency is reduced from the midband value, the effect of C_1 and C_2 becomes more negligible still but the reactance of C_{cc} increases until it no longer acts like a short circuit between plate and grid. Consequently, to determine the low-frequency limit of the amplifier stage in Fig. 3-3, we need consider only the effect of the reactance of C_{cc}. The equivalent circuit, then, for calculating the low-frequency gain is as shown in Fig. 3-4. The voltage gain of this circuit is

$$A(j\omega) = \frac{-g_m R'_L R_g}{R'_L + R_g} \frac{j\omega}{\omega + 1/(R'_L + R_g)C_{cc}} \tag{3-1}$$

where

$$R'_L = \frac{r_p R_L}{r_p + R_L} \tag{3-2}$$

For the usual pentode wideband stage $r_p \gg R_L$ and $R_g \gg R_L$; therefore a good approximation to the above is

$$A(j\omega) = -g_m R_L \frac{j\omega}{j\omega + 1/R_g C_{cc}} = -g_m R_L \frac{j\omega R_g C_{cc}}{j\omega R_g C_{cc} + 1} \tag{3-3}$$

Notice that, if the frequency is high, the gain is constant at the value $-g_m R_L$; this is the midband gain for the complete amplifier, which can be designated A_0. Notice next that when $\omega = 1/R_g C_{cc}$ the real and imaginary terms in the denominator of Eq. (3-3) are equal; this unique frequency will be designated as ω_1, the low-frequency cutoff due to the coupling circuit.[1] Thus in terms of the midband gain A_0 and the cutoff frequency ω_1, Eq. (3-3) can be written as follows:

$$\frac{A[j(\omega/\omega_1)]}{A_0} = \frac{j(\omega/\omega_1)}{j(\omega/\omega_1) + 1} \tag{3-3a}$$

This is a general equation which describes the frequency behavior of any amplifier of this type. The amplitude and phase response are shown in the curves of Fig. 3-5. The amplitude is plotted as $20 \log (A/A_0)$, that is, the so-called decibel; notice that at the cutoff frequency ω_1 the gain is down 3.0 db from the midband value A_0. Similarly the phase shift differs by 45° from the midband value. The curves in Fig. 3-5 may be obtained by inserting different values of ω into Eq. (3-3a) and solving for the magnitude and phase of gain ($|A|$ and $\underline{/A}$); however, a quicker way of finding the approximate curve of $|A|$ is to write the logarithm of the magnitude of $A(j\omega)$,

$$20 \log \left| \frac{A}{A_0} \right| = 20 \log \left| \frac{j\omega}{\omega_1} \right| - 20 \log \left| \frac{j\omega}{\omega_1} + 1 \right| \tag{3-4}$$

---

[1] Note further that at the cutoff frequency $\omega_1 = 1/R_g C_{cc}$ the reactance of C_{cc} is exactly equal to the resistance R_g.

The first term is due to the numerator of Eq. (3-3a), and the second term is due to the denominator of Eq. (3-3a). These terms may now be plotted on semilogarithmic graph paper, as shown in Fig. 3-6. The first term is a straight line rising 20 db/decade (a 10:1 change in frequency) and intersecting the line 0 db at $\omega = \omega_1$. (The slope of this line is also $+6$ db/oc-

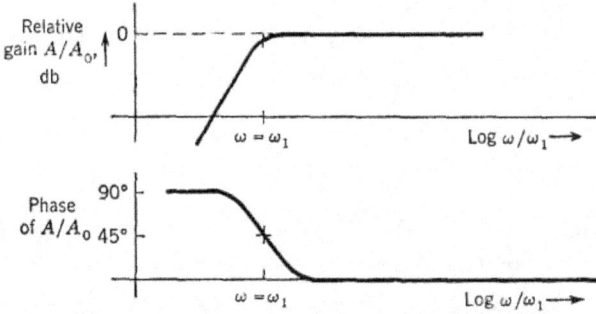

Fig. 3-5 Magnitude and phase of the gain of the circuit shown in Fig. 3-4.

tave, or doubling, of frequency.) The second term is constant at low frequencies, where $\omega R_g C_{cc} \ll 1$, and decreases at a rate of -20 db/decade at high frequencies. The actual curve for the third term is well approximated by using only the two asymptotes which intersect at $\omega = \omega_1$. The maximum error in using the two straight lines occurs at the frequency of their inter-

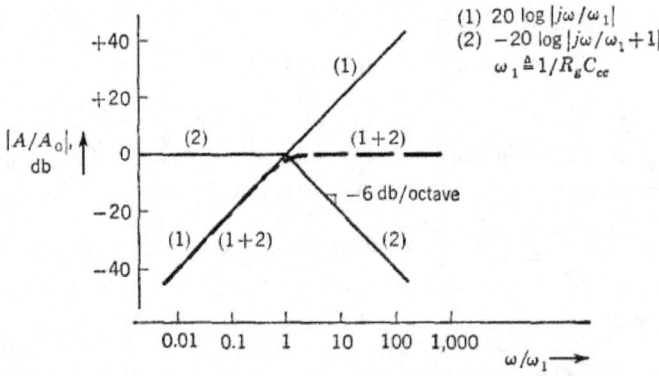

Fig. 3-6 Obtaining the gain characteristic of Eq. (3-3a) by asymptotic construction.

section and is 3 db. Corrections for this will be discussed later. The response due to both terms is simply found by adding the curves together, giving the result shown in Figs. 3-6 and 3-5. Note that the approximate method requires only the calculation of $\omega_1 = 1/R_g C_{cc}$ if only the shape of the gain curve is required.

The phase response may be found in a similar manner,

$$\underline{/A\!\left(j\frac{\omega}{\omega_1}\right)} = \underline{/j\frac{\omega}{\omega_1}} - \underline{/j\frac{\omega}{\omega_1}+1} \qquad (3\text{-}5)$$

The first term is a constant $+90°$ phase shift; the second again may be approximately plotted in terms of the asymptotic value at high frequencies $(-90°)$, the asymptotic value at low frequencies $(0°)$, and a transition

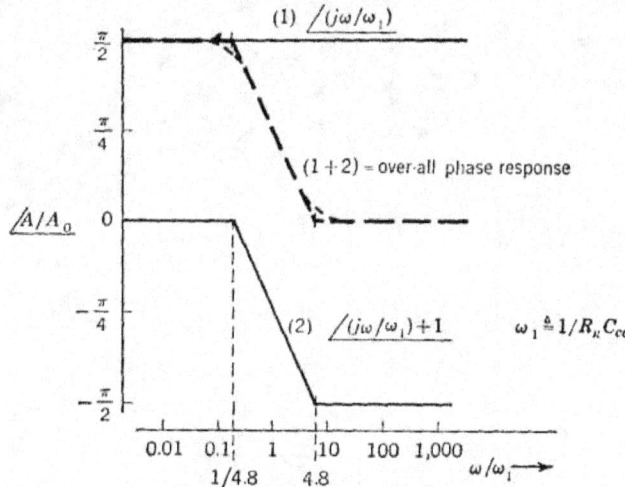

Fig. 3-7 Obtaining the phase characteristic of Eq. (3-3a) by asymptotic construction.

region about the frequency ω_1. The details of the latter construction are shown in Fig. 3-7. Again the phase due to both terms is found by adding the individual curves.

Gain functions more complicated than Eq. (3-3), but of the same general type, can be displayed in the same way, by asymptotic approximations to the frequency response. The general form of Eq. (3-3) is

$$A(j\omega) = K\frac{j\omega}{j\omega + \omega_1} \qquad (3\text{-}6)$$

and is a particular case of a class of gain functions expressed by

$$A(j\omega) = K\frac{(j\omega + \omega_1)(j\omega + \omega_3) \cdots (j\omega + \omega_m)}{(j\omega + \omega_2)(j\omega + \omega_4) \cdots (j\omega + \omega_n)} \qquad (3\text{-}7)$$

The ω_m and ω_n are real numbers and correspond to the critical (also "cutoff" or "corner") frequencies in the asymptotic response graph.[1] As another example, take

$$A(j\omega) = K \frac{j\omega(j\omega + \omega_1)}{(j\omega + \omega_2)^2} \tag{3-8}$$

In logarithmic form, the amplitude response in decibels becomes

$$20 \log |A(j\omega)| = 20 \log K + 20 \log |j\omega| + 20 \log |j\omega + \omega_1|$$
$$- 40 \log |j\omega + \omega_2| \tag{3-8a}$$

$$= 20(\log K + \log \omega_1 - 2 \log \omega_2) + 20 \log |j\omega|$$
$$+ 20 \log \left| j \frac{\omega}{\omega_1} + 1 \right| - 40 \log \left| j \frac{\omega}{\omega_2} + 1 \right|$$

$$\tag{3-8b}$$

The relative gain, relative to the gain A_0 at $\omega = 0$, is

$$20 \log \left| \frac{A}{A_0} \right| = 20 \log |j\omega| + 20 \log \left| j \frac{\omega}{\omega_1} + 1 \right| - 40 \log \left| j \frac{\omega}{\omega_2} + 1 \right|$$

$$\tag{3-8c}$$

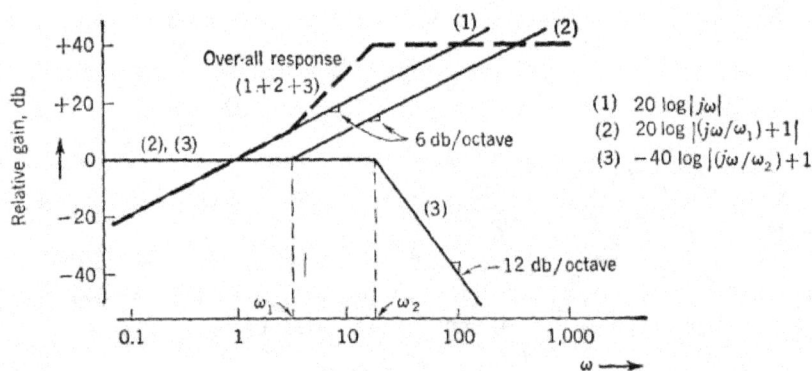

Fig. 3-8 Obtaining the amplitude response of Eq. (3-8c).

Each term can be considered separately and plotted by straight-line asymptotes (see Fig. 3-8). The first term gives a straight line with positive

[1] The most general form of a gain function may include terms such as $(j\omega)^2 + k_1(j\omega) + k_2$ which cannot be factored into the form of Eq. (3-6) with real ω_m or ω_n. Asymptotic plotting of such terms can also be accomplished; see J. L. Bower and P. M. Schultheiss, "Introduction to the Design of Servomechanisms," chap. 3, John Wiley & Sons, Inc., New York, 1958.

slope (20 db/decade, or 6 db/octave) passing through the 0-db level at $\omega = 1$. The second term is constant at 0 db up to the critical frequency $\omega = \omega_1$, whereupon it assumes a "unit" positive slope of 20 db/decade. Finally, the third term is constant at 0 db up to its critical frequency $\omega = \omega_2$, after which it assumes a negative slope of *twice* the unit value, or 40 db/decade, owing to its arising from a second-order factor in the denominator of Eq. (3-8). Adding together the three responses gives the total, as shown in Fig. 3-8.

While the straight-line approximation may suffice for many analyses, it is possible to refine the curves by the generalized correction factors given in Table 3-1. Each straight-line approximation, such as that for the second

Table 3-1

f/f_c (or f_c/f) *	Gain correction, † db	True phase (or complement of true phase)
0.1	0.04	5.7
0.2	0.17	11.3
0.4	0.64	21.8
0.6	1.34	31.0
0.8	2.14	38.6
1.0	3.01	45.0

* f_c is the critical frequency of the numerator or denominator term in question.

† The correction is positive for a numerator term (zero) and negative for a denominator term (pole).

term in Eq. (3-8c), is modified by the tabulated correction terms, and the new curves summed up. In the case of higher-order terms, such as the second-order third term in Eq. (3-8c), the gain correction is multiplied by the order of the term.

The phase shift of the gain function can also be approximated as before,

$$\underline{/A(j\omega)} = \underline{/j\omega} + \underline{/j\frac{\omega}{\omega_1} + 1} - 2\underline{/j\frac{\omega}{\omega_2} + 1} \qquad (3\text{-}8d)$$

The result is shown in Fig. 3-9, using the same principle employed in the construction of Fig. 3-7. Note that numerator terms in the gain function [Eq. (3-8)] contribute a leading phase, while denominator terms contribute a phase lag. A second-order term gives twice the phase shift, as though it were two first-order terms.

A more accurate phase curve can be obtained by substituting for the straight-line plot of Fig. 3-9 the true phase values of Table 3-1 for each term in the gain function. For the second-order term the phase values are merely doubled.

3-2 The Geometric Interpretation of Gain Functions. It is appropriate now to introduce a further generalization which will have great value throughout most of the later portions of the book. It consists

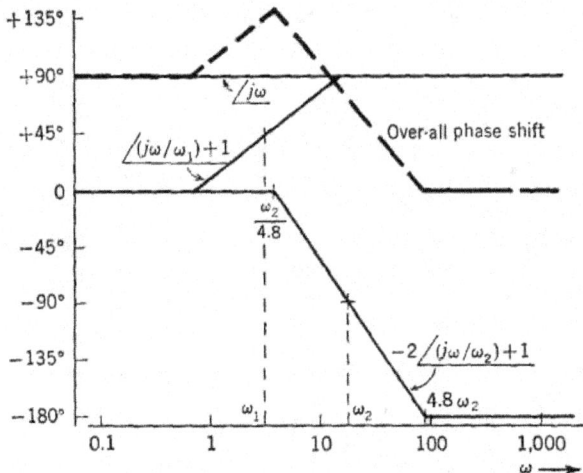

Fig. 3-9 Phase response given by Eq. (3-8d).

first in substituting for $j\omega$ a more generalized frequency variable p, which also is complex but can comprise both real and imaginary components; thus $p \overset{\Delta}{=} \sigma + j\omega$.

It is desirable to have a good understanding of the variable p, where it comes from, and what it signifies.[1] It can be considered, for the present, as originating in either of two ways:

1. A generalization of $\epsilon^{j\omega t}$ into ϵ^{pt}, in the manner of $\epsilon^{j\omega t}$ being a useful generalization of the more elementary $\cos \omega t$, which is a perfectly correct expression for sinusoidally varying currents or voltages but which is more cumbersome analytically than $\epsilon^{j\omega t}$. Note that $\epsilon^{j\omega t} = \cos \omega t + j \sin \omega t$ and thus contains $\cos \omega t$, plus additional information ($j \sin \omega t$) which is not used directly but which is economically carried along because of the greater facility with which $\epsilon^{j\omega t}$ behaves analytically. Similarly ϵ^{pt} has great facility

[1] Further study is recommended, in treatments of network theory such as D. F. Tuttle, "Network Synthesis," vol. I, chap. 3, John Wiley & Sons, Inc., New York, 1958, or in treatments of transform analysis, e.g., M. F. Gardner and J. L. Barnes, "Transients in Linear Systems," John Wiley & Sons, Inc., New York, 1942.

and, more importantly, provides some new insight into the functional behavior of amplifier gain functions.

2. A function deriving from Laplace transform theory, whereby any time-varying current or voltage can be transformed from a time function (a differential equation usually) into an algebraic function of a complex variable p. For example, the gain function in Eq. (3-3) can be rewritten, by substituting p for $j\omega$,

$$A(p) = K \frac{p}{p - p_1} \tag{3-9}$$

where

$$K = -g_m R$$

$$p_1 = \frac{-1}{R_g C_{cc}}$$

Still more generally, any amplifier gain function (where the amplifier uses linear, lumped elements) may be written in the form

$$A(p) = K \frac{\Pi(p - p_m)}{\Pi(p - p_n)} = K \frac{p^m + a_1 p^{m-1} + \cdots + 1}{p^n + b_{1p}^{n-1} + \cdots + 1} \tag{3-10}$$

Equation (3-10) shows that the gain may be written in the form of a constant, or *scale factor*, K, multiplied by the quotient of two polynomials in p, a *rational fraction*. The factors of the numerator give those values of p—the p_m which may or may not be real—for which the gain function is zero, and hence they are called the *zeros* of $A(p)$. The factors of the denominator give the values of $p = p_n$ for which the gain is infinite, and they are the *poles* of $A(p)$; these too can be either real or complex. With gain on a logarithmic basis, the zeros are points of infinite loss, and the poles are points of infinite gain. The poles are also sometimes known as the natural modes. In the example of Eq. (3-7) only one term of the continued product in the numerator and denominator is present. The pole is located at $p = -\omega_1$, and the zero is at $p = 0$.

Thus the gain function of Eq. (3-8) can also be written in terms of p, giving

$$A(p) = K \frac{p(p - p_1)}{(p - p_2)^2} \tag{3-10a}$$

An interesting and useful geometrical interpretation can be given to gain functions of p, such as Eq. (3-10a). Suppose that we plot in the p plane the poles and zeros of Eq. (3-10a); this is shown in Fig. 3-10. Note first that they lie on the negative real (σ) axis for this example, recalling that Eq. (3-10a) reverts to Eq. (3-8) by replacing p by $j\omega$. Note further that the σ coordinate of the zero or pole coincides *numerically* with the critical frequencies ω_1 and ω_2 of the amplitude response. Thus

$$p_1 = \sigma_1 + j0 = -\omega_1 + j0$$

$$p_2 = \sigma_2 + j0 = -\omega_2 + j0 \tag{3-11}$$

This coincidence is restricted to the case of poles and zeros lying solely on the σ axis. In general, poles or zeros may also exist as conjugate pairs with

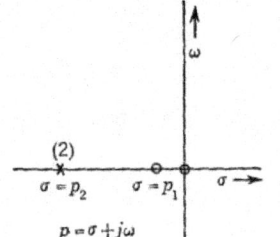
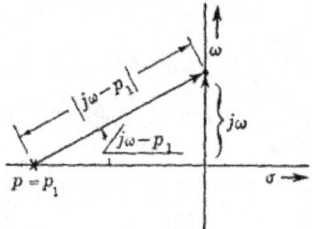

Fig. 3-10   Pole-zero diagram for Eq. (3-10a).

Fig. 3-11   The construction for finding $\underline{/j\omega - p_1}$ and $|j\omega - p_1|$ for one factor of $A(j\omega)$.

respect to the σ axis. In Sec. 8-1 the restrictions on pole-zero locations for physically realizable networks will be discussed.

The geometric interpretation consists in preparing a pole-zero diagram such as Fig. 3-10 and from it determining $A(j\omega)$ with nothing more than a ruler and protractor. For instance, Eq. (3-10a) shows $A(p)$ to be simply a scale factor times the quotient of some vector distances on the p plane, and when we let $p = j\omega$, we can find the magnitude and phase angle of $A(j\omega)$ as follows:

$$|A(j\omega)| = K \frac{|j\omega| \, |j\omega - p_1|}{|j\omega - p_2|^2} \tag{3-12}$$

Each of these factors may be measured with a ruler from the point corresponding to the frequency in question, ω, to the appropriate singularity. Hence $|j\omega|$ is measured from the origin to ω; $|j\omega - p_1|$ is measured from p_1 to ω. The resulting distances are substituted into Eq. (3-12) to find $|A(j\omega)|$.

For the phase we write

$$\underline{/A(j\omega)} = \underline{/j\omega} + \underline{/j\omega - p_1} - 2\,\underline{/j\omega - p_2} \tag{3-13}$$

The construction for finding the angle is shown in Fig. 3-11 for a single pole (or zero) singularity; the angle indicated as $\underline{/jω - p_1}$ is simply measured with a protractor. The geometric method is equally applicable to functions containing complex poles and zeros.

3-3 High-frequency Response of Pentode Stage. With the methods just described for quickly determining the important steady-state properties of a gain function, we may now go on and briefly investigate the high-frequency cutoff of an audio or video amplifier. Referring again to Fig. 3-3, which is the complete pentode equivalent circuit omitting the effects of the bias impedances, we remember that in the calculation of the low-frequency behavior we ignored the effect of C_1 and C_2 because their shunting effect became less and less as the frequency was reduced; at the same time, C_{cc} became more and more important because its reactance is in *series* with the signal. Considering the high-frequency end of the passband, however, the effect of C_{cc} becomes negligible because it acts more like the desired short circuit than at midband. The reactance of the *shunting* capacitance ($C_1 + C_2 \stackrel{\Delta}{=} C$) sooner or later begins to divert some of the signal current g_mV_g to ground as the frequency is increased. Hence an equivalent circuit to represent the pentode amplifier at high frequencies can regard C_{cc} as a short circuit but must retain C; the resulting equivalent circuit is shown in Fig. 3-12a. The gain of the stage is thus

$$A(p) = \frac{-g_m}{C} \frac{1}{p + 1/RC} \tag{3-14}$$

where

$$R \stackrel{\Delta}{=} \frac{1}{1/r_p + 1/R_L + 1/R_g} \cong R_L \qquad r_p \gg R_L \ll R_g \tag{3-15}$$

$$C \stackrel{\Delta}{=} C_1 + C_2$$

The gain function thus has a simple pole at $p = -1/RC \cong -1/R_LC$, giving the pole-zero diagram shown in Fig. 3-12b.

The amplitude response as a function of $jω$ can be determined from Eq. (3-14) by substituting $p = jω$,

$$A(jω) = \frac{-g_m}{C} \frac{1}{jω + 1/RC} = \frac{-g_mR}{jωRC + 1} \tag{3-16}$$

As in the case of the low-frequency response [Eq. (3-3)] there is a cutoff frequency, in this case $ω_2 = 1/RC$, and also at frequencies much below this there is the same limiting gain $A_0 = -g_mR \cong -g_mR_L$. Hence a normalized version of Eq. (3-16) can be written as follows:

$$\frac{A[j(ω/ω_2)]}{A_0} = \frac{1}{j(ω/ω_2) + 1} \tag{3-16a}$$

The amplitude response can be plotted readily on a logarithmic basis by means of the straight-line asymptotes shown in Fig. 3-12c. At low frequencies the gain is constant, and at high frequencies it falls away at 20 db/decade. The asymptotes intersect at the cutoff frequency ω_2. At this cutoff frequency the reactance of the shunting capacitance C is equal to the resistance R in parallel with it; thus the magnitude of the parallel impedance

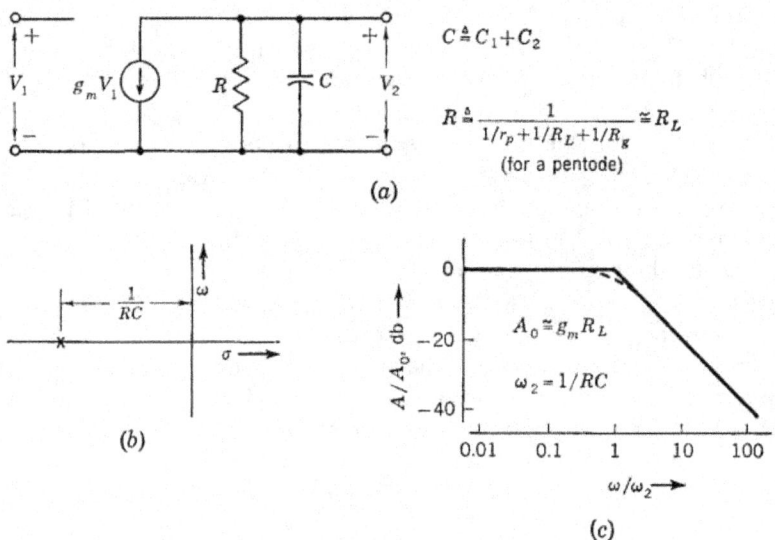

$$C \triangleq C_1 + C_2$$

$$R \triangleq \frac{1}{1/r_p + 1/R_L + 1/R_g} \cong R_L$$

(for a pentode)

(a)

$$\frac{1}{RC}$$

(b)

$$A_0 \cong g_m R_L$$

$$\omega_2 = 1/RC$$

(c)

Fig. 3-12 (a) An equivalent circuit for computing the h-f response of a pentode stage. (b) The pole-zero diagram for (a). (c) The resulting amplitude response.

of the combination is $0.707R$ (or -3 db in logarithmic units), and the phase angle is $45°$.

Note that the gain as ω becomes much less than $1/R_L C$ (the upper band-edge frequency) is the same as that given by Eq. (3-1) for ω much greater than $1/R_g C_{cc}$ (the lower band-edge frequency). Hence, if $1/R_L C \gg 1/R_g C_{cc}$, as is always the case in a wideband amplifier, the gain is constant over a large region and is given as a limiting case of either Eq. (3-1) or Eq. (3-14). The fact that the high-frequency solution indicates constant gain in the frequency region of the low-frequency cutoff (and vice versa) is actually the proof that the two solutions may be made separately. To find the over-all gain function of the stage, the two individual solutions [Eqs. (3-1) and (3-14)] are multiplied together. (The constant must be such as to give the correct midband gain.)

$$A(p) = \frac{-g_m}{C} \frac{1}{p + 1/R_L C} \frac{p}{p + 1/R_g C_{cc}} \qquad (3\text{-}17)$$

The pole-zero diagram and the over-all response are shown in Fig. 3-13. The pole far out on the σ axis is seen to be the controlling factor in the high-frequency behavior of the amplifier, while the pole-zero combination near the origin determines the low-frequency behavior. Note that nothing on the pole-zero diagram indicates the absolute *magnitude* of the gain at any frequency but that all the information determining the shape of the gain and phase response is present. Also note that another way of stating the condition for separation of the low- and high-frequency equivalent circuits is that the singularities determining the high-frequency response (the pole at

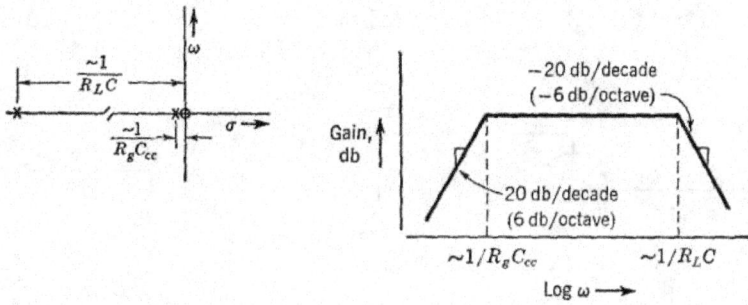

Fig. 3-13 The response of a complete pentode amplifier stage (exclusive of biasing impedances).

$1/R_LC$) must be far away (in terms of the *ratio* of frequencies) from the singularities determining the low-frequency response.

Notice once more that poles or zeros at the origin or on the real σ axis, as in Figs. 3-10, 3-12, and 3-13, contribute in similar fashion to the amplitude response if one works with the asymptotic approximation. A pole provides an amplitude response which *decreases* at a rate of 20 db/decade at high frequencies; a zero provides an *increase* of 20 db/decade. Each of these high-frequency asymptotes intersects the 0-db reference at the appropriate "cutoff" frequency, which is equal numerically to the σ coordinate of the pole or zero. The individual asymptotes add on a logarithmic plot to provide the over-all response of a given gain function, as in Figs. 3-6, 3-8, 3-12, and 3-13. The phase responses add in a similar fashion, as in Figs. 3-7 and 3-9.

The essence of the preceding discussion is to divide the calculation of the response of a lowpass amplifier into three parts, one part determining the low-frequency cutoff and one part the high-frequency cutoff. A third part, which we shall consider next, determines the gain in the region in which reactive effects are negligible. This is the so-called midband region.

3-4 Midband Properties of the Common Amplifier Configurations. Since a vacuum tube or transistor is usually considered to have three useful terminals (control grid, cathode, and plate, or base, emitter, and collector), three configurations are possible with one terminal grounded. Taking the vacuum tube first, these are shown in Fig. 3-14 together with their equivalent circuits. (The tube may be considered to be either a triode or a pentode with suitable values of r_p and μ.) The important properties of these circuits, which are tabulated in Table 3-2 and should be well known, are the voltage gain, the output impedance (the impedance looking into the terminals to which R_L is connected), and the input impedance. The equivalent circuits shown omit all reactive elements; hence the values in Table 3-2 are valid only at midband. Several useful points should be noted in this table. The grounded-plate stage gives a voltage gain of less than 1 but a power gain greater than 1. The voltage gain of both the

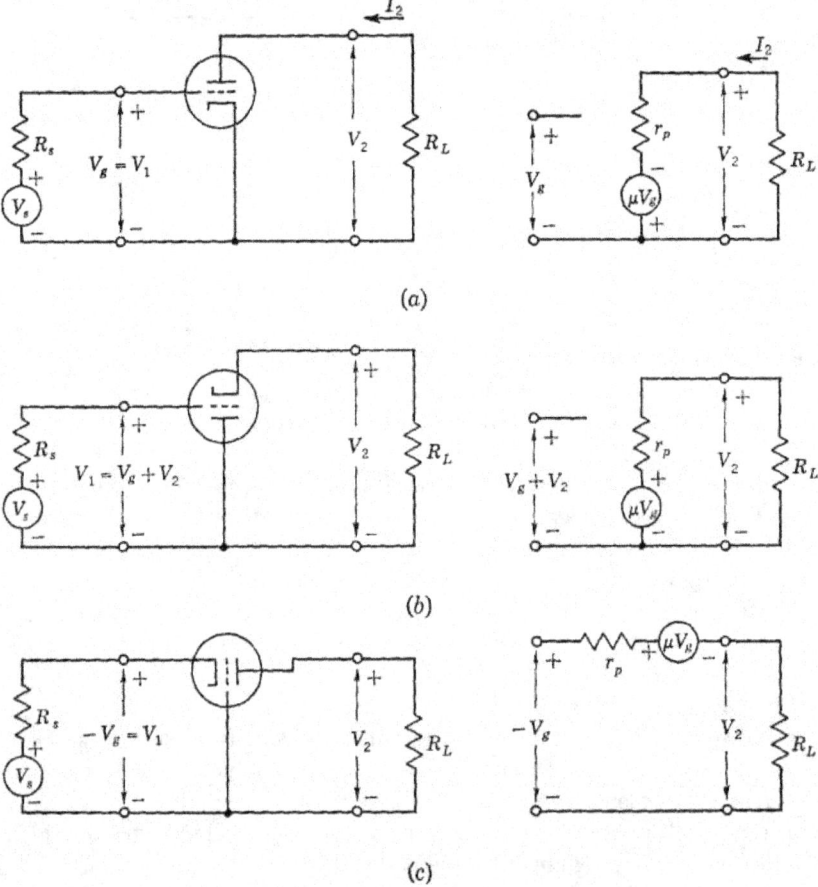

Fig. 3-14   The three basic vacuum-tube amplifier configurations.   (a) Grounded cathode.   (b) Grounded plate (cathode follower).   (c) Grounded grid.

Table 3-2 *

	Grounded cathode	Grounded grid	Grounded plate
$VG \triangleq \dfrac{V_2}{V_1}$	$\dfrac{\mu R_L}{r_p + R_L} = \dfrac{g_m R_L}{1 + R_L/r_p}$ $\cong g_m R_L \quad (R_L \ll r_p)$	$\dfrac{(\mu + 1) R_L}{r_p + R_L}$	$\dfrac{\mu R_L}{(\mu + 1) R_L + r_p}$ $\cong \dfrac{g_m R_L}{1 + g_m R_L} \quad (\mu \gg 1)$
R_{in}	∞	$\dfrac{r_p + R_L}{\mu + 1}$ $\cong \dfrac{1}{g_m} \quad (R_L \ll r_p)$	∞
R_{out}	r_p	$(\mu + 1) R_s + r_p$	$\dfrac{r_p}{\mu + 1}$ $\cong \dfrac{1}{g_m} \quad (\mu \gg 1)$
Transducer gain (power)	$\dfrac{4\mu^2 R_L R_s}{(r_p + R_L)^2}$ $\cong 4 g_m{}^2 R_L R_s \quad (R_L \ll r_p)$	$\cong \dfrac{4 R_L/R_s}{[1 + (r_p + R_L)/\mu R_s]^2} \; (\mu \gg 1)$ $\cong \dfrac{4 R_L/R_s}{(1 + 1/g_m R_s)^2} \quad (R_L \ll r_p)$	$\dfrac{4\mu^2 R_L R_s}{[(\mu + 1) R_L + r_p]^2}$ $\cong \dfrac{4 g_m{}^2 R_L R_s}{1 + g_m R_L} \quad (\mu \gg 1)$

* See Fig. 3-14.

grounded-grid and grounded-cathode stages is the same, but because of the relatively low input impedance of the former, the grounded-grid stage gives less power gain. Finally, if a low output impedance is desired, the grounded-plate stage (cathode follower) is best and the grounded-grid stage is worst. Note also the approximate formulas and the limits on their validity. The formulas requiring $R_L \ll r_p$ are almost always accurate in practice when the tube is a pentode.

The equivalent-circuit representation for the transistor is somewhat more difficult than for the vacuum tube, and so the various quantities of interest will be developed here in an approximate manner. The exact equations may be developed by a straightforward loop or nodal analysis, but the resulting equations are frequently so cumbersome as to be useless in increasing one's understanding of the fundamentals of transistor amplifier operation. However, use of the approximate relations can be had only if one thoroughly understands the approximations which have been made.

Consider first the *common-base* equivalent circuit shown in Fig. 3-15. The various known currents are drawn in the diagram, on the assumption that the voltage across r_b is much less than that across R_L. This is very reasonable if the stage is an amplifier stage. The equation for V_1 may be written by inspection once the currents are known.

$$V_1 \cong r_e I_1 + r_b I_1 \left[(1 - \alpha) + \frac{\alpha R_L}{r_c + R_L} \right] \tag{3-18}$$

Since $R_L \ll r_c$ usually, the equation for R_{in} may be written

$$R_{in} = \frac{V_1}{I_1} \cong r_e + r_b(1 - \alpha) \qquad \frac{R_L}{r_c + R_L} \ll (1 - \alpha) \qquad (3\text{-}19)$$

For the same conditions, the current gain is

$$\frac{I_2}{I_1} \cong \frac{-\alpha r_c}{r_c + R_L} \cong -\alpha \qquad \text{if } R_L \ll r_c \qquad (3\text{-}20)$$

The voltage gain A is

$$A \cong \frac{\alpha r_c R_L}{[r_e + r_b(1 - \alpha)](r_c + R_L)} \qquad (3\text{-}21)$$

$$\cong \frac{\alpha R_L}{r_e + r_b(1 - \alpha)} \qquad \text{if } R_L \ll r_c \qquad (3\text{-}22)$$

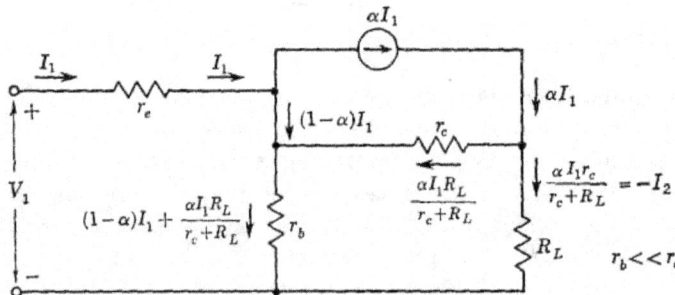

Fig. 3-15 Approximate currents in a common-base (CB) amplifier.

In the common-base connection the voltage gain per stage is approximately α, if stages are cascaded without transformers so that $R_L = R_{in}$. Consequently common-base stages are rarely used in wideband amplifiers because of the difficulty of making the necessary wideband transformer.

The power gain of the transistor itself is

$$PG = \frac{P_o}{P_{in}} = \left(\frac{I_2}{I_1}\right)^2 \frac{R_L}{R_{in}} \cong \frac{\alpha^2 R_L}{r_e + r_b(1 - \alpha)} \qquad (3\text{-}23)$$

Again this quantity will be less than 1 if $R_L = R_{in}$.

A similar analysis may be made for the *common-emitter* circuit with the aid of Fig. 3-16. In this figure I_1 is assumed; the other currents are found by simply applying Kirchhoff's law to each node in turn and assuming that the voltage across r_e is much less than that across R_L (which is always true if the stage amplifies). From the known currents the following equations may be written by inspection:

$$R_{\text{in}} = \frac{V_1}{I_1} \cong r_b + \frac{r_e}{1 - \alpha} - \frac{r_e R_L \beta}{r_c(1 - \alpha) + R_L} \qquad \text{if } R_L \ll r_c(1 - \alpha) \qquad (3\text{-}24)$$

$$\cong r_b + \frac{r_e}{1 - \alpha} \tag{3-25}$$

$$I_2 \cong \beta I_1 \tag{3-26}$$

$$PG = \left(\frac{I_2}{I_1}\right)^2 \frac{R_L}{R_{\text{in}}} \cong \frac{\beta^2 R_L}{r_b + r_e/(1 - \alpha)} \tag{3-27}$$

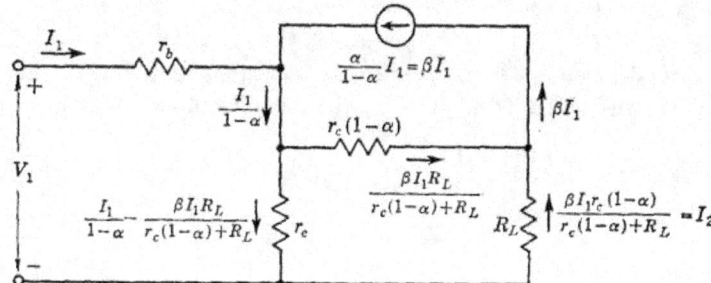

Fig. 3-16 Approximate currents in a common-emitter (CE) amplifier.

Since R_{in} is usually much less than $r_c(1 - \alpha)$, the above approximations are fulfilled in a cascade of common-emitter stages, except possibly for the last stage, where R_L may be fairly large. Such a cascade is shown in Fig. 3-17, where any resistors associated with the collector or base are assumed large compared with R_{in}. The power gain of the circuit in Fig. 3-17 is

$$PG = \left(\frac{\beta_1 \beta_2 \beta_3 I_1}{I_1}\right)^2 \frac{R_L}{R_{\text{in}}} = \frac{\beta_1{}^2 \beta_2{}^2 \beta_3{}^2 R_L}{R_{\text{in}}} \tag{3-28}$$

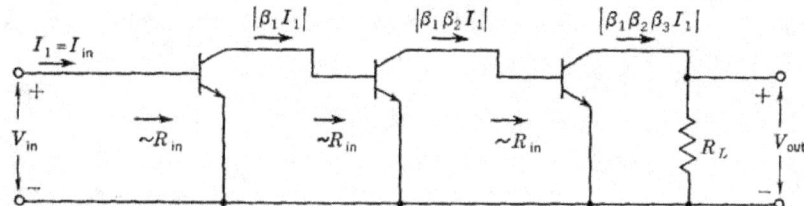

Fig. 3-17 Elementary amplifier made up of common-emitter amplifiers.

This is actually the maximum possible gain for the combination since in practice some of the collector current will flow into the collector load resistances and the biasing resistances. However, the preceding equation gives the order of magnitude of the over-all gain and is therefore quite useful. The common-emitter amplifier is probably the configuration most often used since it gives a large gain without the use of transformers.

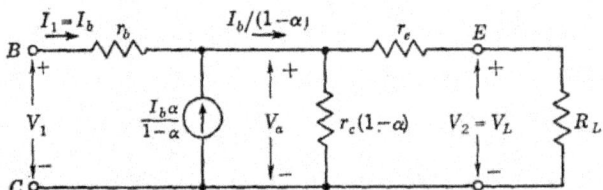

Fig. 3-18 Equivalent circuit for a common-collector (CC) stage.

The equivalent circuit for a *common-collector* stage including R_L may be drawn as shown in Fig. 3-18. The input resistance of the stage is

$$R_{in} = \frac{V_1}{I_b} = (r_b I_b + V_a) \frac{1}{I_b} \tag{3-29}$$

$$V_a = \frac{I_b}{1 - \alpha} \frac{r_c(1 - \alpha)(r_e + R_L)}{r_c(1 - \alpha) + r_e + R_L} \tag{3-30}$$

$$R_{in} = r_b + \frac{1}{1 - \alpha} \frac{r_c(1 - \alpha)(r_e + R_L)}{r_c(1 - \alpha) + r_e + R_L}$$

$$\cong \frac{r_c(r_e + R_L)}{r_c(1 - \alpha) + R_L} + r_b \tag{3-31}$$

If $r_c(1 - \alpha) \gg R_L$, as is frequently the case, then

$$R_{in} \cong \frac{r_e + R_L}{1 - \alpha} + r_b \cong \frac{R_L}{1 - \alpha} \qquad R_L \gg r_e \tag{3-32}$$

Hence, the input impedance may be made quite large. The voltage gain is

$$A = \frac{V_2}{V_1} = V_a \frac{R_L}{r_e + R_L} \frac{1}{V_1} \tag{3-33}$$

$$\cong \frac{r_e + R_L}{1 - \alpha} \frac{R_L}{r_e + R_L} \frac{1}{r_b + \dfrac{r_e + R_L}{1 - \alpha}} \qquad \text{if } r_c(1 - \alpha) \gg R_L \tag{3-34}$$

$$A \cong \frac{R_L}{(1 - \alpha)r_b + r_e + R_L} \tag{3-35}$$

$$\cong 1 \quad \text{if } r_b(1 - \alpha) + r_e \ll R_L \ll r_c(1 - \alpha) \tag{3-36}$$

In some applications the precise difference between the true voltage gain A and unity is of importance (as, for example, in the case of some active filters). A quite exact expression for this case is

$$1 - A \cong \frac{1}{1 + \dfrac{R_L}{r_b(1 - \alpha) + r_e + \dfrac{r_b R_L}{r_c}}} \quad \text{good for any } R_L \tag{3-37}$$

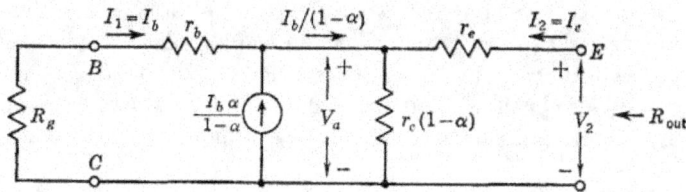

Fig. 3-19 A common-collector stage driven from source impedance of R_g.

The maximum value of voltage gain obtained with infinite R_L is thus $A = r_c/(r_b + r_c)$, which for a typical transistor is about $0.999+$. The power gain is

$$PG = \frac{V_2{}^2/R_L}{V_1{}^2/R_{\text{in}}} \cong \frac{R_L}{[(1 - \alpha)r_b + r_e + R_L](1 - \alpha)} \tag{3-38}$$

$$\cong \frac{1}{1 - \alpha} \quad \text{if } r_c(1 - \alpha) \gg R_L \gg r_e \tag{3-39}$$

Another quantity of interest is the impedance seen by the load when the common-collector stage is driven by a given source impedance, as in Fig. 3-19.

If we ignore $r_c(1 - \alpha)$ because it is usually large compared with $R_g + r_b$, we get

$$V_a = -I_b(R_g + r_b) \tag{3-40}$$

$$I_2 \cong \frac{-I_b}{1 - \alpha} \tag{3-41}$$

$$V_2 \cong V_a - \frac{I_b}{1-\alpha} r_e \cong -I_b(R_g + r_b) - \frac{r_e I_b}{1-\alpha} \qquad (3\text{-}42)$$

$$R_{\text{out}} \cong \frac{V_2}{I_2} = \frac{-I_b(R_g + r_b) - r_e I_b/(1-\alpha)}{-I_b/(1-\alpha)}$$

$$\cong (R_g + r_b)(1-\alpha) + r_e$$

$$\cong \frac{R_g}{\beta} \qquad \text{for } R_g \gg r_b \qquad (3\text{-}43)$$

The common-collector stage has the properties of an impedance transformer; i.e., the input impedance is roughly βR_L, and the output impedance is roughly R_g/β. However, this is accomplished with a voltage gain of almost unity and a considerable power gain. Thus, the circuit operates like a cathode follower but has a gain A nearer unity. For this reason the common collector is often called an *emitter follower*. Similarly, the circuit is useful for coupling a high-impedance source to a lower-impedance load. It may also be noted that the common-emitter and common-cathode stages are somewhat analogous, as are the common-base and common-grid stages.

3-5 D-C Bias Considerations. Proper d-c biasing of tubes and transistors is important because the bias determines the operating point of the active device. A good bias circuit is one which maintains the proper operating point during the life of the device and when the device is replaced. From the standpoint of amplifier theory, biasing is important because the bias circuit almost inevitably introduces parasitic elements which affect the signal response of the amplifier.

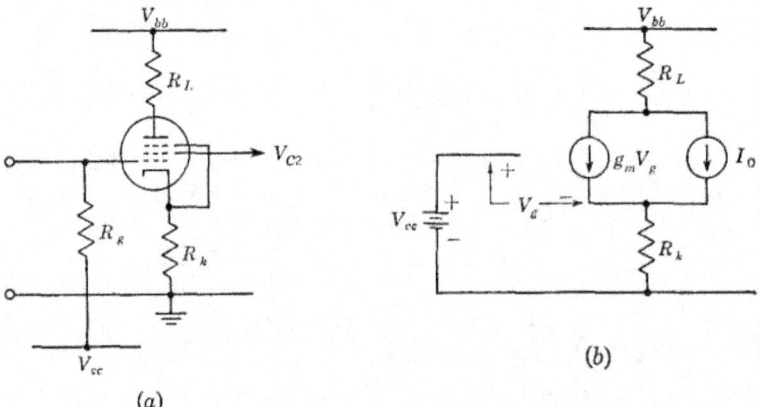

(a)   (b)

Fig. 3-20 Vacuum-tube cathode bias circuit.

Pentode Amplifier. To take first the simpler case of the vacuum tube, a basic and almost universally used biasing scheme is shown in Fig. 3-20a. The voltage V_{cc} may be of either polarity or zero; R_k is the cathode bias resistor. A somewhat simplified equivalent circuit is shown in Fig. 3-20b, where the pentode is represented by two current generators: the $g_m V_g$ generator is the ordinary small-signal current generator, and the generator I_0 is an additional current generator such that the d-c plate current I_b $= g_m V_g + I_0$. (These quantities may be found from the static plate curves of a tube, $g_m = \Delta I_b / \Delta V_g$, taken near the operating point I_b and V_g. From the calculated value of g_m and the known values of V_g and I_b, I_0 may be calculated with the preceding equation.) For this discussion r_p is considered to have negligible effect. The circuit is solved for I_b, the d-c plate current, resulting in

$$I_b = \frac{g_m V_{cc} + I_0}{1 + g_m R_k} = \frac{V_{cc}/R_k + I_0/g_m R_k}{1 + 1/g_m R_k} \qquad (3\text{-}44)$$

The plate current can be made most stable by making $g_m R_k$ large so that $I_b \cong V_{cc}/R_k$. This usually necessitates making V_{cc} a few volts positive so that the desired I_b is obtained. The resulting plate current is very stable with respect to changes in both I_0 and g_m, making this method of operation desirable with high-performance pentodes, where both these quantities are likely to be highly variable. In the simplest and usual case, $V_{cc} = 0$, and a moderate degree of bias stability is obtained because both I_0 and g_m tend to change together. If $R_k = 0$ and $V_{cc} < 0$, so-called "fixed bias" results and the plate current is extremely dependent upon both g_m and I_0 since both tend to change I_b in the same direction. Consequently this method of operation should not be used where much reliability or stability is required.

Because adequate bypassing of R_k down to very low frequencies is sometimes very difficult, the feedback biasing circuit of Fig. 3-21 is a practical alternative if both positive and negative supplies are available. In practice, the current through the R_1 branch is much smaller than the current through R_L, and V_g is only a very few volts; hence the approximate plate current and voltage are

$$V_b \cong \frac{-V_{cc}(R_1 + R_2)}{R_3} \qquad (3\text{-}45)$$

$$I_b \cong \frac{V_{bb} - V_b}{R_L} \qquad (3\text{-}46)$$

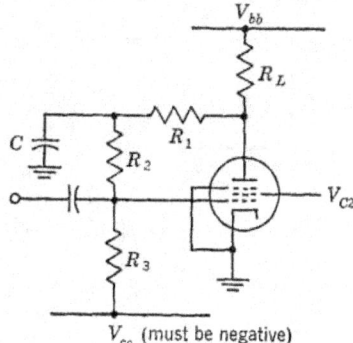

Fig. 3-21 Feedback biasing circuit.

The capacitor C is added to make the feedback inoperative at signal frequencies in the passband of the amplifier. Omitting C reduces the input impedance of the stage. This particular circuit is exactly the same in operation as the transistor circuit shown in Fig. 3-26.

Transistor Amplifier. In the practical realization of a transistor amplifier with proper d-c biasing the amplifier may be expected to operate satisfactorily over a wide range of temperatures, and a number of different transistors may be used interchangeably. On the other hand, improper biasing may cause the amplifier to become inoperative at elevated temperatures or in some cases may even destroy the transistor.

Three factors cause difficulty in transistor biasing: variation of β_{DC} (ratio of d-c collector current to d-c base current) from unit to unit and with temperature, variations in the temperature-sensitive current I_{C0} (I_{C0} is the collector current with open emitter, $I_E = 0$), and changes in V_{BE} due to temperature. These factors all cause changes in the base current I_B for a given collector current. Since a resistance is usually necessary in the base circuit, variation of base current leads to a change in base voltage, thereby changing the operating point of the transistor. Consequently a circuit is desired which produces little change in operating point (I_C and V_{CB}) with changes in base current. A change in V_{BE} has relatively little effect on the bias circuits, because V_{BE} is small compared with the voltage drop in the resistor which determines the emitter current.

In the following discussion only changes induced by I_{C0} will be discussed since these are most important in germanium transistors. In silicon transistors I_{C0} is not important in most circuits until temperatures of 100°C or higher are reached, but changes in β_{DC} *are* important. However, a circuit which has low sensitivity to I_{C0} is also insensitive to changes in β_{DC}, owing either to temperature or to replacing transistors.

The current I_{C0} is a result of the thermal generation of minority carriers in the collector and base regions. These minority carriers are drawn across the reverse-biased collector junction because the junction field is in the direction to attract them across the junction. The current I_{C0} is thus a d-c component which flows from collector to base, as in Fig. 3-22. The current gain α_{DC} is the current transfer ratio (sometimes called h_{FB} or α_{FB}) involving the total emitter and collector currents. If we neglect the collector resistance, which is usually quite high in a junction transistor, we may write the following

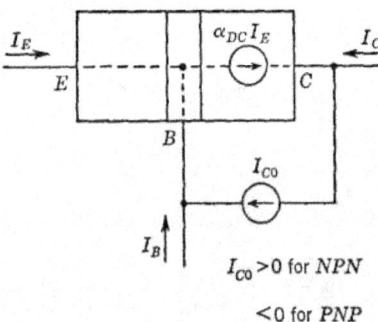

Fig. 3-22 Simple model for I_{C0} calculations.

equation for I_C (this equation may also be used as the defining equation for α_{DC}):[1]

$$I_C = I_{C0} - \alpha_{DC} I_E \qquad (3\text{-}47)$$

This assumes that α_{DC} is constant over the operating range and that I_C is independent of V_C.

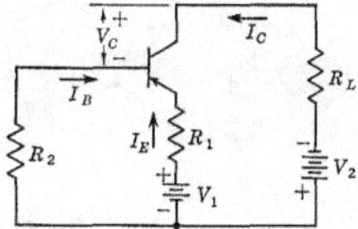

Fig. 3-23 A general transistor biasing circuit.

Let us now investigate the "d-c" characteristics of a basic biasing circuit, as in Fig. 3-23. This circuit is sufficiently general to enable us to analyze the common biasing schemes. Let us also assume that the emitter-to-base voltage is zero; then we can write

$$V_1 = R_1 I_E - R_2 I_B \qquad (3\text{-}48)$$

$$V_2 = -V_C + R_2 I_B - R_L I_C \qquad (3\text{-}49)$$

$$I_C = -I_B - I_E \qquad (3\text{-}50)$$

$$I_C = I_{C0} - \alpha_{DC} I_E \qquad (3\text{-}51)$$

Expressing I_E and I_B in terms of I_C, I_{C0}, and α, we get

$$I_E = \frac{I_{C0} - I_C}{\alpha_{DC}} \qquad (3\text{-}52)$$

$$I_B = -I_C - I_E = \frac{I_C(1 - \alpha_{DC}) - I_{C0}}{\alpha_{DC}} \qquad (3\text{-}53)$$

Substitute these values in Eq. (3-48), and solve for I_C.

$$I_C = \frac{I_{C0}(R_1 + R_2)}{R_1 + R_2(1 - \alpha_{DC})} - \frac{\alpha_{DC} V_1}{R_1 + R_2(1 - \alpha_{DC})} \qquad (3\text{-}54)$$

By differentiating I_C with respect to I_{C0} we get a measure of the sensitivity of I_C to changes in I_{C0}.

$$\frac{\partial I_C}{\partial I_{C0}} = \frac{R_1 + R_2}{R_1 + R_2(1 - \alpha_{DC})} \overset{\Delta}{=} S \qquad (3\text{-}55)$$

Equation (3-55) tells us the magnification of changes in I_{C0} which appear

[1] The following is adapted from chap. 3 of R. F. Shea (ed.), "Transistor Circuit Engineering," John Wiley & Sons, Inc., New York, 1957.

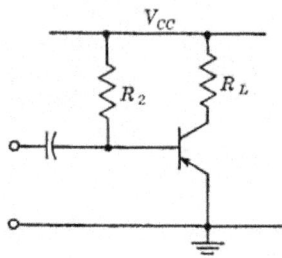

Fig. 3-24 An amplifier with "fixed-base bias."

in I_C. The best condition is to make $R_2 = 0$; then $S = 1$. This is obvious, since then I_{C0} cannot change I_E; hence the component $I_C = -\alpha_{DC}I_E$ is a constant. The worst situation is to make $R_1 = 0$. Such an amplifier is shown in Fig. 3-24. In such an amplifier I_B is constant as I_{C0} changes, but I_C changes by $I_{C0}/(1 - \alpha_{DC})$. In a typical transistor I_{C0} at 25°C might be -10 μa. Since I_{C0} doubles, approximately, for each 10°C temperature rise, at 65°C $I_{C0} = -(2^4) \times 10 = -160$ μa. Let us assume that $\alpha_{DC} = 0.98$ and $I_C = -1$ ma (at 25°C).

From Eq. (3-53) I_B at 25°C is

$$I_B = \frac{I_C(1 - \alpha_{DC}) - I_{C0}}{\alpha_{DC}} = \frac{-(1)(0.02) + 0.01}{0.98}$$

$$= -0.0102 \text{ ma} \tag{3-56}$$

The collector current at $T = 65°C$ may be found by using the above value of I_B, and I_{C0} as found previously.

$$I_C = \frac{\alpha_{DC}I_B + I_{C0}}{1 - \alpha_{DC}} = \frac{(0.98)(-0.0102) + (-0.160)}{0.02}$$

$$= -8.5 \text{ ma} \tag{3-57}$$

This amount of collector current could very easily cause the transistor to saturate ($V_{CE} \cong 0$) and make the amplifier completely inoperative.

A defect of the basic circuit in Fig. 3-23 is that two batteries are required. A simple transformation (Fig. 3-25) will take us to a more usual circuit. For this circuit S is obtained by substituting $R_2 = R_aR_b/(R_a + R_b)$ into Eq. (3-55).

$$S = \frac{R_1/R_a + R_1/R_b + 1}{R_1/R_a + R_1/R_b + (1 - \alpha_{DC})} \tag{3-58}$$

Note that for small S the resistance $R_a \parallel R_b$ should be small. The circuit in Fig. 3-25 is useful in the design of an amplifier if the desired operating point and S are known.

Other schemes for stable biasing may be developed. The one shown in Fig. 3-26 is a feedback method of decreasing the effect of I_{C0}. The details are given in the reference below and in Prob. 3-7.[1]

[1] A. W. Lo et al., "Transistor Electronics," pp. 134ff., Prentice-Hall, Inc., Englewood Cliffs, N.J., 1955.

If I_C tends to increase, the action of this circuit is to decrease the base current, thus tending to hold I_C constant.   The circuit is useful where a moderate degree of stabilization is required and where a minimum number

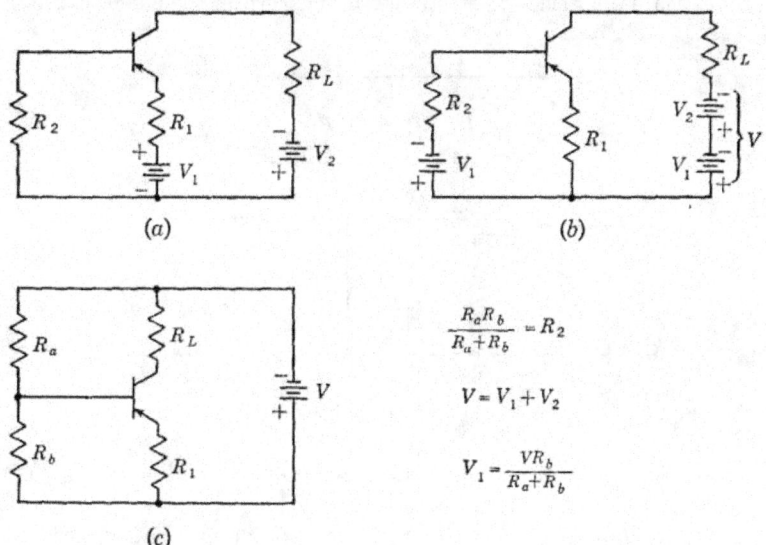

(a)   (b)

(c)

$$\frac{R_a R_b}{R_a + R_b} = R_2$$

$$V = V_1 + V_2$$

$$V_1 = \frac{V R_b}{R_a + R_b}$$

Fig. 3-25 Development of single-battery biasing from Fig. 3-23.

of components must be used.   Note that the circuit works only when $I_{C0} < I_C(1 - \alpha_{DC})$ [see Eq. (3-53)].

Biasing circuits which have small S also tend to reduce changes in I_C due to substituting transistors with different β_{DC}.   This is quite important

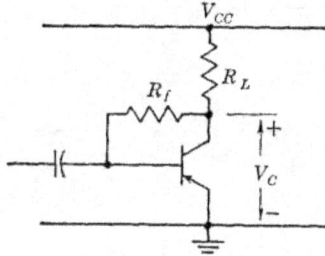

Fig. 3-26 A method of feedback biasing (cf. Fig. 3-21).

since different transistors of the same type may easily have values of β_{DC} differing by two or three times.

3-6 Determination of the Low-frequency Response Due to Bias Impedances. Considering the complete pentode amplifier stage, shown in Fig. 3-27 without the components affecting only the high-frequency response, the reactance of capacitors C_{cc}, C_s, and C_k will decrease the low-frequency gain and introduce phase shift. The effect of C_{cc} has already been investigated in Sec. 3-1. Here we shall ascertain the effect of the bias

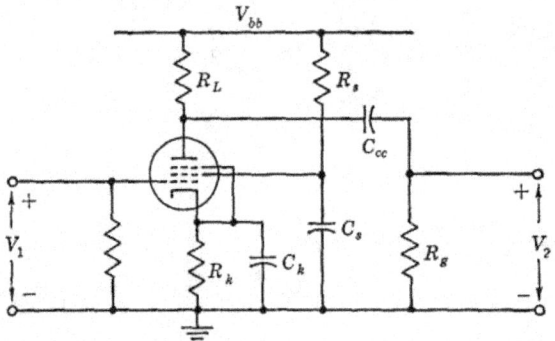

Fig. 3-27 A complete pentode stage for low-frequency analysis.

impedances separately and then combine the effects to find the response due to all three.

Cathode Impedance. Considering first the cathode bias impedance alone, the equivalent circuit shown in Fig. 3-28 may be drawn. A simple and useful way to investigate this circuit is to determine the y parameters for

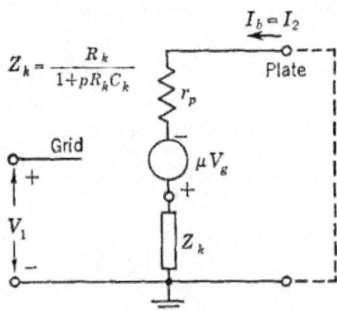

Fig. 3-28 Equivalent circuit for determining the effect of Z_k.

the tube and Z_k. Both y_{11} and y_{12} are zero, since $I_1 = 0$ at low frequencies. The transfer admittance y_{21} is found by determining the short-circuit current $I_b = I_2$,

$$I_2 = \frac{\mu V_g}{r_p + Z_k} \tag{3-59}$$

$$V_g = V_1 - V_k = V_1 - I_2 Z_k \tag{3-60}$$

Therefore

$$I_2 = \frac{\mu(V_1 - I_2 Z_k)}{r_p + Z_k} \tag{3-61}$$

$$\frac{I_2}{V_1} = y_{21} = \frac{1}{\dfrac{r_p}{\mu} + \dfrac{(\mu+1)Z_k}{\mu}} \cong \frac{g_m}{1 + g_m Z_k} \qquad \text{if } \mu + 1 \cong \mu \quad (3\text{-}62)$$

The approximate equation is excellent for pentodes since μ is then very large. The impedance looking into the output terminals may be found by

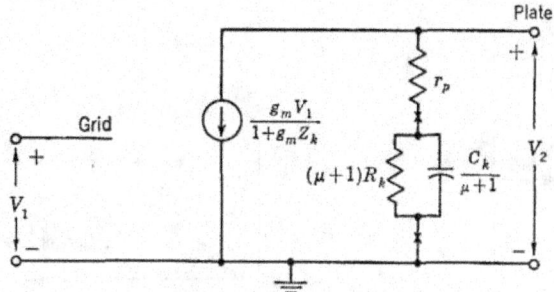

Fig. 3-29 Equivalent circuit for pentode or triode with nonzero cathode bias impedance.

dividing the open-circuit output voltage by the short-circuit current,

$$V_{2oc} = -\mu V_1 \tag{3-63}$$

$$\frac{1}{y_{22}} = \frac{-V_{2oc}}{I_2} = r_p + (\mu+1)Z \tag{3-64}$$

The new equivalent circuit thus becomes that shown in Fig. 3-29. In the pentode, r_p is usually so large that it is ignored in comparison with the load resistor, and so the equivalent circuit is effectively just a modified current generator $g_m/(1 + g_m Z_k)$. Note that, if Z_k is negligible, the equivalent circuit becomes the usual one for the tube alone. By substituting the value of $Z_k = R_k/(1 + pR_kC_k)$ into Eq. (3-62) a modified value of g_m is found,

$$y_{21} = g'_m = \frac{g_m}{1 + g_m R_k} \frac{pR_kC_k + 1}{p[R_kC_k/(1 + g_m R_k)] + 1} \tag{3-65}$$

$$= g_m \frac{p + 1/R_kC_k}{p + [(1 + g_m R_k)/R_kC_k]} \tag{3-65a}$$

At high frequencies $g'_m \to g_m$, but for very low frequencies the effective transconductance is reduced to $g'_m \to g_m/(1 + g_m R_k)$. The pole-zero diagram for Eq. (3-65a) is shown in Fig. 3-30a. The effect of the pole and zero is to cause the gain to decrease for $\omega < (1 + g_mR_k)/R_kC_k$, as is shown in Fig. 3-30b. Two important points may be gained from Fig. 3-30b; the first is that the gain begins to fall at a frequency considerably higher than the natural frequency of the cathode circuit, $1/R_kC_k$. The second is that, since the frequency $(1 + g_mR_k)/R_kC_k$ is strongly dependent upon g_m if the

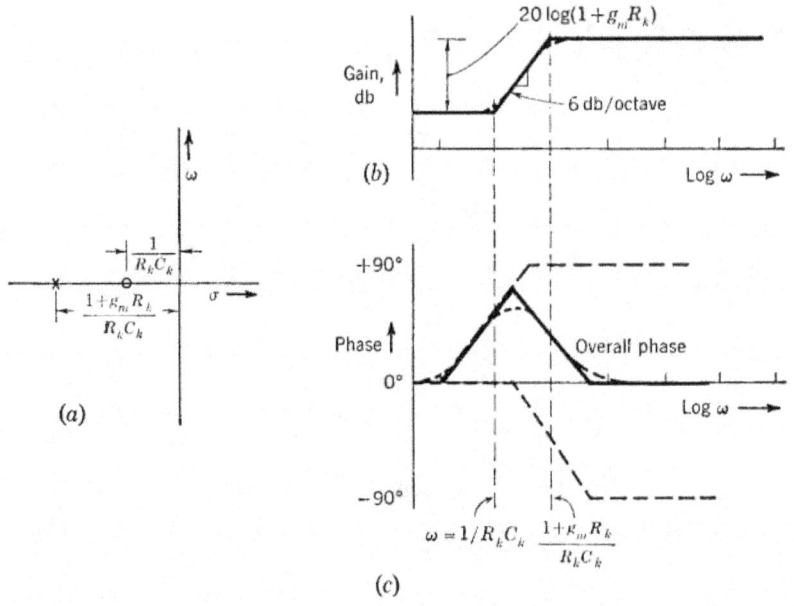

Fig. 3-30 Gain of a stage with cathode bias (no effects due to screen or coupling circuit). (a) Pole-zero diagram. (b) Gain magnitude. (c) Phase of output relative to the input.

stage is stably biased, the point at which the gain begins to decrease is not a particularly constant one because of the variability of g_m; hence there is little point in making elaborate calculations to find the exact low-frequency cutoff.

The phase shift caused by the cathode impedance is also of importance, and the contribution to the total phase due to each term in Eq. (3-65a) is

$$\underline{/A(j\omega)} = \underline{/g_m'} = \underline{/j\omega + 1/R_kC_k} - \underline{/j\omega + (1 + g_mR_k)/R_kC_k} \quad (3\text{-}66)$$

The resulting phase response is shown in Fig. 3-30c.

One final point remains to be clarified. In the preceding it has been assumed either that the screen is operated from a zero-impedance supply or that it has been bypassed by an infinitely large capacitor. In either case the effect of the cathode circuit is dependent upon the return path for the screen current. As shown in Fig. 3-31, the screen supply (or bypass capacitor) may be returned either to the cathode or to ground. In the former case (Fig. 3-31a) only the plate current flows through Z_k, and the normal pentode g_m is the correct transconductance to use in calculating the pole position. In the latter case (Fig. 3-31b), both the screen and plate current flow through Z_k, and the correct transconductance to use is the g_m of the tube connected as a triode (g_{mT}), which is approximately equal to

$$g_{mT} = \frac{I_b + I_{c2}}{I_b} g_m \qquad (3\text{-}67)$$

where I_b and I_{c2} are the d-c plate and screen currents, respectively. In

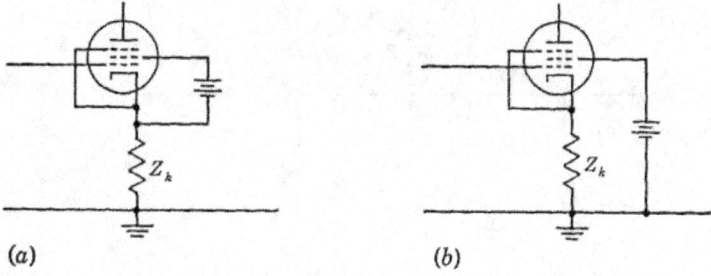

(a)   (b)

Fig. 3-31 Two methods of screen connection in pentode amplifiers.

either case, the effective g_m should reduce to the pentode g_m at high frequencies.

Emitter Bias Impedance. For the moment we shall digress from computing the effect of impedance in the screen of a pentode and find instead the effect of impedances in series with the emitter of a common-emitter transistor amplifier. We shall find that the effect is very similar to the cathode bias impedance so that we may draw upon the preceding discussion.

The transistor case is complicated by the fact that changes in the collector impedance have some small effect on the input to the transistor. These effects are usually of little consequence at low frequencies if the collector is loaded by at least the base circuit of a following transistor, as is the case in a chain of common-emitter stages. Consequently, in most cases the resistor $r_c(1 - \alpha)$ may be omitted and the equivalent circuit for the bias problem drawn as shown in Fig. 3-32.[1] All the impedances to the

---

[1] The resulting equivalent circuit for the transistor is the same as that using the h parameters and neglecting h_{22} and h_{12}. If necessary the effect of h_{22} may be approximated by adding a resistor of $1/h_{22}$ in shunt with the output current generator. The effect of h_{12} is usually quite small.

left of the base (including R_a and R_b in Fig. 3-25) are lumped into one
Thévenin equivalent R_g. It is desirable to split the equivalent circuit into
two separate loops, as shown in Fig. 3-32b. That the two circuits are
identical may be shown by writing the equation for the input loop in both
cases,

$$V_1 = I_1 R_g + I_1 r_b + \frac{r_e I_1}{1 - \alpha} + \frac{Z_e I_1}{1 - \alpha} \tag{3-68}$$

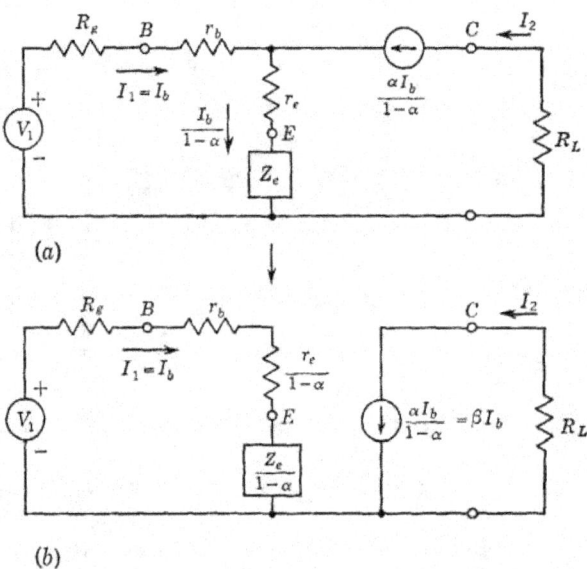

(a)

(b)

Fig. 3-32 Circuit for calculating the low-frequency cutoff due to the emitter biasing
circuit.

This equation is the same for both figures. By substituting Z_e we get

$$V_1 = I_1 \left[R_g + r_b + \frac{r_e}{1 - \alpha} + \frac{R_e/(1 - \alpha)}{1 + p R_e C_e} \right] \tag{3-69}$$

$$\underbrace{\phantom{R_g + r_b + \frac{r_e}{1 - \alpha} + \frac{R_e/(1 - \alpha)}{1 + p R_e C_e}}}_{R_s}$$

$$\frac{I_1}{V_1} = \frac{1}{R_s} \frac{p + 1/R_e C_e}{p + \dfrac{R_s + R_e/(1 - \alpha)}{R_s R_e C_e}} \tag{3-70}$$

Since the voltage gain of the amplifier is directly proportional to I_1, the above expression is directly proportional to the gain of the amplifier. Thus the presence of Z_e causes the gain of the stage to have a zero at $p = -1/R_eC_e$ and a pole at $p = -[R_s + R_e/(1 - \alpha)]/R_sR_eC_e$. This is the frequency at which the reactance of the effective emitter bypass capacitance $C_e(1 - \alpha)$ is equal to the parallel combination of R_s and $R_e/(1 - \alpha)$. Equation (3-70) has exactly the same pole-zero configuration as the equation describing the response due to a cathode bias impedance [Eq. (3-65)]; consequently the shape of the response is the same as shown in Fig. 3-30 with the new pole and zero frequencies inserted. Again note that the gain begins to drop at

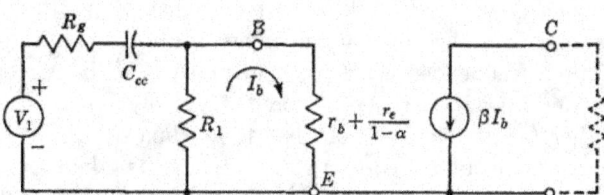

Fig. 3-33 Circuit for calculating the low-frequency cutoff due to the coupling circuit.

a frequency considerably higher than $\omega = 1/R_eC_e$, the natural frequency of the bias circuit itself. In the case of the transistor bias circuit, the source impedance driving the stage in question has an important effect on the frequency of the pole—the higher the source impedance, the lower the frequency at which the gain begins to fall. Notice that the source impedance usually cannot be increased without limit because of the necessity of having resistance from the base to ground to give bias stability, as shown in the preceding section.

Because the source impedance does affect the low-frequency cutoff, the effects of the coupling capacitor (C_{cc} in Fig. 3-2) and the bias impedance Z_e interact to some extent. Consequently the exact low-frequency cutoff cannot be found by considering the two circuits separately, as may be done in the case of the pentode. However, if the cutoff frequencies of the coupling and bias circuit are quite different, the approximate low-frequency cutoff may be found by taking the higher of the two frequencies. That is, the cutoff frequency of the bias circuit is found by considering C_{cc} to be infinite; then the coupling-circuit frequency is found by regarding C_e as infinite. The circuit for computing the latter frequency is shown in Fig. 3-33, where the transistor is represented as in Fig. 3-32b with $Z_e = 0$. The cutoff frequency for the coupling circuit is that for which the reactance of C_{cc} equals the generator resistance plus the parallel combination of R_1 and the input resistance of the transistor. The transfer admittance may be written as

$$\frac{I_b}{V_1} \triangleq Y_t = \frac{Kp}{p + 1/T} \tag{3-71}$$

$$T = \left\{ R_g + \frac{[r_b + r_e/(1 - \alpha)]R_1}{r_b + r_e/(1 - \alpha) + R_1} \right\} C_{cc} \tag{3-72}$$

$$K = \frac{R_1}{R_g[R_1 + r_b + r_e/(1 - \alpha)] + [r_b + r_e/(1 - \alpha)]R_1} \tag{3-73}$$

If the cutoff frequency $\omega = 1/T$ is considerably different from that computed for the bias circuit (say at least four times), then the cutoff frequency of the entire stage may be taken as the higher of the two cutoff frequencies. If the two cutoff frequencies nearly coincide, curves of the response due to the coupling and bypass circuits considered separately may be sketched and added together (on the assumption that the ordinates are in decibels) to find the approximate cutoff frequency for both circuits taken together. The cutoff frequency found in this manner in a typical case is about 20 per cent lower than the actual cutoff frequency.

3-7 Calculation of Low-frequency Cutoff Due to the Screen-grid Circuit. The calculation of the effect of impedance in the screen circuit requires knowledge of how changes in the screen voltage affect the plate current. Several methods may be used to calculate the effect of the screen, but a most straightforward method, which also brings out quantities which are usually neglected, is to write a set of three simultaneous equations which relate to Fig. 3-34.

$$I_1 = y_{11}V_1 + y_{12}V_2 + y_{13}V_3 \tag{3-74}$$

$$I_2 = y_{21}V_1 + y_{22}V_2 + y_{23}V_3 \tag{3-75}$$

$$I_3 = y_{31}V_1 + y_{32}V_2 + y_{33}V_3 \tag{3-76}$$

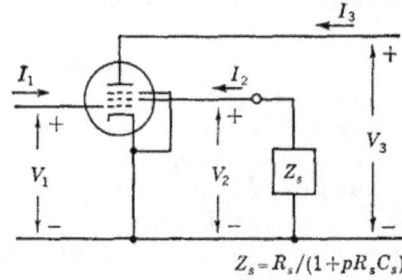

$$Z_s = R_s/(1 + pR_sC_s)$$

Fig. 3-34 Reference voltages and currents for pentode operation with ungrounded screen.

Each of the y's can be defined by measuring an appropriate current with all but one V equal to zero; for example, $y_{33} = I_3/V_3$ with V_1 and V_2 equal to zero. Hence $y_{33} = 1/r_p$, the normal plate resistance. For the negative-grid pentode at low frequencies $I_1 = 0$; hence $y_{11} = y_{12} = y_{13} = 0$. The normal pentode transconductance is $y_{31} = g_m$; y_{21} is the transconductance from the control grid to the screen grid and is given by the relation $y_{21} \cong g_m I_{c2}/I_b$, where I_{c2} and I_b are the screen and plate d-c currents, respectively. The dynamic resistance of the screen grid with all other electrodes grounded is $1/y_{22} = r_{p2}$. An approximate value for r_{p2} may be obtained if the plate resistance of the tube connected as a triode (screen and suppressor connected to the plate) is known: $r_{p2} \cong r_{pt} I_{bt}/I_{c2}$, where r_{pt} is the triode plate resistance, I_{bt} is the d-c current to the triode plate, and I_{c2} is the d-c current to the screen. The calculation of these approximate quantities is based upon the fact that the ratio of plate to screen current is a fixed quantity dependent upon the geometry of the tube and relatively independent of the plate and screen voltages as long as the tube is operated above the knee of the pentode plate characteristics.[1] Because of this y_{23} is approximately zero and can usually be neglected. The remaining admittance y_{32} represents the change in plate current caused by a change in screen voltage and is $y_{32} \cong (1/r_{p2})I_b/I_{c2}$. If necessary, all these quantities can be measured with an ordinary vacuum-tube bridge.

[1] An example of the calculation of these y's is

$$y_{21} = \frac{I_2}{V_1} = \frac{I_3}{V_1}\frac{I_2}{I_3} = g_m \frac{I_2}{I_3} \cong g_m \frac{I_{c2}}{I}$$

As an example of the quantities involved, the data below refer to a 6AU6 operated at $E_b = 250$ volts, $E_{c2} = 125$ volts, $I_b = 7.6$ ma, $I_{c2} = 3.0$ ma. The measured quantities are obtained with a tube bridge; the calculated quantities are determined as described in the text. Note the good agreement.

	Measured	Calculated
g_m	5,050 μmhos	
r_p	1.41 megohms	
r_{pt}	6.41 kilohms	
	($E_b = 125$ volts)	
y_{21}	2,030 μmhos	1,992 μmhos
y_{22}	1/23,300 mho	1/22,700 mho
y_{23}	0.68 μmho	0
y_{32}	107 μmhos	112 μmhos

To solve for the effects of Z_s, we write

$$I_2 = y_{21}V_1 - y_{22}I_2Z_s \qquad \text{since } V_2 = -I_2Z_s \text{ and } y_{23} \cong 0 \quad (3\text{-}77)$$

$$= \frac{y_{21}V_1}{1 + y_{22}Z_s} \qquad\qquad (3\text{-}78)$$

$$I_3 = y_{31}V_1 - y_{32}I_2Z_s + y_{33}V_3 \qquad\qquad (3\text{-}79)$$

$$= y_{31}V_1 - \frac{y_{32}y_{21}V_1Z_s}{1 + y_{22}Z_s} + y_{33}V_3 \qquad\qquad (3\text{-}80)$$

The first two terms in the last equation give the effect of control-grid voltage V_1 on the plate current; hence we may obtain a modified transconductance which takes into account the effect of Z_s,

$$\frac{I_3}{V_1} = y_{31} - \frac{y_{32}y_{21}Z_s}{1 + y_{22}Z_s} \qquad\qquad (3\text{-}81)$$

If the values previously found for the y's are substituted into the above equation, the similarity to a modified g_m is easier to see,

$$\frac{I_3}{V_1} = g_m - \frac{g_m Z_s/r_{p2}}{1 + Z_s/r_{p2}} = g_m\left(1 - \frac{Z_s}{r_{p2} + Z_s}\right) \qquad (3\text{-}82)$$

Note that, if $Z_s = 0$, the result is the normal transconductance found with V_2 equal to zero. Substituting the value of Z_s for the screen circuit param-

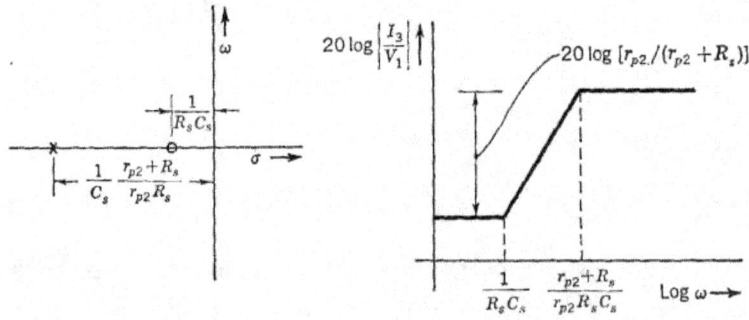

Fig. 3-35 Response due to the screen circuit alone.

eters given in Fig. 3-27 gives

$$Z_s = \frac{R_s}{1 + pR_sC_s} \tag{3-83}$$

$$\frac{I_3}{V_1} = g_m\left[1 - \frac{R_s/(1 + pR_sC_s)}{r_{p2} + R_s/(1 + pR_sC_s)}\right] \tag{3-84}$$

$$= \frac{g_m(p + 1/R_sC_s)}{p + [(r_{p2} + R_s)/r_{p2}R_s](1/C_s)} \tag{3-85}$$

The pole in Eq. (3-85) corresponds to the frequency at which the reactance of the bypass capacitor C_s equals the parallel combination of the dynamic screen resistance and the screen dropping resistor R_s. The resulting pole-zero diagram and frequency response shown in Fig. 3-35 are very similar to those for the cathode circuit shown in Fig. 3-30.

3-8 Low-frequency Response of Cascaded Stages. In the preceding discussion each of the low-frequency cutoffs was discussed separately. To combine the effects due to coupling, cathode, and screen circuits or the effect due to several stages, one must know how the functions describing the effects combine. As an example, consider a pentode stage with a coupling circuit and a cathode bias circuit. In this case the over-all equivalent circuit using the equivalent for the tube and Z_k as shown in Fig. 3-29 is that shown in Fig. 3-36. In almost all cases the parallel combination of R_L and the branch containing r_p is essentially equal to R_L. Therefore, the equation for voltage gain, referring to Eq. (3-65) for the coupling circuit, is the *product* of the functions describing the cathode circuit and the coupling circuit,

$$A(p) = \frac{g_m}{1 + g_mZ_k}\frac{R_LR_g}{R_L + R_g}\frac{p}{p + 1/(R_L + R_g)C_{cc}} \tag{3-86}$$

$$\cong g_mR_L\left(\frac{p + 1/R_kC_k}{p + \dfrac{1 + g_mR_k}{R_kC_k}}\frac{p}{p + 1/R_gC_{cc}}\right)\quad R_L \ll R_g \tag{3-87}$$

Because the over-all response is the product of the individual responses, the over-all pole-zero diagram is merely the superposition of those describing the behavior of the cathode and coupling circuits individually, and the amplitude (on a logarithmic scale) and phase responses can be added together to give the correct over-all response, as shown in Fig. 3-36. Note that the necessary requirement for the responses to be multiplied in this case was the neglect of a reverse coupling, i.e., output to input; in this case the neglected element was r_p, which is indeed a good approximation for a pentode, but less so in the case of a triode because r_p may not be much greater than R_L. A situation similar to the triode case arises when both cathode and screen effects in a pentode must be included: the screen acts much like the plate of a triode as far as the cathode is concerned. Hence the effects of the cathode and screen circuit do not exactly add; since, however, an exact solution of the low-frequency cutoff is not usually desired, the approximate solution obtained by ignoring interaction is frequently employed. A similar

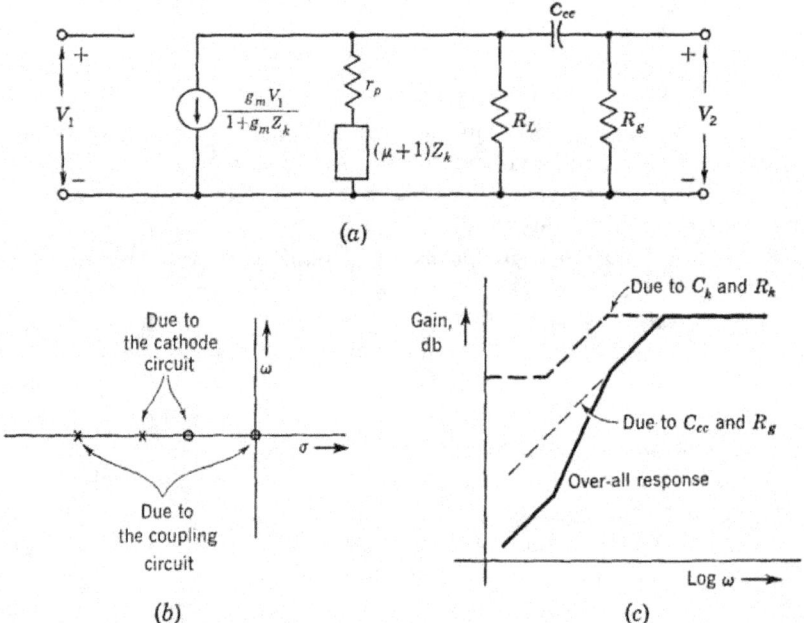

Fig. 3-36 Low-frequency characteristics of a vacuum-tube stage including the effects due to the coupling and cathode biasing circuits. (*a*) Equivalent circuit. (*b*) Pole-zero diagram. (*c*) Response.

effect exists in a cascade of transistor common-emitter stages because the transistor is not quite a unilateral device even at low frequencies; hence the effects of successive stages cannot exactly be added, but for most practical purposes addition suffices. One can thus conclude that the responses of the individual stages of a cascade of grounded-cathode stages may be exactly added. If other types of stages are employed, some care should be used in combining the responses.

3-9 Determination of the Amplifier High-frequency Response. The determination of the high-frequency cutoff of a pentode stage [1] is much easier than the determination of the low-frequency cutoff because the only

---

[1] The triode is little used in wideband amplifiers because of the excessive shunting capacitance. For example, see Prob. 2-7.

significant elements are the shunt resistances and capacitances in the interstage, as shown in Fig. 3-12. The transfer function for high frequencies written in terms of the variable p was given in Eq. (3-14) and is repeated here for convenience,

$$A(p) = \frac{-g_m}{C} \frac{1}{p + 1/RC} \qquad (3\text{-}88)$$

where

$$C = C_1 + C_2$$

$$\frac{1}{R} = \frac{1}{r_p} + \frac{1}{R_L} + \frac{1}{R_g} \cong \frac{1}{R_L} \qquad \text{for } r_p \gg R_L \ll R_g$$

The bandwidth of the stage is $B = 1/2\pi RC$, and the midband gain of the stage $A = -g_m R$. The product of gain and bandwidth is

$$\text{Gain} \times \text{bandwidth} = GB = \frac{g_m}{2\pi C} \qquad (3\text{-}89)$$

and is independent of either the gain or the bandwidth. Thus one can attain more gain at the expense of bandwidth, or vice versa. Note that the capacitance C will be largely the input and output capacitances of the tube plus the additional stray wiring capacitance. Hence the gain-bandwidth product is largely dependent upon the tube, since the wiring capacity is usually kept to a minimum. More will be said concerning this when we try to improve the gain-bandwidth product (or, as we shall see, improve the gain/rise-time *quotient*).

Let us for the moment leave the problem of improvement to the next chapter and find the analogous h-f behavior for a transistor amplifier. Here the situation is more complicated because of the more complex equivalent circuit for the device. Let us first examine the case of a common-base stage

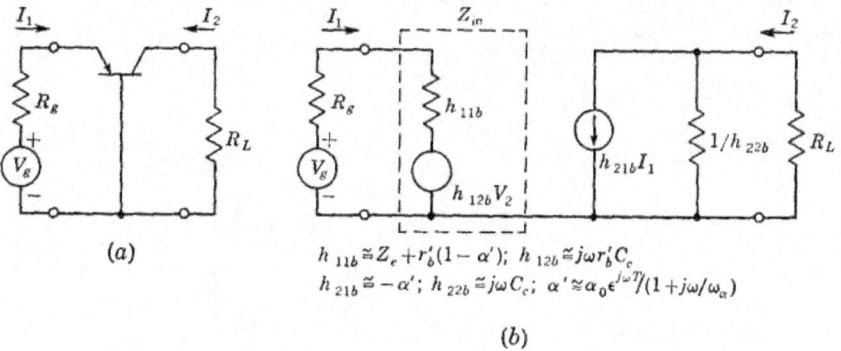

$$h_{11b} \cong Z_e + r_b'(1 - \alpha'); \quad h_{12b} \cong j\omega r_b' C_c$$
$$h_{21b} \cong -\alpha'; \quad h_{22b} \cong j\omega C_c; \quad \alpha' \cong \alpha_0 \epsilon^{j\omega T}/(1 + j\omega/\omega_\alpha)$$

(b)

Fig. 3-37 Common-base amplifier stage. (a) Actual circuit. (b) Equivalent circuit using high-frequency h parameters.

driven from a resistive source R_g and coupled to a load R_L, as shown in Fig. 3-37. Such a situation might arise in the last stage of an amplifier which must drive a rather large resistive load over a relatively large bandwidth. Since simplifications to the equivalent circuit for the transistor were given in terms of the h parameters in Sec. 2-6, the simplest course at this time is to use the h-parameter representation as shown in Fig. 3-37b. The approximate values of the h's are repeated in the figure. Note that the output is effectively shunted by C_c since $h_{22} = pC_c$; therefore one reason for the gain falling off at high frequencies is the shunting effect of the collector capacitance. However, even if R_L were made very small to make the effect of C_c small, two other effects would still cause an h-f cutoff. One is caused by the decrease in current gain as the frequency ω_α is approached; the other is caused by the increase in input impedance Z_{in} with increasing frequency. The following should serve to bring out these three effects analytically:

First let us calculate the bandwidth of the output circuit alone, assuming the input current I_1 to be constant.

$$V_2 = -I_2 R_L = \frac{-h_{21} I_1 R_L}{1 + h_{22} R_L} = \frac{I_1 \alpha_0 \epsilon^{-pT}}{1 + p/\omega_\alpha} \frac{R_L}{1 + pR_L C_c} \tag{3-90}$$

The bandwidth for constant I_1 is dependent upon both the alpha-cutoff frequency and the "output circuit bandwidth" $1/R_L C_c$. For our example we shall assume that the load is a high resistance so that $\omega_\alpha \gg 1/R_L C_c$, thus making the latter factor dominant.

To find the effect of the input impedance, we may write Z_{in} in terms of the h parameters,

$$Z_{\text{in}} = h_{11} - \frac{h_{21} h_{12} Z_L}{1 + h_{22} Z_L} \tag{3-91}$$

Substituting for the h's and retaining only the first two terms of the series expansion for ϵ^{-pT}, we obtain

$$Z_{\text{in}} = Z_e + r_b'(1 - \alpha_0) \frac{1 + p(1/\omega_\alpha - \alpha_0 T)/(1 - \alpha_0)}{1 + p/\omega_\alpha}$$

$$+ \frac{\alpha_0(1 - pT)pr_b' R_L C_c}{(1 + p/\omega_\alpha)(1 + pR_L C_c)} \tag{3-92}$$

This expression can be simplified by combining the terms and discarding terms in p^2, which is a reasonable approximation for the frequency range which will be discussed.

$$Z_{\text{in}} \cong Z_e + r_b'(1 - \alpha_0) \frac{1 + p(1/\omega_\alpha + \alpha_0 T + R_L C_c)/(1 - \alpha_0)}{(1 + p/\omega_\alpha)(1 + pR_L C_c)} \tag{3-93}$$

To simplify things further, let us first assume that the source impedance is large compared with the l-f transistor input impedance, which is true if the input circuit is not to limit the bandwidth to a very small value; second, let us assume that the load impedance and collector capacitance are sufficiently large to limit the bandwidth. These assumptions are thus

$$R_g \gg r_e' + r_b'(1 - \alpha_0) \tag{3-94}$$

and
$$R_L C_c \gg \frac{1}{\omega_\alpha} + \alpha_0 T \tag{3-95}$$

Note that R_g does not need to be very large for the inequality to hold because $r_e' + r_b'(1 - \alpha_0)$ is typically in the range 10 to 50 ohms. With these inequalities satisfied, the approximate value of input current I_1 is

$$I_1 = \frac{V_g}{R_g + Z_{\text{in}}} \cong \frac{V_g}{R_g + r_b'[pR_L C_c/(1 + pR_L C_c)]}$$
$$= \frac{V_g(1 + pR_L C_c)}{R_g + (R_g + r_b')pR_L C_c} \tag{3-96}$$

The voltage gain for these conditions may also be found by using the previous expression [Eq. (3-90)] for V_2/I_1,

$$\frac{V_2}{V_g} = \frac{1 + pR_L C_c}{R_g + (R_g + r_b')pR_L C_c} \frac{\alpha_0 R_L}{1 + pR_L C_c}$$
$$= \frac{\alpha_0 R_L}{R_g} \frac{1}{1 + p(1 + r_b'/R_g)R_L C_c} \tag{3-97}$$

From this rather drastically simplified equation we may now find the stage bandwidth, voltage gain, and gain-bandwidth product.

$$B_r = \frac{1}{(1 + r_b'/R_g)R_L C_c} \qquad \text{bandwidth in radians/sec} \tag{3-98}$$

$$A = \frac{\alpha_0 R_L}{R_g} \tag{3-99}$$

$$GB = (A)(B_r) = \frac{\alpha_0}{(R_g + r_b')C_c} \tag{3-100}$$

The largest bandwidth is obtained for a given R_L when the source resistance is infinite, i.e., when the stage is driven by a current source. However, the gain-bandwidth product is then zero; it is increased by decreasing R_g, but R_g cannot be decreased too much, or the preceding approximations become invalid.

This illustrative calculation for a particular type of grounded-base stage illustrates the typical complexity of even a simple transistor circuit and further shows that useful results, which may be easily interpreted, can be obtained by restricting the solution to a particular range of some variables. This technique of restricting the useful range of a solution to obtain simpler equations is a powerful and necessary tool on many occasions, and particularly in transistor circuits.

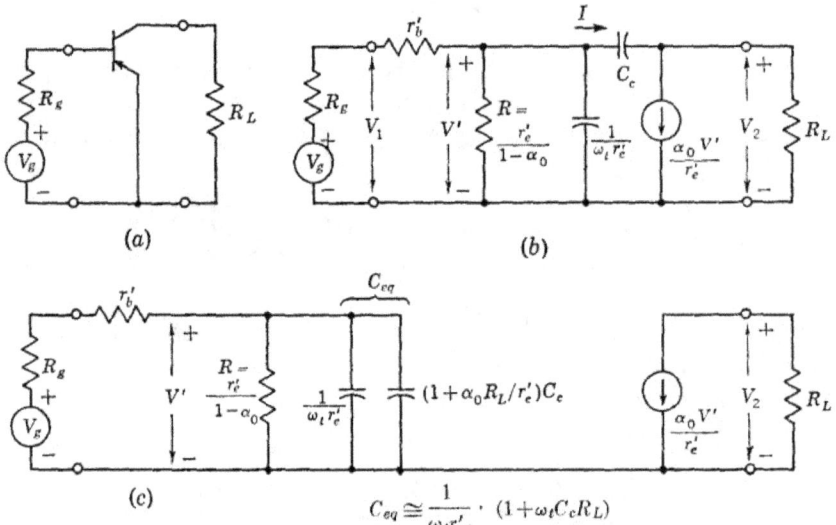

Fig. 3-38 Common-emitter amplifier stage. (*a*) Actual circuit. (*b*) High-frequency equivalent circuit. (*c*) Circuit with C_c represented by a Miller capacitance.

A similar analysis for a common-emitter stage, as shown in Fig. 3-38, proceeds most easily from the hybrid-pi equivalent circuit of Fig. 2-24*c*, repeated in Fig. 3-38*b*. Note that stray capacitances have not been included because their effect is usually small compared with the effects due to the transistor. One should not forget the presence of parasitic capacitances, however, because their effect may not always be negligible, particularly with vhf transistors.

One difficulty with the circuit of Fig. 3-38*b* is the capacitor C_c, which prevents the circuit from being unilateral; i.e., the input circuit is affected by changes in the output circuit. The effect of C_c may be approximated by noting that the current flowing in it is

$$I = (V' - V_2)j\omega C_c \qquad (3\text{-}101)$$

If we neglect the reactance of C_c in shunt with R_L, which is a reasonable thing to do in the usual range of values of R_L used in common-emitter amplifiers, then V_2 is

$$V_2 \cong \frac{-\alpha_0 V'}{r_e'} R_L \qquad\qquad (3\text{-}102)$$

and the current through C_c is approximately

$$I \cong \left(V' + \frac{\alpha_0 V'}{r_e'} R_L \right) j\omega C_c \cong V' \left[j\omega C_c \left(1 + \frac{\alpha_0 R_L}{r_e'} \right) \right] \qquad (3\text{-}103)$$

Consequently the effect of C_c is about the same as connecting a "Miller-effect" capacitance of value $C_c(1 + \alpha_0 R_L/r_e')$ from V' to ground and omitting the capacitor C_c entirely. The resulting equivalent circuit is shown in Fig. 3-38c. Since usually $1/\omega_t r_e' \gg C_c$ and $\alpha_0 \cong 1$, the value for C_{eq} is conveniently approximated by

$$C_{eq} \cong \frac{1}{\omega_t r_e'} + \frac{C_c R_L}{r_e'} = \frac{1}{\omega_t r_e'} (1 + \omega_t C_c R_L) \qquad (3\text{-}104)$$

With the aid of this simplified circuit, the gain and bandwidth are very easily found. The voltage gain is

$$\frac{V_2}{V_g} = A(p) = \frac{\alpha_0 R_L}{(R_g + r_b')(1 - \alpha_0) + r_e'} \frac{1}{1 + \dfrac{p(R_g + r_b')r_e'C_{eq}}{(1 - \alpha_0)(R_g + r_b') + r_e'}}$$

$$(3\text{-}105)$$

$$\frac{V_2}{V_g} \cong \frac{\alpha_0 R_L}{r_e'} \frac{1}{1 + p(R_g + r_b')C_{eq}} \qquad \text{if } \frac{r_e'}{1 - \alpha_0} \gg R_g + r_b' \qquad (3\text{-}106)$$

The final approximate equation is usually valid in a video-type amplifier because R_g is of necessity small. For the common-emitter amplifier the bandwidth

$$B_r \cong \frac{1}{(R_g + r_b')C_{eq}} \cong \omega_t \frac{r_e'}{(R_g + r_b')(1 + \omega_t C_c R_L)} \qquad (3\text{-}107)$$

is increased by *decreasing* the source impedance R_g. The maximum obtainable bandwidth, which is obtained by using a voltage source ($R_g = 0$), is $1/r_b'C_{eq}$. From this it is seen that a transistor suitable for wide bandwidths has a high ω_t, a low collector capacitance C_c, and low base resistance r_b'.

The bandwidth with very large R_g may be obtained from Eq. (3-105) by finding the limit of B_r as R_g approaches infinity.

$$\lim_{R_g \to \infty} B_r = \omega_t \frac{1 - \alpha_0}{1 + \omega_t R_L C_c}$$

This limiting bandwidth (if R_L is small) is the so-called beta-cutoff frequency, $\omega_\beta \triangleq (1 - \alpha_0)\omega_t$, diminished by the effect of C_c. This bandwidth is approximately that of a resistance-coupled stage, designed to give high gain as in an audio amplifier.

Since a single stage rarely produces enough gain, the multistage situation depicted in Fig. 3-39a is of particular interest. Because the output circuit

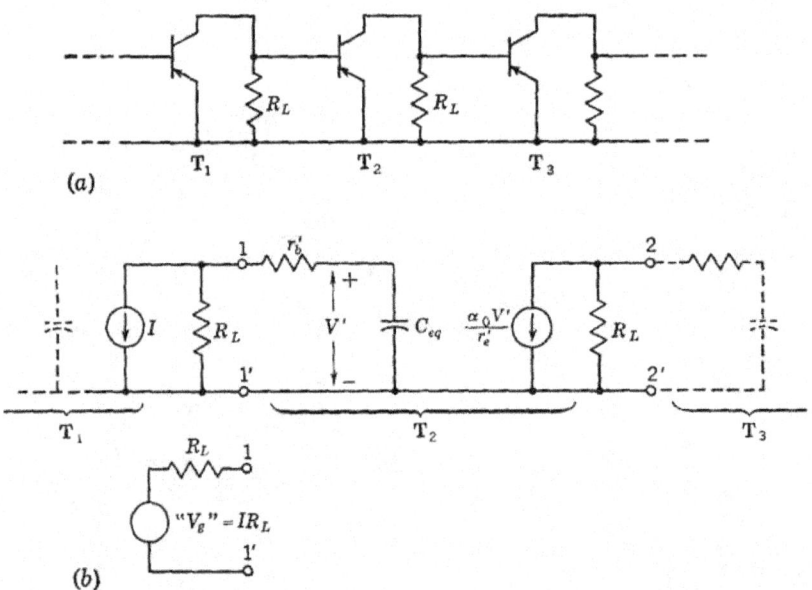

(a)

(b)

Fig. 3-39 Cascaded common-emitter stages. (a) Actual circuit. (b) Simplified equivalent circuit of an "interior" stage.

of the transistor circuit appears essentially like a current generator, an interior stage has a generator impedance of R_L and a load impedance of R_L also. (Note that, in calculating the gain and bandwidth of T_2, the stage is opened at 2-2'. The bandwidth reduction due to the input capacity of T_3 is taken into account in computing the bandwidth of the third stage with R_L regarded as its source impedance.) The bandwidth and gain-bandwidth product [both on the assumption that $r_e'/(1 - \alpha_0) \gg R_g + r_b'$] are

$$B \cong \omega_t \frac{r_e'}{(R_L + r_b')(1 + \omega_t R_L C_c)} \tag{3-108}$$

$$GB \cong \omega_t \frac{\alpha_0 R_L}{(R_L + r_b')(1 + \omega_t R_L C_c)} \tag{3-109}$$

This time, the gain-bandwidth product is a function of the load resistance

R_L and can in fact be maximized by the proper choice of R_L. The value of R_L which maximizes GB is

$$R_{L,\text{opt}} \cong \sqrt{\frac{r_b'}{\omega_t C_c}} \qquad (3\text{-}110)$$

The resulting maximum GB is

$$GB_{\max} \cong \frac{\omega_t}{(1 + \sqrt{\omega_t r_b' C_c})^2} \qquad (3\text{-}111)$$

One way of designing a stage is to give R_L its optimum value first and then adjust the d-c emitter current for the desired gain and bandwidth ($A = \alpha_0 R_L / r_e'$ and $r_e' \cong kT/qI_E$). This method of optimization is not applicable over a very wide range and with many transistors is only of academic interest because ω_t often depends quite strongly on I_E, particularly in drift transistors. Consequently one does better by choosing I_E to maximize ω_t.

For estimation purposes the inequality

$$GB \lesseqgtr \omega_t \qquad (3\text{-}112)$$

tells us that a rough estimate for the gain of a stage is obtained by dividing ω_t by the desired bandwidth.

It is interesting to compare the three amplifier examples discussed in this section. In the vacuum-tube amplifier the high-frequency limit is imposed by the tube and stray circuit capacitances. Both sources of capacitance are of the same order of magnitude. The gain of a stage is inversely proportional to the bandwidth. In the common-emitter transistor stage the high-frequency limit arises because of the effects of r_b', ω_t, and C_c; stray capacitance has relatively small effect (on the assumption of normal wiring). The gain of a stage is *not* exactly inversely proportional to the bandwidth. In both types of amplifiers the bandwidth is increased by decreasing the load and source resistance; in the common-base stage, on the other hand, we found that the bandwidth increases with increasing source resistance and decreasing load resistance. Numerous other important examples could be worked out for the high-frequency response; however, we shall stop at this point in the discussion of steady-state response and go on to the discussion of amplifiers with transient signals, since it is this type of signal that the majority of wideband amplifiers must handle.

PROBLEMS

3-1. Write the equations for V_2/V_1 for the circuits shown in Fig. P3-1, and sketch straight-line approximations of the amplitude and phase response (V_2/V_1 versus frequency, etc.). Give the values of the corner frequencies in terms of the appropriate R's and C's.

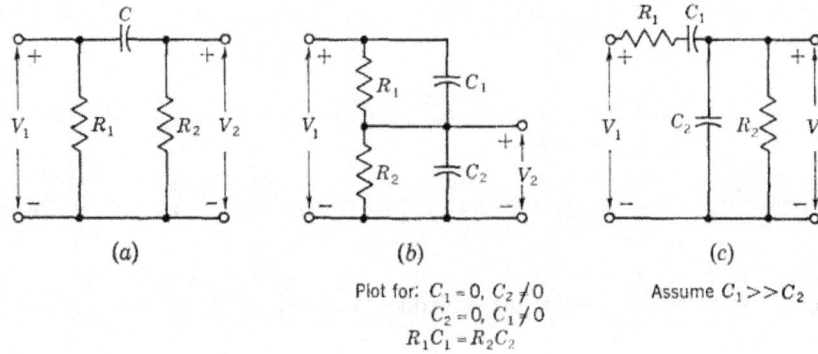

(a)   (b)   (c)

Plot for: $C_1 = 0$, $C_2 \neq 0$
$C_2 = 0$, $C_1 \neq 0$
$R_1 C_1 = R_2 C_2$

Assume $C_1 >> C_2$

Fig. P3-1

3-2. Find the y parameters in terms of r_p and μ for a grounded-grid triode stage. Discuss the conditions under which such a stage might be useful.

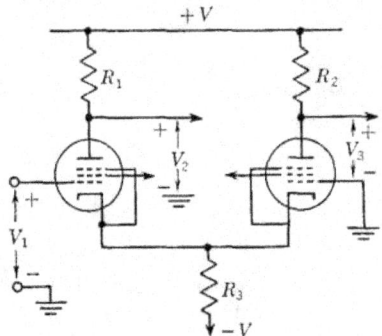

Fig. P3-3

3-3. A circuit commonly used as a phase inverter, especially in d-c amplifiers, is the so-called paraphase inverter shown in Fig. P3-3, where the tubes may be either triodes or pentodes.

a. For two triodes derive an expression giving the voltage gain from the input grid to each plate.

b. Derive an expression similar to that of (a), but using the pentode equivalent circuit neglecting r_p. Under what conditions might this equivalent circuit be valid for triodes as well?

c. For the following circuit values and the 12AX7 triode characteristics determine the value of R_2 so that the gain to the two plates is identical. $R_1 = 68$ kilohms, $R_3 = 220$ kilohms, $\mu = 100$, $r_p = 60$ kilohms.

3-4a. Find the exact value of transducer gain for the amplifier in Fig. P3-4a. Assume that $r_e = 15$ ohms, $r_b = 250$ ohms, $r_c = 2$ megohms, and $\alpha = 0.98$.

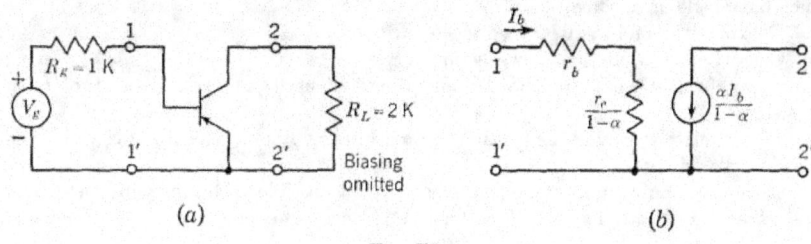

Fig. P3-4

b. Find an approximate value for the transducer gain by replacing the T equivalent with the simpler circuit of Fig. P3-4b, which omits r_c. Compare the two values of gain you obtain, and comment on the circuit conditions which are necessary for approximate equivalence between the two values of gain.

3-5. Find the value of C_{cc} and C_e to give a low-frequency cutoff of 100 cps due to each individual capacitor in the transistor amplifier of Fig. P3-5 (i.e., make both cutoff frequencies coincide). Use the data of Table 2-1.

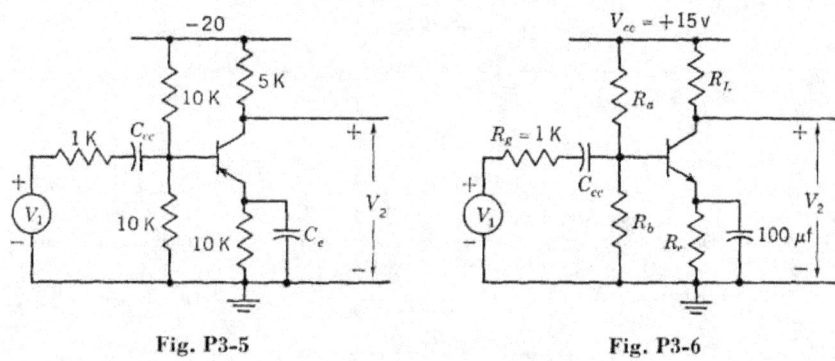

Fig. P3-5   Fig. P3-6

3-6. The single-stage transistor amplifier shown in Fig. P3-6 is to have a voltage gain V_2/V_1 of 50 and a stability factor of 5. The known circuit values are given in the diagram. The transistor is to operate with $I_E = -0.5$ ma and $V_{CE} = +5$ volts. (Assume that $V_{BE} \cong 0$.) At this operating point the transistor parameters are $h_{11e} = 5{,}000$ ohms; $h_{12e} = 2.5 \times 10^{-3}$; $h_{21e} = 100$; $h_{22e} = 50$ μmhos; $I_{C0} = 2$ μa.

a. Give the values for R_L, R_a, R_b, and R_e. (There is a unique value for each element.)

b. How low can the low-frequency cutoff be made by increasing C_{cc}?

3-7. Show that the feedback biasing circuit shown in Fig. 3-26 is actually the same as the single-battery biasing circuit shown in Fig. 3-25 by redrawing the circuit. (Some of the resistors in the latter circuit will be missing, of course.) Then by substituting the appropriate resistors into Eq. (3-58) show that the sensitivity factor is

$$S = \frac{R_L + R_f}{R_L + R_f(1 - \alpha_{DC})}$$

3-8. The amplifier in Fig. P3-8 is composed of a common-emitter stage driving a common-collector stage. Such an amplifier combination might be used to drive a low-impedance load. The transistor parameters are $r_c = 1$ megohm, $\alpha = 0.98$, $r_e = 6$ ohms, and $r_b = 500$ ohms. (In all calculations it is important to disregard elements which have little or no effect.)

a. Calculate the mid-frequency voltage gain V_L/V_g.

b. Calculate the transducer power gain.

c. *Approximately* what are the d-c operating potentials at each element of both transistors?

d. Approximately what is the maximum rms voltage without peak clipping which the amplifier will deliver?

e. Calculate the maximum transducer power gain possible with any source and load impedances. What are the source and load impedances which give this maximum gain?

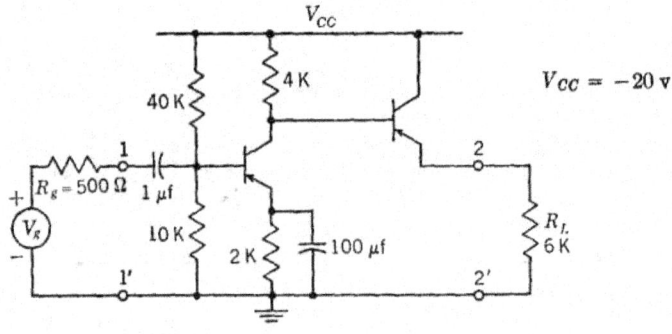

Fig. P3-8

3-9. A pentode stage using the 6AU6 (use the measured characteristics given in Sec. 3-7) is to have a low-frequency cutoff of 20 cps and a high-frequency cutoff of at least 2 Mc. Design the stage so that all the low-frequency cutoffs occur at the same frequency. Find all the missing values in the circuit of Fig. P3-9. What is the maximum gain that may be obtained?

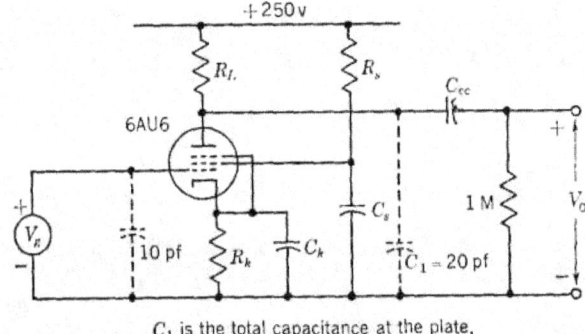

C_1 is the total capacitance at the plate.

Fig. P3-9

3-10. The amplifier shown in Fig. P3-10 is a common-collector stage driving a common-emitter stage. The purpose of the common-collector stage is to provide a relatively high input impedance compared with the 1-kilohm source. The two transistors are identical and have the following low-frequency parameters: $r_e = 15$ ohms; $r_b = 500$ ohms; $\alpha/(1 - \alpha) = 50$; $r_c = 1$ megohm. The high-frequency parameters are $r_e' = 25$ ohms; $r_b' = 100$ ohms; $f_t = 20$ Mc; $C_c = 10$ pf. Use any reasonable approximations for computing the following:

a. Find the mid-frequency voltage gain, that is, V_2/V_1.

b. Find the value of the capacitors C_1 and C_2 to give a lower cutoff frequency of 20 cps. Make $C_1 = (1 - \alpha)C_2$ so that the capacitors have equal effect.

c. Find the input and output resistance of the amplifier in the mid-frequency range.

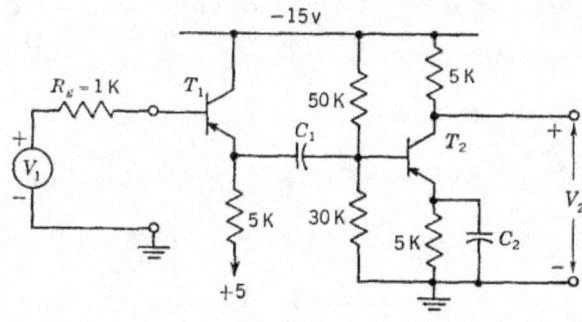

Fig. P3-10

d. As a start on computing the high-frequency cutoff of the amplifier find the Thévenin equivalent circuit of the first stage (viewed into the emitter and ground terminals). Pick a suitable equivalent circuit with an eye to simple analysis. Neglect the effect of C_c, which is usually a reasonable approximation, and assume that α can be adequately represented by

$$\alpha = \frac{\alpha_0}{1 + j\omega/\omega}$$

e. What would the high-frequency cutoff of T_1 be if it were feeding a 100-ohm load? (The capacitance $1/\omega_\alpha r_e'$ may be neglected here.)

3-11. An unbypassed cathode resistance is often used to decrease the distortion introduced by a stage. Compute the output resistance of such a stage as shown in Fig. P3-11, and compare it with the output resistance of the stage with the cathode bypassed.

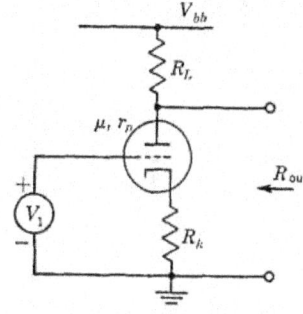

Fig. P3-11

3-12. It was shown in Sec. 3-9 that the high-frequency cutoff of a pentode amplifier stage is $\omega_1 \triangleq 1/RC$, where R and C are as defined in the section. For n stages in cascade call the high-frequency cutoff ω_n. Show that

$$\frac{\omega_n}{\omega_1} = \sqrt{2^{1/n} - 1}$$

$$\cong \frac{0.833}{\sqrt{n}} \qquad n \geqq 2$$

Is this expression also good for cascaded resistance-coupled common-emitter stages? Why?

3-13. Show that the reciprocal of the expression derived in Prob. 3-12 gives the relation between the low-frequency cutoff of one stage and n identical stages, assuming that the cutoffs are due only to the coupling circuits.

4

Step Response of Lowpass (Video) Amplifiers: Speed of Rise

Most lowpass amplifiers are required to handle transient signals rather than steady-state sine-wave signals. Hence what one would really like to know is the behavior of the amplifier with an input similar to that which the amplifier must actually amplify. Test data obtained with actual signals may be difficult to interpret or to generalize from, although this method of testing is used with video television amplifiers. Here the test signal might be obtained from a special test pattern, passed through the amplifier, and displayed on a picture tube. More usually, however, the transient test signal is a step voltage or low-frequency square wave. Such a signal would occur in a television system at an abrupt transition from white to black. Since the use of an amplifier is most often to amplify a transient signal, it is logical to design the amplifier on the basis of its transient response rather than its steady-state response. Consequently, we shall consider improvements on the high-frequency response obtained in the analysis of the last chapter on the basis of the transient behavior, rather than merely trying to extend the bandwidth. In some ways the transient behavior is more difficult to deal with, and in very complicated cases we may have to fall back upon steady-state analysis; therefore an additional point of interest is the relationship of the transient response to the steady-state response.

4-1 Pentode Stage—Choice of Tube. Let us begin the transient study with the simplest case: the pentode stage shown in Fig. 3-1 with the equivalent circuit shown in Fig. 3-12a. The response of this stage to a unit step voltage input [$u(t)$] may be found by assuming $V_1(p) = \mathcal{L}[v_1(t)] = \mathcal{L}[u(t)] = 1/p$; then the output voltage of one stage is [see Eq. (3-88)]

$$V_2(p) = V_1(p)A(p) = \frac{-g_m}{pC}\frac{1}{p + 1/RC} \tag{4-1}$$

$$v_2(t) = -g_mR(1 - \epsilon^{-t/RC}) \tag{4-2}$$

The 10 to 90 per cent rise time T_R and ultimate (same as mid-frequency) gain A are

$$T_R = 2.2RC \tag{4-3}$$

$$|A| = g_mR \tag{4-4}$$

If one sets out to design the resistance-coupled amplifier stage to give both high gain and a small rise time, one is confronted with a contradiction. From Eqs. (4-3) and (4-4) it is seen that for a given tube, i.e., a given g_m and C, one can increase R_L to raise the gain but in doing so one lengthens the rise time. This proportionality of gain and rise time can be expressed as a quotient whose magnitude is independent of R and depends primarily upon the tube,

$$\frac{\text{Gain}}{\text{Rise time}} = \frac{A}{T_R} = \frac{g_m}{2.2C} \tag{4-4a}$$

The capacitance C includes both input and output capacitances of the tube (on the assumption that both tubes associated with the interstage network are of the same type), together with the stray wiring capacitance. The latter can usually be made small compared with the tube capacitance, and in any case it is apparent that, if two tubes have equal g_m but different C, the one with the smaller C will be better. In Table 4-1 are listed the transconductances g_m, the total capacitance C (which includes 4 pf for stray

Table 4-1 *

Tube	C_i, pf	C_o, pf	C, pf	g_m, μmhos	Gain/rise time, μsec
6AK5	4.0	2.8	10.8	5,000	210
6AG5	6.5	1.8	12.3	5,000	185
6AH6	10.0	2.0	16.0	9,000	256
6AU6	5.5	5.0	14.5	5,200	163
6EW6	9.9	2.5	16.4	14,500	402
5847	7.1	2.9	14.0	12,500	405
6688	7.5 ($\pm$0.9)	3.0 ($\pm$0.5)	14.5	16,500 ($\pm$2,300)	517

* All capacitances measured with tube cold (cf. Table 10-7).

wiring capacitance), and the A/T_R quotient, which becomes a sort of "figure of merit" for the tube in a resistance-coupled amplifier circuit. A graphical tabulation of many tubes is shown in Fig. 4-1.[1]

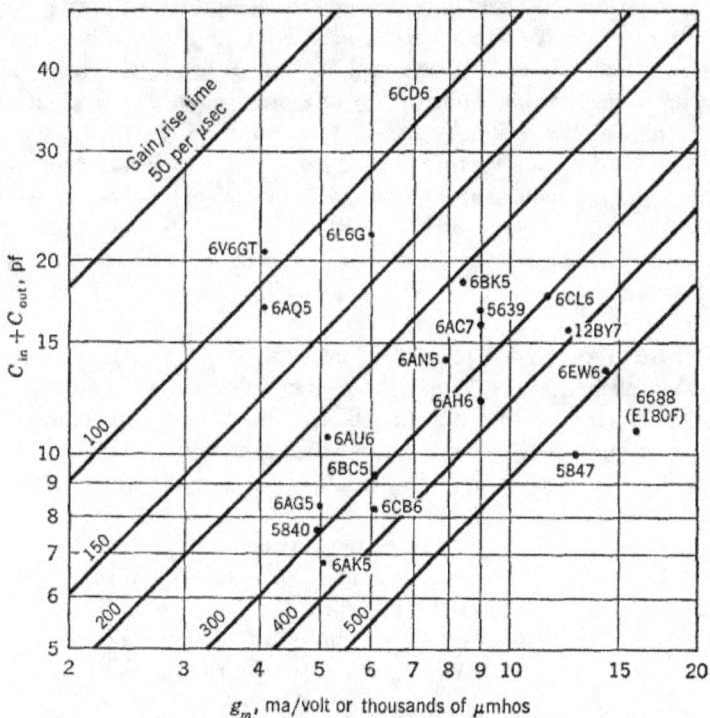

Fig. 4-1   Comparison of tube types for gain/rise time. No allowance has been made for wiring capacitance; so any point should be moved vertically by the proper amount. Note that points for low-C tubes move farther vertically for a given added wiring capacitance. Graphical presentation suggested by a similar plot from J. R. Whyte, *Electronics*, April, 1952.

Although Eq. (4-4a) and Table 4-1 are derived from the properties of the elementary resistance-coupled amplifier stage, it will turn out that with more complicated networks the same figure of merit for the tube will apply. Better circuits will give more gain for the same rise time, but $g_m/2.2C$ is still a common multiplier for all.

4-2   Pentode Stage—Choice of Circuit.   It might well be expected that by going to a more sophisticated circuit than the elementary resistance-

---

[1] For actual computation, the values of capacitance used should be those with the tube operating normally, i.e., the "hot" capacitance. For comparison of measured values of hot and cold capacitance, see Table 10-7.

coupled interstage network one could achieve better performance in terms of more gain for a given rise time, i.e., a higher gain/rise-time quotient. The g_m/C ratio of the tube is a universal factor that appears in the gain/rise-time quotient for all the circuits; so one should eliminate this as a factor in comparing alternative circuits to be used with the same tube. Since the resistance-coupled circuit is the simplest, we can use it as a reference and divide the gain/rise-time quotient for any other circuit by $g_m/2.2C$. This will then give a *relative speed*, or figure of merit, for the *circuit*.[1] There are several basic network structures which can be used; the more complicated ones give greater speed, but the designer must decide where to draw the line, for beyond some point the added complexity does not give enough improvement to justify the added difficulty of design and adjustment (particularly in a manufacturing or field servicing situation, where unskilled workers are involved).

4-3 Shunt-peaked Circuit.   The first step in circuit refinement beyond the elementary resistance-coupled circuit leads one to the so-called shunt-peaked circuit,[2] shown in Fig. 4-2a. This circuit provides a substantial increase in speed relative to the resistance-coupled circuit, and with little increase in complexity. It is probably the most widely used of the circuits discussed here.

It is necessary to specify a parameter that defines the value of L relative to R_L and C. Notice that L is a variable to be adjusted after R_L has been chosen to provide the desired gain, since the final value of gain for the circuit is $g_m R_L$ as before and C is fixed by the tube. Let this factor be called m, after Valley and Wallman,[3] defined by the relationship

$$m \triangleq \frac{L}{R_L{}^2 C} \qquad R_L \ll R_g,\, r_p \qquad (4\text{-}5)$$

[1] A more complicated figure of merit which employs a factor for the overshoot introduced by many circuits is given by R. C. Palmer and L. Mautner, A New Figure of Merit for the Transient Response of Video Amplifiers, *Proc. IRE*, vol. 37, pp. 1073–1077, September, 1949. To use this figure, it is necessary to know the acceptable limit of overshoot for the service intended, e.g., perhaps 2 per cent for television, as suggested by these writers.

[2] The name is derived from steady-state considerations, where the parallel resonance of L and C tends to produce a peak in the curve of amplification vs. frequency. By way of contrast, the *series*-peaking circuit, which enjoyed popularity for a time, employs an inductance in series with the coupling capacitor C_c.

[3] G. E. Valley, Jr., and H. Wallman (eds.), "Vacuum Tube Amplifiers" (vol. 18, M.I.T. Radiation Laboratory Series), p. 73, McGraw-Hill Book Company, Inc., New York, 1948. Other writers use the same or similar factors; F. E. Terman, "Electronic and Radio Engineering," 4th ed., pp. 292–296, McGraw-Hill Book Company, Inc., New York, 1955, calls the factor Q_2, since m is the Q of the circuit at the frequency f_2, where $X_c = R_L$. Also $m = Q_0{}^2$, where Q_0 is the circuit Q at the resonant frequency $f_0 = 1/2\pi\sqrt{LC}$.

As the factor m is increased from zero (corresponding to the simple resistance-coupled case) to a value of 0.6 by increasing L for a given combination of R_L and C, the step-response curves that result are as shown in Fig. 4-2b. It can be seen that as m is increased the rise time decreases, with overshoot appearing for values of m greater than 0.25. A physical interpretation of the effect of adding the inductance is that, in order to charge the capacitor C as rapidly as possible, the maximum current should flow from the gen-

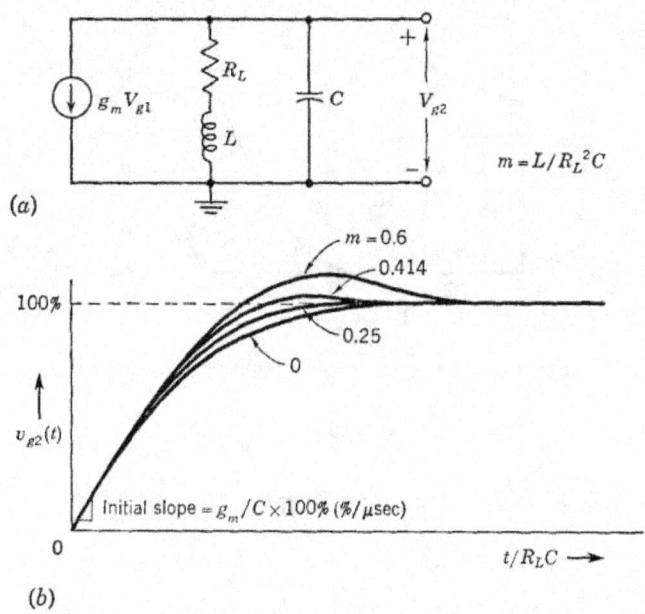

Fig. 4-2 A shunt-peaked stage. (a) Circuit. (b) Step response.

erator into C. Without L the initial current does flow entirely in C, but as the voltage builds up, more of the generator current is bypassed into the resistor R_L. Adding L slows up the increasing current in the R_L branch. Eventually, with L made larger, the current build-up in R_L is slowed so much that the capacitor voltage overshoots its final value. (Note that if L were infinite the R_L branch would be an open circuit and the capacitor voltage could increase indefinitely at a rate $I/C = g_m V_{g1}/C$.)

The figure of merit η of the shunt-peaked circuit, i.e., its speed (10 to 90 per cent rise) relative to the resistance-coupled circuit, increases with m, as illustrated in Fig. 4-3. Also shown is the curve of overshoot, which is zero up to $m = 0.25$ and then increases with m.

From Fig. 4-3 it can be seen that the most beneficial range of m is from 0 to 0.25, where the relative speed η increases from 1.0 to 1.4 without any overshoot appearing.   Beyond an m of 0.25 the overshoot commences, with both η and the overshoot increasing, but the latter more rapidly.[1] The way in which values of m greater than 0.25 come to produce overshoot can be illustrated with the aid of plots of the poles and zeros of the transfer

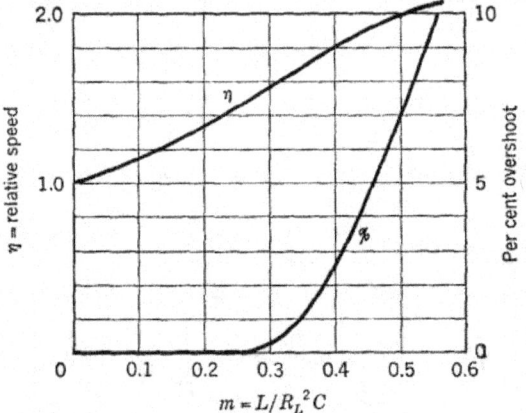

Fig. 4-3 Characteristics of a shunt-peaked stage.

function (gain) of p, as in Fig. 4-4; this function is

$$A(p) = \frac{-g_m}{C} \frac{p + R/L}{p^2 + p(R/L) + 1/LC} \tag{4-6a}$$

$$= \frac{-g_m}{C} \frac{p - p_0}{(p - p_1)(p - p_2)} \tag{4-6b}$$

A study of a table of Laplace transforms, or indeed a more thorough analysis,[2] will show that, when system or transfer functions of the kinds in (a), (b), or (c) of Fig. 4-4 are driven by a step function V_1/p, the time responses contain only damped exponentials which combine to give a monotonic rise to the final value.   On the other hand, when the system function contains conjugate pairs of complex poles (off the negative real axis), sine and cosine terms can appear in the solution, giving rise to the oscillatory overshoot.

[1] Discussions based upon steady-state analysis sometimes mention a value of $m =$ 0.414 as "critical peaking," or "critical compensation." It turns out that this value of m is the crossover point between a frequency-response curve that falls off uniformly at the high-frequency end—as does the resistance-coupled case—and one that has a peak or hump. But from the standpoint of the transient response to a step, this value of m has no particular significance (Fig. 4-2b).

[2] J. H. Mulligan, Jr., Effect of Pole and Zero Locations on Transient Response, *Proc. IRE*, vol. 37, pp. 516–529, May, 1949.

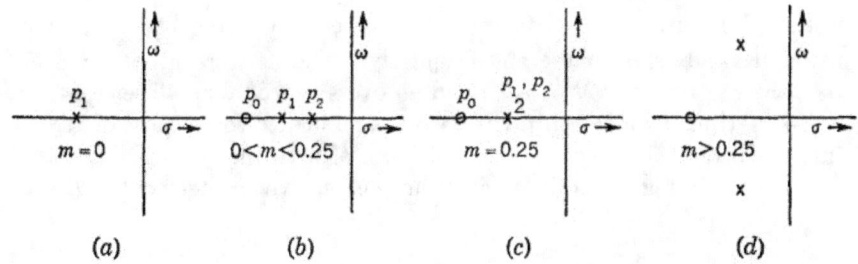

Fig. 4-4 Pole-zero diagrams for shunt-peaked stages with different damping (m).

4-4 Improved Shunt-peaked Circuits. Properly proportioning the capacitance shown as C_a in Figs. 4-5 and 4-6 can provide improved performance compared with the simple shunt-peaked circuit. Two parameters describe the relations of the elements in Figs. 4-5 and 4-6. One is m

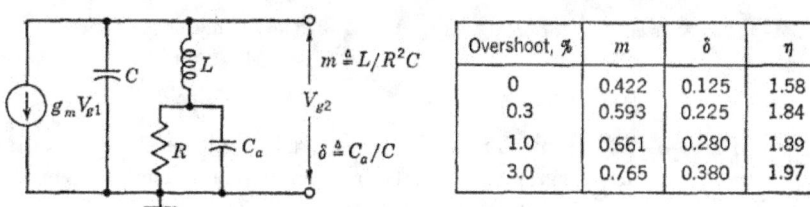

Overshoot, %	m	δ	η
0	0.422	0.125	1.58
0.3	0.593	0.225	1.84
1.0	0.661	0.280	1.89
3.0	0.765	0.380	1.97

Fig. 4-5 An improved two-terminal network.

$= L/R^2 C$; the second is $\delta = C_a/C$. By analytically studying the step response of the network for different values of δ and m, optimum values for these parameters may be found for a desired value of overshoot. This is a numerical process requiring a great deal of calculation. Values obtained in

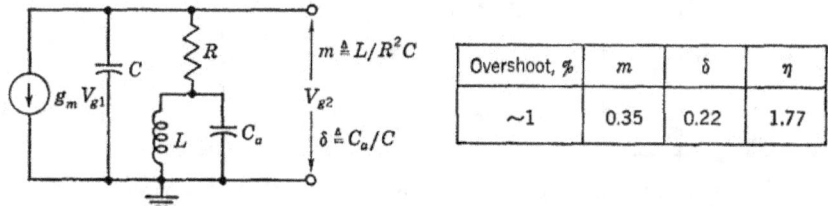

Overshoot, %	m	δ	η
~1	0.35	0.22	1.77

Fig. 4-6 Doba linear-phase network.

this way by F. A. Muller are tabulated in Fig. 4-5.[1] A similar network shown in Fig. 4-6 with a value of $m = 0.35$ and $\delta = 0.22$ gives a network credited to S. Doba of the Bell Telephone Laboratories. This latter net-

[1] F. A. Muller, High-frequency Compensation of RC Amplifiers, *Proc. IRE*, vol. 42, no. 8, pp. 1271–1276, August, 1954.

work is known as a linear-phase network because of the linear relation between steady-state phase and frequency. The network gives $\eta = 1.77$ (i.e., the rise time is $2.2RC/1.77$) and an overshoot of about 1 per cent.

4-5 Other Two-terminal Networks. Further complexity of structure can be added to the two-terminal form of network by considering the reactance X in Fig. 4-7 to be an arbitrarily extensive arrangement of L and

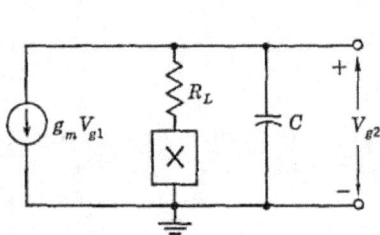
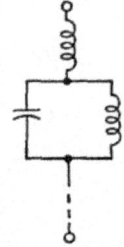

Fig. 4-7 Generalized two-terminal network.

Fig. 4-8 The arrangement of L and C for X in Fig. 4-7 as investigated by Elmore.

C, as in Fig. 4-8. This situation has been explored by Elmore,[1] with results as follows: (1) the improvement in speed of rise, as each new element (L or C) is added, diminishes rapidly after the first one or two; (2) an ultimate improvement factor of 2.12 for an infinite number of elements is indicated by the analysis, though not proved conclusively.

4-6 Four-terminal Networks. A whole family of four-terminal networks can be devised which give substantial improvement over the two-

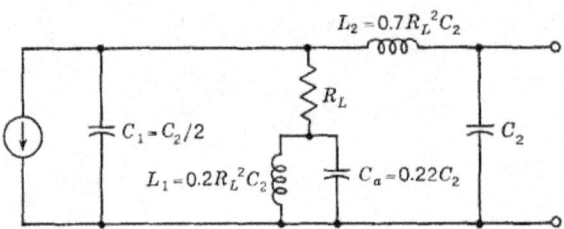

Fig. 4-9 Four-terminal "linear-phase" network for $C_1 = C_2/2$.

terminal variety. Added complexity results, however, and the response of the networks to a step is sensitive to the ratio of C_1/C_2.

Two typical examples are illustrated in Figs. 4-9 and 4-10. The former is called the "four-terminal linear-phase network" and provides a relative

[1] W. C. Elmore, The Transient Response of Damped Linear Networks with Particular Regard to Wideband Amplifiers, *J. Appl. Phys.*, vol. 19, pp. 55–63, January, 1948.

speed of 2.48 over the resistance-coupled counterpart. A 1:2 ratio of $C_1:C_2$ is assumed in the design.

The circuit of Fig. 4-10 is commonly called the "series-shunt-peaked network." It assumes a 1:1 capacitance ratio but is less sensitive to deviation from this value than the network of Fig. 4-9.

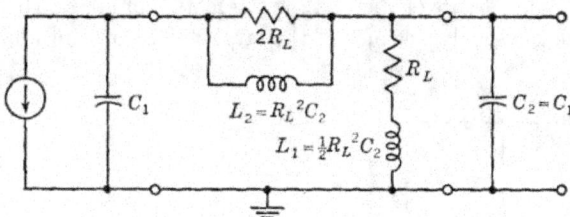

Fig. 4-10 Four-terminal network for $C_1 = C_2$.

In Fig. 4-11a is a third circuit, this one for a 1:1 ratio of capacitance. Failure to realize this ratio results in response waveforms as shown in Fig. 4-11b.

A network which is faster and more complicated than the preceding because of the utilization of mutual inductance is shown in Fig. 4-12. This

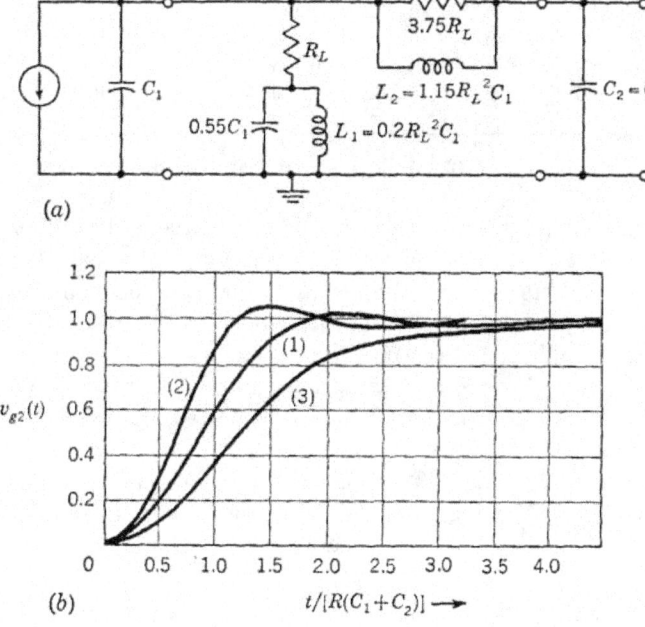

Fig. 4-11 Four-terminal network designed for $C_1 = C_2$. (a) Circuit. (b) Step response for (1) $C_2 = C_1$, (2) $C_2 = C_1/2$, (3) $C_2 = 2C_1$.

network has been investigated by F. A. Muller for a range of capacitance values. A part of his data are reproduced in the table in Fig. 4-12.[1]

The relative speed of the four-terminal networks is substantially greater than that of the two-terminal forms. The circuit of Fig. 4-9 provides $\eta = 2.48$, while the circuits of Figs. 4-10 and 4-11 give values of 2.06 and 2.1, respectively. The circuit of Fig. 4-12 is the fastest of all, but it may not be easy to realize physically because of the large values of coupling coefficient k required.

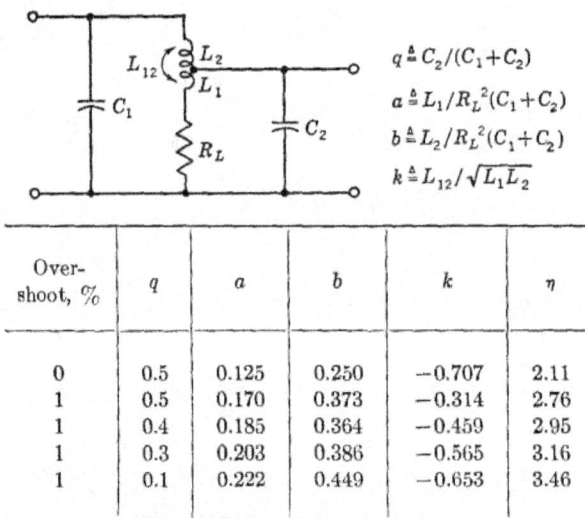

$$q \triangleq C_2/(C_1 + C_2)$$
$$a \triangleq L_1/R_L{}^2(C_1 + C_2)$$
$$b \triangleq L_2/R_L{}^2(C_1 + C_2)$$
$$k \triangleq L_{12}/\sqrt{L_1 L_2}$$

Over-shoot, %	q	a	b	k	η
0	0.5	0.125	0.250	−0.707	2.11
1	0.5	0.170	0.373	−0.314	2.76
1	0.4	0.185	0.364	−0.459	2.95
1	0.3	0.203	0.386	−0.565	3.16
1	0.1	0.222	0.449	−0.653	3.46

Fig. 4-12 Muller four-terminal network.

The maximum speed of a four-terminal network has not been conclusively established [2] but is probably in the neighborhood of 4. As in the case of two-terminal circuits, the actual networks in practical use fall short of the maximum but provide much improvement over the resistance-coupled form. More complicated structures than those shown here can be devised, but the added speed comes slowly with the extra elements required.[3]

4-7   Transient Response of Transistor Video Stages—The Series-peaked Stage. The equivalent circuit employed for analysis of transistor video stages is that of Fig. 3-38c. One of the two simple interstages is the series-peaked circuit of Fig. 4-13. This circuit derives its name from the fact that the inductor is placed in series with the input of the following

[1] Muller, *op. cit.*

[2] Valley and Wallman, *op. cit.*, pp. 81–82; also Ph.D. dissertation by George W. C. Mathers, Stanford University, 1951.

[3] A class of interstage networks based upon filter theory is presented in a classic paper by Harold Wheeler, Wideband Amplifiers for Television, *Proc. IRE*, vol. 27, p. 429, 1939.

stage. The second type of interstage employs an inductor in series with R_L, that is, in shunt with the second transistor (Fig. 4-17). This is shunt peaking and analogous to the pentode case already discussed. The two cases will be discussed in some detail, for the useful relations are again approximations, and the details of the approximations must be known for the results to be applied properly.

In discussing either case, the voltage gain taken as V_2'/V_1' will be used, although these two voltages are internal node voltages not actually appearing in the circuit. The voltage gain between any two identical points in adjacent stages will be the same as V_2'/V_1', however.

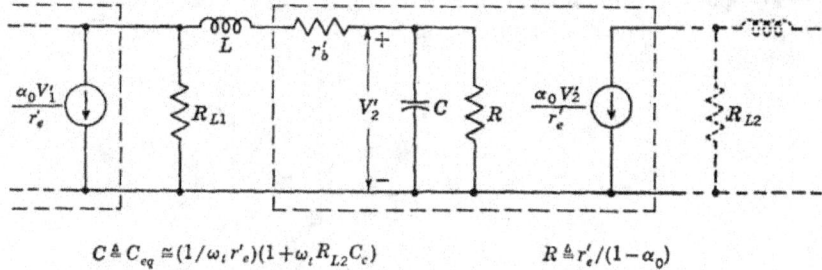

$$C \triangleq C_{eq} \cong (1/\omega_t r_e')(1 + \omega_t R_{L2} C_c) \qquad\qquad R \triangleq r_e'/(1 - \alpha_0)$$

Fig. 4-13 Transistor equivalent circuit and series-peaked interstage.

The voltage gain of the series-peaking circuit will be considered first because the results are somewhat simpler and are also useful in designing an input stage where R_L is then the source resistance in series with a voltage generator V_g. The transfer impedance Z_{21} of the interstage shown in Fig. 4-13 is

$$Z_{21} \triangleq \frac{V_2'}{I_1} = \frac{R_{L1}R/(r_b' + R_{L1} + R)}{p^2 \dfrac{LCR}{r_b' + R_{L1} + R} + p\left[\dfrac{L}{r_b' + R_{L1} + R} + \dfrac{R(r_b' + R_{L1})C}{r_b' + R_{L1} + R}\right] + 1}$$

(4-7)

For convenience define

$$R_1 \triangleq r_b' + R_{L1} + R \qquad\qquad (4\text{-}8)$$

$$R_{eq} \triangleq \frac{R(r_b' + R_{L1})}{r_b' + R_{L1} + R} \qquad\qquad (4\text{-}9)$$

Note that R_{eq} is the resistance in parallel with C. With these substitutions Z_{21} may be written as

$$Z_{21} = \frac{R_{L1}R/R_1}{p^2(LCR/R_1) + p(R_{eq}C + L/R_1) + 1} \qquad (4\text{-}10)$$

For $L = 0$ the voltage gain is

$$A = \frac{V_2'}{V_1'} = \frac{-\alpha_0 Z_{21}}{r_e'} = \frac{-\alpha_0 R_{L1} R}{R_1 r_e'} \frac{1}{p R_{eq} C + 1} \tag{4-11}$$

The form of the step response, the constant in the numerator of Eq. (4-11) being disregarded, is

$$v_2'(t) \approx (1 - \epsilon^{-t/R_{eq}C}) \tag{4-12}$$

The rise time of the uncompensated stage, as for Eq. (4-2), is thus $2.2 R_{eq} C$. If L is increased from zero, the rise time decreases without any overshoot appearing until a critical value of L is reached. To facilitate understanding the effect of changing L, write Z_{21} in factored form, assuming that L/R_1

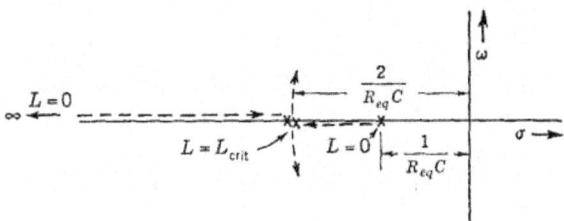

Fig. 4-14   Locus of the poles of a series-peaked stage as L is increased from zero.

$\ll R_{eq} C$. This will be shown to be usually true in a video amplifier. The two poles of Z_{21}, if L/R_1 is neglected, are

$$p_1, p_2 = \frac{-R_{eq} C \pm \sqrt{R_{eq}^2 C^2 - 4 L C R / R_1}}{2 L C R / R_1} \tag{4-13}$$

The locus of the poles p_1, p_2 as L varies is shown in Fig. 4-14. The maximum value of L without having complex p_1 and p_2 is that which makes the radical vanish. This value of L gives critical damping. Making L larger introduces oscillatory terms which cause overshoot in the output waveform.

$$R_{eq}^2 C^2 = \frac{4 L C R}{R_1} \tag{4-14}$$

$$L_{crit} = \frac{R_1 R_{eq}^2 C}{4R} = \frac{R C (r_b' + R_{L1})^2}{4(R_{L1} + r_b' + R)} \tag{4-15}$$

Substituting this value of L into Eq. (4-10) gives

$$A(p) = \frac{-\alpha_0 R_{L1} R}{R_1 r_e'} \frac{1}{[p(R_{eq}C/2) + 1]^2} \tag{4-15a}$$

The output voltage $v_2'(t)$ for a step input is

$$v_2'(t) \approx 1 - \left(1 + \frac{2t}{R_{eq}C}\right)\epsilon^{-2t/R_{eq}C} \qquad (4\text{-}16)$$

The 10 to 90 per cent rise time T_R of $v_2'(t)$ is then found by numerical substitution in Eq. (4-16),

$$T_R = 1.7 R_{eq}C \qquad (4\text{-}17)$$

The inductor causes an improvement in rise time of

$$\eta \triangleq \frac{\text{rise time of simple resistance interstage}}{\text{rise time of interstage in question}} \qquad (4\text{-}18)$$

$$\eta = \frac{2.2 R_{eq}C}{1.7 R_{eq}C} = 1.3 \qquad (4\text{-}19)$$

The addition of enough inductance to give critical damping thus materially improves the rise time, but not quite as much as in the pentode case, where $m = 0.25$ gives $\eta = 1.4$.

The inequality which was used to give these results may now be investigated.

$$R_{eq}C \gg \frac{L}{R_1} = \frac{R_{eq}{}^2 C}{4R} \qquad (4\text{-}20)$$

This may be written in the form

$$1 \ll 4\left(1 + \frac{R}{R_{L1} + r_b'}\right) \qquad (4\text{-}21)$$

The inequality is sufficiently well satisfied if the stage bandwidth is at least a few times the beta-cutoff frequency. Reducing R_{L1} for shorter rise times makes the inequality still more valid.

It is instructive to form the gain/rise-time quotient for a transistor stage, as was done for the pentode amplifier. Using Eqs. (4-11) and (4-12), we obtain

$$A = \frac{\alpha_0 R_{L1} R}{R_1 r_e'} \qquad (4\text{-}22)$$

$$\frac{A}{T_R} = \frac{\eta \alpha_0 R_{L1} R}{r_e'(R + R_{L1} + r_b')} \frac{1}{2.2 R_{eq}C} \qquad (4\text{-}23)$$

This may also be written in a normalized form on the assumption that the stage in question is one of a series of identical stages, $R_{L1} = R_{L2} = R_L$.

Note that R_{L2} enters into the determining of the value of C [see Eq. (3-104)].

$$\frac{A}{\omega_t T_R} = \frac{\eta \alpha_0}{2.2(1 + r_b'/R_L)(1 + \omega_t R_L C_c)}$$

$$= \frac{\eta \alpha_0}{2.2(1 + r_b'/R_L)(1 + a R_L/r_b')} \tag{4-24}$$

where $\qquad\qquad a \triangleq \omega_t C_c r_b'$

Note that, in contrast to the pentode amplifier, the gain/rise time is not a constant but depends upon the ratio of load resistance R_L to base resistance r_b' and also upon $a = \omega_t C_c r_b'$, which is a constant for any given transistor.

Decreasing the rise time of a stage by decreasing R_L increases the first term of the denominator of Eq. (4-24) but decreases the second term. Hence there is an optimum R_L which gives the greatest gain/rise time. This optimum is the same R_L which gave the maximum gain-bandwidth product in Eq. (3-110), which is repeated here,

$$R_{L,\text{opt}} = \sqrt{\frac{r_b'}{\omega_t C_c}} \qquad \text{or} \qquad \frac{r_b'}{R_{L,\text{opt}}} = \sqrt{a} \tag{4-25}$$

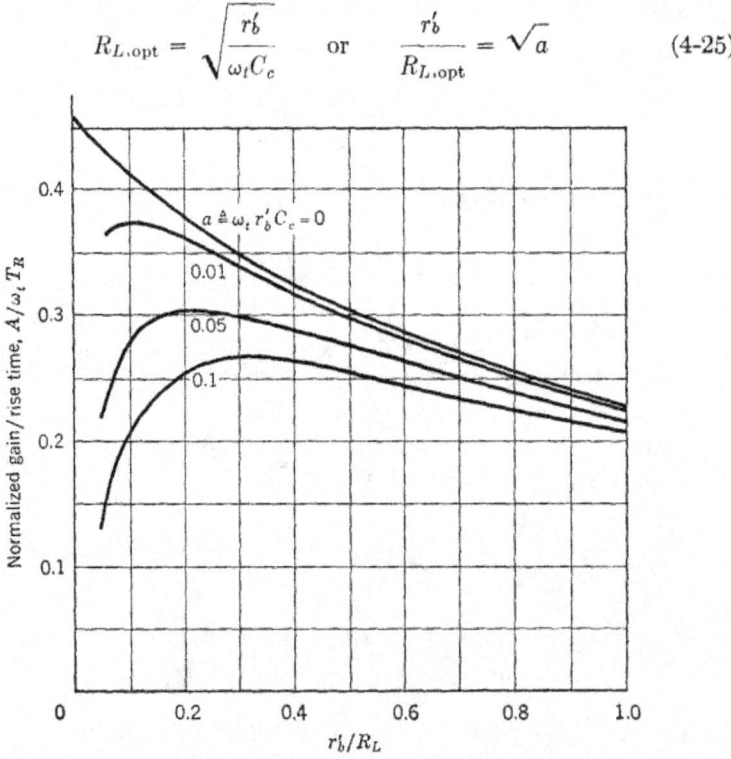

Fig. 4-15 Normalized gain/rise time for an iterative ($R_{L1} = R_{L2}$) transistor stage with no peaking.

The effect of changing R_L on the gain/rise time is easily seen in the graph of Fig. 4-15. The gain/rise time is reduced at high values of R_L by the effect of the feedback capacitance C_c and at low values of R_L by the effect of r_b'. The curves are not exactly correct for very large values of R_L, however, because of the approximations in deriving the effect of the feedback capacitor (see Fig. 3-38c). The effect of r_b' and C_c is serious in a fast amplifier, as can be seen from the value of $a = \omega_t r_b' C_c = 0.06$ for a typical high-frequency transistor with $f_t = 100$ Mc, $r_b' = 50$ ohms, and $C_c = 2$ pf. Figure 4-15 also shows the value of gain/rise time which is obtained for a series-peaked stage if the value of $A/\omega_t T_R$ is multiplied by $\eta = 1.3$.

The calculation of the required load resistance R_L for a given stage rise time in a chain of iterative stages involves the solution of a quadratic; however, a graph can be prepared which simplifies the calculation. The rise time for the series-peaked stage (after substituting for R_{eq} and C) is

$$T_R = \frac{2.2}{\eta} \frac{R(R_L + r_b')}{R + R_L + r_b'} \frac{1}{\omega_t r_e'} (1 + \omega_t R_L C_c)$$

$$= \frac{2.2}{\eta} \frac{R_L + r_b'}{1 + (R_L + r_b')/R} \frac{1}{\omega_t r_e'} \left(1 + \omega_t r_b' C_c \frac{R_L}{r_b'}\right) \tag{4-26}$$

$$\underbrace{T_R \left(1 + \frac{R_L + r_b'}{R}\right)}_{= T_R'} = \underbrace{\frac{2.2 r_b'}{\eta \omega_t r_e'}}_{= T_1} \left(\frac{R_L}{r_b'} + 1\right) \left(1 + \underbrace{\omega_t r_b' C_c}_{= a} \frac{R_L}{r_b'}\right) \tag{4-27}$$

In the usual case T_R is approximately equal to T_R', although the former is always somewhat smaller. The modified rise time T_R' is now only a function of R_L/r_b', which is what we seek, and of two quantities depending only upon the transistor characteristics: $a \triangleq \omega_t r_b' C_c$ and $T_1 \triangleq 2.2 r_b'/\eta \omega_t r_e'$. Hence we may plot T_R'/T_1 as a function of R_L/r_b' for some typical values of a. This is done in Fig. 4-16, which is useful for both the uncompensated stage and the series-peaked stage by taking η to be 1.0 and 1.3, respectively. Although it might appear from an inspection of Fig. 4-16 that the load resistance is proportional to r_b' for a given T_R, this is not true, for the rise time is also normalized by r_b'. Hence, although r_b' is important in determining R_L, a small error in the measurement of r_b' will not lead to a proportional error in R_L or rise time.

To find the exact R_L for a desired rise time, an iterative process must be used, since T_R is not exactly equal to T_R'. As an example, assume that the characteristics of a transistor are

$$r_b' = 50 \text{ ohms} \qquad\qquad C_c = 5 \text{ pf}$$

$$r_e' = 10 \text{ ohms} \qquad\qquad \frac{1}{1 - \alpha_0} = 50$$

$$\omega_t = 2 \times 10^8 \text{ radians/sec}$$

Then $a = \omega_t r_b' C_c = 0.05$, and $T_1 = 2.2 r_b'/\eta \omega_t r_e' = 4.23 \times 10^{-8}$ sec (for $\eta = 1.3$).

Suppose that the load resistance R_L for a stage rise time of 0.15 μsec is to be found. For a first trial take $T_R' = 0.15$ μsec; then $T_R'/T_1 = 3.54$.

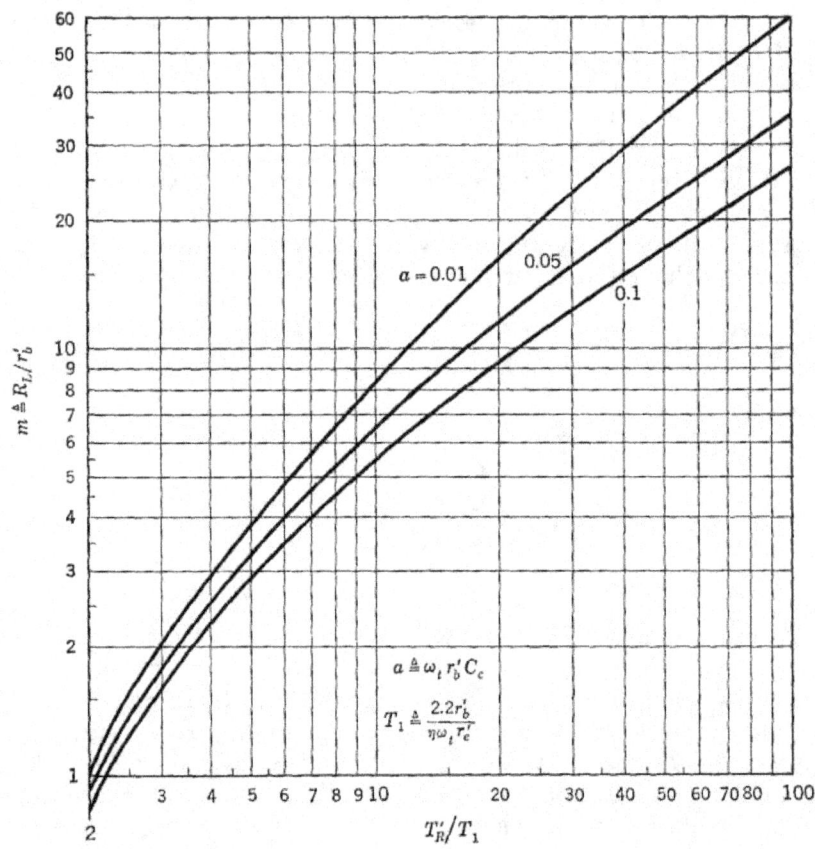

Fig. 4-16 Load resistance as a function of stage rise time (for iterative stages, that is, $R_{L1} = R_{L2}$).

Figure 4-16 gives $R_L/r_b' = 2.2$ or $R_L = 110$ ohms. The ratio $T_R'/T_R = 1 + (R_L + r_b')/R = 1 + (110 + 50)/500 = 1.32$. Hence the actual rise time obtained with the 110-ohm load resistor is $0.15/1.32 = 0.114$ μsec. To obtain a more accurate value, a modified value of T_R' may be used, which is $(0.15)(1.32) = 0.198 \cong 0.2$ μsec. This value of T_R' leads to $R_L = 155$ ohms and to a calculated $T_R = 0.142$ μsec. If this value is not close enough to the desired T_R, the process may be repeated; however, the approximations in the equivalent circuit make the use of extreme precision in these calculations of doubtful value.

4-8 The Shunt-peaked Transistor Interstage. The shunt-peaked interstage shown in Fig. 4-17 has no more elements than the previously described series-peaked stage but has considerably improved performance. The transfer impedance of the shunt-peaked stage is

$$Z_{21} = \frac{[RR_{L1}/(R_{L1} + r_b' + R)](pL/R_{L1} + 1)}{\dfrac{p^2 RLC}{R_{L1} + r_b' + R} + p\left[\dfrac{(r_b' + R_{L1})RC}{R_{L1} + r_b' + R} + \dfrac{L}{R_{L1} + r_b' + R}\right] + 1} \tag{4-28}$$

This equation has the same denominator as the equation for the series-peaked circuit; however, there is now a zero in Z_{21} at $p = -R_{L1}/L \triangleq -z_0$.

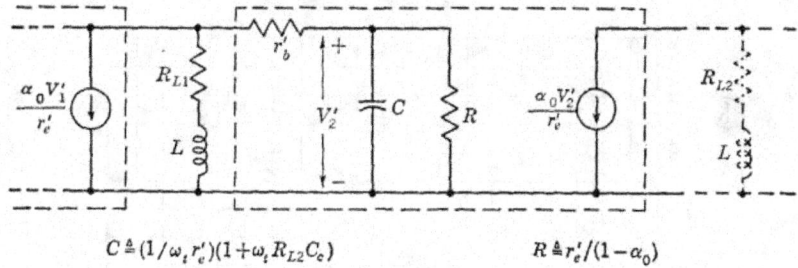

$$C \triangleq (1/\omega_t r_e')(1 + \omega_t R_{L2} C_c) \qquad\qquad R \triangleq r_e'/(1 - \alpha_0)$$

Fig. 4-17 A shunt-peaked transistor interstage.

Again the value of L may be chosen so that critical damping is obtained—this will require the same value of L as before,

$$L_{\text{crit}} = \frac{(r_b' + R_{L1})^2 RC}{4(R_{L1} + r_b' + R)} \tag{4-29}$$

For this value of L the poles of Z_{21} are in the same position as for the series-peaked circuit,

$$p_1 = p_2 = \frac{2(R_{L1} + r'_1 + R)}{(R_{L1} + r_b')RC} = \frac{2}{R_{\text{eq}}C} \tag{4-30}$$

To compute the rise time for the interstage, we must also consider the effect of the zero in Z_{21}; consequently the form of gain function we must consider is

$$A(p) = \frac{K(p + z_0)}{(p + p_1)^2} = \frac{K'(s + z_0/p_1)}{(s + 1)^2} \tag{4-31}$$

where $$s = \frac{p}{p_1}$$

The rise time may be found from Eq. (4-31) by calculating the response to a unit step for various values of z_0/p_1. To find the range of values for the latter, we find

$$\frac{z_0}{p_1} = \frac{2}{1 + r_b'/R_{L1}} \tag{4-32}$$

Hence the maximum value of z_0/p_1 is 2; the minimum value that is useful is probably in the neighborhood of unity since $R_{L1} < r_b'$ leads to very low gain. From calculations of rise time for various values of z_0/p_1 the rise-time improvement factor η may be calculated as a function of z_0/p_1. For the shunt-peaked interstage, η as a function of r_b'/R_{L1} is shown in Fig. 4-18.

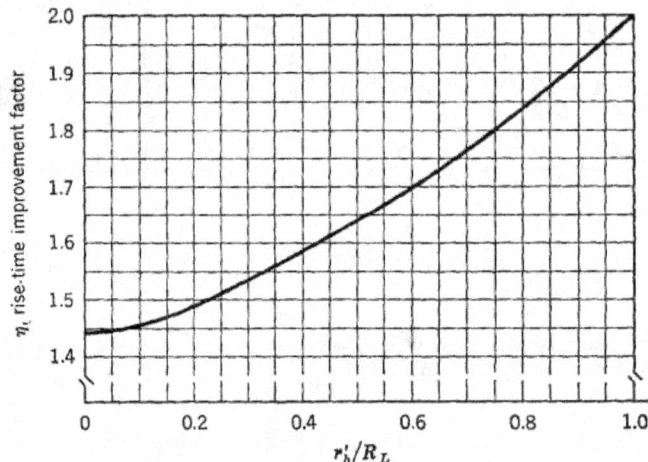

Fig. 4-18 Rise-time-improvement factor for a critically damped, shunt-peaked transistor stage.

The value $\eta = 1.44$ for large values of R_{L1} is somewhat better than that obtained for the series-peaked interstage, but in the range of r_b'/R_{L1} frequently encountered in a fast amplifier the value of η obtained for the shunt-peaked stage is very considerably better. The improvement obtained by using shunt peaking is graphically shown in Fig. 4-19, where normalized gain/rise time is shown as a function of r_b'/R_{L1} for an interstage with no peaking, with series peaking, and with shunt peaking. These curves are for the iterative situation, $R_{L1} = R_{L2} = R_L$, as in a chain of identical stages.

The rise time of the shunt-peaked stage is given by Eq. (4-26) with appropriate values of η; however, the equation must be treated differently to find R_L for a required rise time in a chain of identical stages, because η is now a function of r_b'/R_L. Equation (4-26) may be written in the form

$$\underbrace{T_R\left(1 + \frac{R_L + r_b'}{R}\right)}_{= T_R'} = \underbrace{\frac{2.2r_b'}{\omega_t r_e'}}_{= T_2}\left[\frac{1}{\eta}\left(\frac{R_L}{r_b'} + 1\right)\left(1 + \underbrace{\omega_t r_b' C_c \frac{R_L}{r_b'}}_{= a}\right)\right] \quad (4\text{-}33)$$

The quotient T_R'/T_2 is now a function of R_L/r_b' and transistor parameters

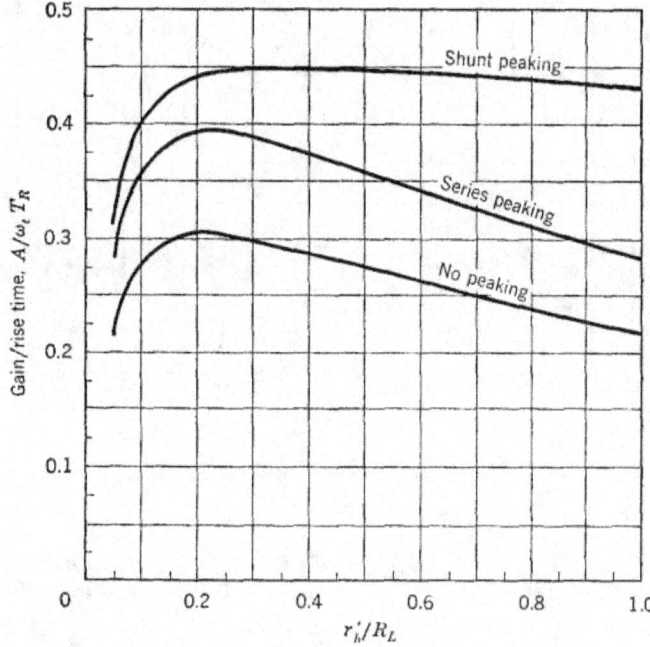

Fig. 4-19 Comparison of normalized rise times for three interstage networks using a transistor having $a \approx 0.05$.

since $\eta = f(R_L/r_b')$. A graph of R_L/r_b' as a function of T_R'/T_R for various values of $a = \omega_t r_b' C_c$ is shown in Fig. 4-20. The use of Fig. 4-20 and Fig. 4-16 is identical, but Fig. 4-20 is valid only for shunt-peaked stages, since the value of η is "built in" instead of appearing in T_1.

The step responses of the three types of transistor video interstage are compared in Fig. 4-21. These curves may be considered to be the response obtained by keeping R_L fixed and changing L to the proper position in the circuit and the proper value for each curve. Note that, although the rise time is reduced by series peaking, the time delay is increased. This results because dV/dt is zero at $t = 0^+$ in the series-peaked stages, whereas dV/dt is maximum at $t = 0^+$ for both the unpeaked and shunt-peaked stages. Although the series-peaked circuit is inferior to the shunt-peaked with regard to rise time, series peaking is useful at the input of a transistor amplifier driven from a resistive source.

A comparison of the results obtained by peaking circuits for both the vacuum tube and transistor shows that comparable values of η may be

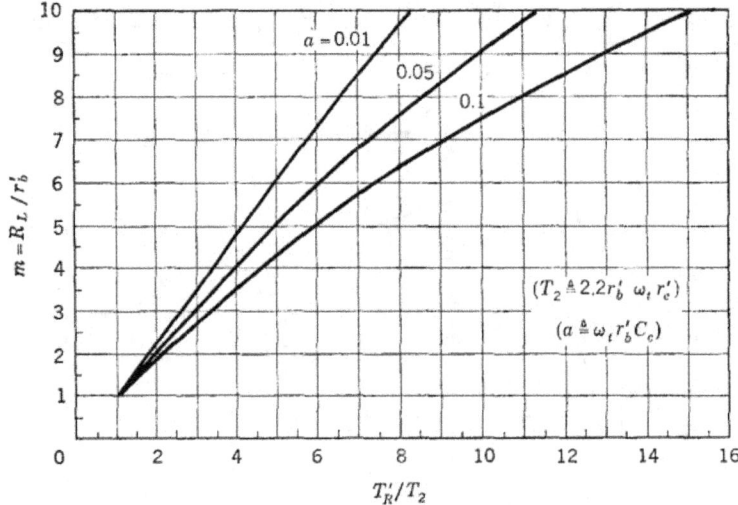

Fig. 4-20 Normalized load resistance as a function of normalized rise time for a shunt-peaked stage ($R_{L1} = R_{L2}$).

obtained for both cases, with the same degree of circuit complexity. The solution of the transistor case is more complicated, however, primarily because of the presence of r_b' and C_c. Four-terminal interstages can be devised for transistor use also, but they have one limitation not present in pentode-tube use: the higher gain is obtained by an impedance-transforming action at the higher frequencies; i.e., the input impedance of the interstage tends to become large at high frequencies. For transistor use this action tends to accentuate the effect of C_c in the transistor driving the interstage: thus the bandwidth of the previous stage is reduced by more than the Miller-effect capacitance calculations of Eq. (3-104) would indicate. Hence the increased gain/rise-time quotient obtained by the four-terminal interstage is somewhat offset

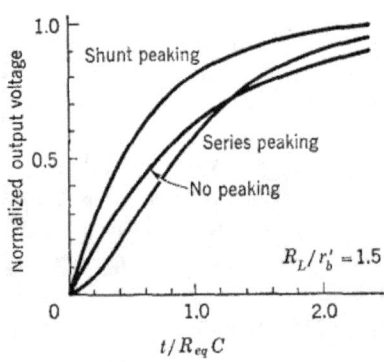

Fig. 4-21 Step response of a transistor stage.

by the decrease in the bandwidth of the driving stage. The effect is, of course, to some degree present even in the transistor peaking circuits discussed herein.

4-9 Amplifier Stages in Cascade. In general, a single stage of amplification is not adequate to meet the gain requirements of a system, and hence several stages will be connected in cascade. The question arises

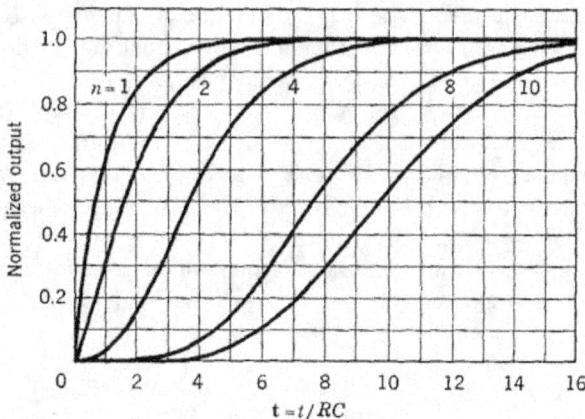

Fig. 4-22 Step response of an n-stage RC amplifier.

as to the over-all response of the system when the responses of the individual stages are known. Of course, the entire system could be analyzed as a formal circuit problem, but there are some simple rules of great general value.

The nature of the transient response of several pentode stages (identical) in cascade is illustrated in Figs. 4-22 and 4-23; both curves are from Valley

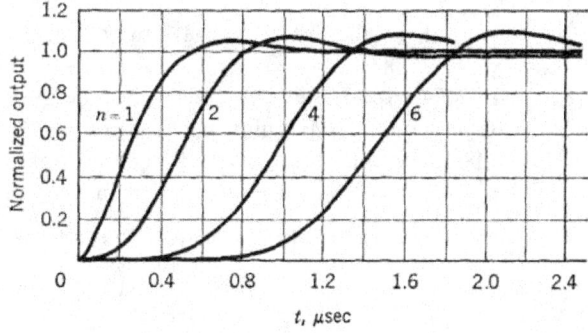

Fig. 4-23 Step response of n stages each having $|A| \approx 1/\sqrt{1 + (f/10^6)^4}$ (1 Mc bandwidth).

and Wallman.[1] The stages shown in Fig. 4-22 have no peaking and have no overshoot. The response of the stages in Fig. 4-23, on the other hand, is the step response of a rather complicated interstage adjusted to have a very flat, steady-state amplitude response, with substantial transient overshoot in even a single stage.

The two figures show the general properties of the increasing delay time and rise time resulting as the number of stages is increased. Moreover, if there is overshoot in a single stage, the amount of this increases as stages are added. On the other hand, if there is no overshoot in a single stage, the adding of stages will not introduce overshoot (Fig. 4-22).

The quantitative details are of interest. Various attempts have been made to analyze the problem, some based upon an empirical study of cascades of particular circuits.[2] The analysis is quite laborious, even if the network functions and their inverse transforms are simple, because of the arbitrary definition of the rise time in terms of the 10 and 90 per cent levels. For instance, the network function for a cascade of resistance-coupled pentode stages, whose response in time is given in Fig. 4-22, is simply

$$V_n(p) = \frac{V_{\text{output for } n \text{ stages}}}{(-g_m R_L)^n} = \left(\frac{1}{p R_L C + 1}\right)^n V_{\text{in}}(p) \qquad (4\text{-}34)$$

To get an answer in terms of normalized time $\mathbf{t} = t/R_L C$, take $s = p R_L C$ as the variable. Assume a unit step as the input, $V_{\text{in}} = 1/s$,

$$V_n(s) = \frac{1}{s(s+1)^n} \qquad (4\text{-}35)$$

Inverse transform $\qquad v_n(\mathbf{t}) = 1 - \epsilon^{-\mathbf{t}} \sum_{r=0}^{r=n-1} \frac{\mathbf{t}^r}{r!} \qquad (4\text{-}36)$

(NOTE: $0! = 1$.) Equation (4-36) describes the family of curves in Fig. 4-22. Unfortunately there is no correspondingly simple expression to describe the rise time T_R as a function of the number of stages; one must compute the function $v_n(\mathbf{t})$ for each value of n, determining the time between the 10 and 90 per cent levels. Valley and Wallman give the results for values of n up to 10. These are given here in Table 4-2.

[1] See also A. V. Bedford and G. L. Fredendall, Transient Response of Multi-stage Video-frequency Amplifiers, *Proc. IRE*, vol. 27, p. 277, April, 1939.

[2] Valley and Wallman, *op. cit.*, pp. 65–66, 77–78; Bedford and Fredendall, *ibid.*; H. E. Kallman, R. E. Spencer, and C. P. Singer, Transient Response, *Proc. IRE*, vol. 33, pp. 169–195, March, p. 482, July, 1945; D. G. Tucker, Bandwidth and Speed of Build-up as Performance Criteria for Pulse and Television Amplifiers, *J. IEE*, vol. 94, pp. 218–226, May, 1947.

Table 4-2

| | No. of stages, n | | | | | | | | | |
	1	2	3	4	5	6	7	8	9	10
*	1.0	1.5	1.9	2.2	2.5	2.8	2.9	3.14	3.27	3.54
†	...	1.56	1.91	2.2	2.46	2.7	2.91	3.11	3.3	3.48

* Rise time T_R divided by $2.2RC$.
† Approximation by $1.1\sqrt{n}$ as in Eq. (4-38).

The results of empirical studies such as these can be summarized into several working rules, as formulated by Valley and Wallman:

1. In circuits having little or no overshoot, the over-all rise time T_{RN} is approximately given by [1]

$$T_{RN} = \sqrt{T_{R1}^2 + T_{R2}^2 + T_{R3}^2 + \cdots + T_{Rn}^2} \qquad (4\text{-}37)$$

1*a*. A better approximation for identical stages (Table 4-2) is

$$T_{RN} = 1.1\sqrt{n}\,T_{R1} \qquad (4\text{-}38)$$

2. In circuits having little or no overshoot, the total overshoot for n stages is essentially that of a single stage.

3. If the overshoot of a single stage is in the order of 5 or 10 per cent, then the total overshoot goes as the square root of the number of stages.

3*a*. When the stage overshoot is 5 or 10 per cent, the total rise time is somewhat less than that given by rule 1.

Of the above rules, the first is perhaps the most important. There are a large number of cases in which the overshoot must be kept small and one is concerned with rise time alone. A valuable contribution to the analysis of this case has been made by Elmore,[2] putting rule 1 on a somewhat more rigorous basis.

To follow Elmore's analysis, we first redefine the delay time and the rise time to be functions which are more readily expressible yet which give essentially the same numerical results in typical circuits as the previous

[1] Elmore, *op. cit.* See also related papers: W. C. Elmore, Electronics for the Nuclear Physicist, *Nucleonics*, vol. 2, pp. 4–17, February, pp. 16–36, March, pp. 43–55, April, pp. 50–58, May, 1948; W. C. Elmore and M. Sands, "Electronics: Experimental Techniques," chap. 3, National Nuclear Energy Series, McGraw-Hill Book Company, Inc., New York, 1949.

[2] The Transient Response of Damped Linear Networks, *loc. cit.*

definitions. The delay time T_D is defined in terms of the impulse response, or, what is the same thing, the derivative of the step response. In Fig. 4-24 is shown a step response $v(t)$ and in Fig. 4-25 its corresponding impulse response $v'(t)$. The 50 per cent level of $v(t)$ was the previous definition of T_D; the maximum of $v'(t)$ might have been equally good and would give the

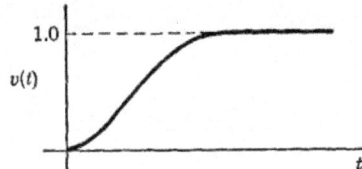

Fig. 4-24   A typical step response.   **Fig. 4-25** Impulse response corresponding to the step response of Fig. 4-24.

same value of T_D in many cases. Instead of either of these, however, Elmore uses the centroid of the area under $v'(t)$,

$$T_D = \frac{\displaystyle\int_0^\infty t v'(t)\, dt}{\displaystyle\int_0^\infty v'(t)\, dt} \qquad (4\text{-}39)$$

or, for a normalized step response $v(t)$ of height 1.0 as in Fig. 4-24,

$$T_D = \int_0^\infty t v'(t)\, dt \qquad (4\text{-}39a)$$

It should be noted that this definition is not well suited to functions $v(t)$ that have overshoot, for in such cases there would be negative areas occurring in $v'(t)$ and these would improperly alter the value of T_D. Hence this definition and the derivations to follow should be restricted to monotonic or non-overshoot responses. But to judge by the results of the empirical studies referred to previously, the results can probably be applied to cases of small overshoot, say less than 5 per cent.

The new definition of rise time T_R is based upon a quantity known as the "radius of gyration" in kinematics or as the "standard deviation" in statistics,

$$T_R{}^2 = \text{const} \times \int_0^\infty (t - T_D)^2 v'(t)\, dt \qquad (4\text{-}40)$$

[Note that $\int_0^\infty v'(t)\, dt = 1$ is implicit in the denominator.] The value of

the constant can be chosen arbitrarily to make the numerical values come
out right. Elmore uses 2π, because this makes the rise time computed for
an infinite number of resistance-coupled stages [1] agree with the value
obtained for the rise time in terms of the maximum slope, which is

$$T_R = \frac{\text{area under } v'(t)}{\text{max value of } v'(t)} \tag{4-41}$$

An alternative form of Eq. (4-40) with the constant inserted is

$$T_R{}^2 = 2\pi \left[\int_0^\infty t^2 v'(t)\, dt - T_D{}^2 \right] \tag{4-42}$$

Before proceeding further with the analysis, it might be well to see what
kind of numerical values come out of these definitions for T_D and T_R and
compare them with previous definitions:

	RC stage	Shunt-peaked stage, $m = 0.25$
T_D:		
Defined as t/RC to 50% level........	0.7	0.6
Defined by Eq. (4-39)...............	1.0	0.75
T_R:		
Defined as t/RC between 10 and		
90% levels......................	2.2	1.57 *
Defined by Eq. (4-42)..............	2.5 ($\sqrt{2\pi}$)	1.66 †

* Improvement over RC is 1.4.
† Improvement over RC is 1.51.

As can be seen, the two definitions of T_R agree moderately well. The
agreement would be even closer for a large number of resistance-coupled
stages because of the basis for selecting the constant 2π.

Now we come to the problem of determining the rise time and delay time
for a cascade of stages. We do not yet know how the time responses for
several stages combine, but we do know that the voltage gains $A_i(p)$ combine as a continued product in a cascade of amplifier stages consisting of
ideal pentodes as unilateral coupling elements separating the interstage

[1] In this case the response $v'(t)$, as well as $v(t)$ and the steady-state amplitude response,
approaches a gaussian error function; see Valley and Wallman, op. cit., pp. 723–724.

networks. Thus, the gain function for a cascade of n stages (see Fig. 4-26) as a function of the complex frequency variable p is given by

$$A_n(p) = \frac{V_n(p)}{V_0(p)} = \prod_{i=1}^{n} A_i(p) \tag{4-43}$$

where $V_0(p)$ = voltage at input of 1st stage

$V_n(p)$ = voltage at output of nth stage

$A_i(p)$ = gain function of ith stage = $V_i(p)/V_{i-1}(p)$

The next step is to normalize each of the $A_i(p)$ by dividing by $-g_m R_L$, so that the normalized gain for each stage is unity. Then we drive the circuit

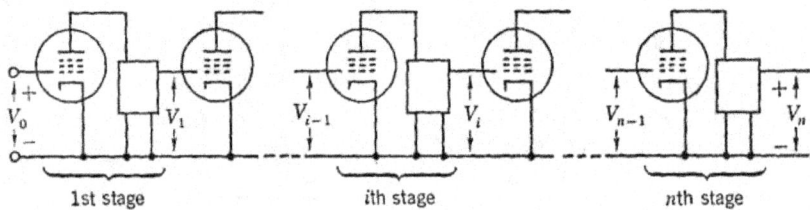

Fig. 4-26 Nomenclature for the stages in a multistage amplifier.

with a unit impulse, i.e., the time derivative of a unit step, so that $V_0(p)$ = 1.0 (which is the Laplace transform of the unit impulse).

$$a_n(p) = V_n(p) = \prod_{1}^{n} a_i(p) = \mathcal{L}[v_n'(t)] \tag{4-44}$$

Normalized
over-all gain
function

Normalized
gain function
of the ith
stage

We take next the general expression for the Laplace transform, expand ϵ^{-pt} in a power series, integrate term by term, and gather and identify terms with the definitions of Eqs. (4-39) and (4-42).

$$a_n(p) = V_n(p) = \int_0^\infty v_n'(t)\epsilon^{-pt}\,dt$$

$$= \int_0^\infty v_n'(t)\left[1 - pt + \frac{(pt)^2}{2!} - \cdots\right]dt$$

$$= 1 - p\int_0^\infty tv_n'(t)\,dt + \frac{p^2}{2!}\int_0^\infty t^2 v_n'(t)\,dt - \cdots$$

$$= 1 - pT_{DN} + \frac{p^2}{2!}\left(\frac{T_{RN}^2}{2\pi} + T_{DN}^2\right) - \cdots \tag{4-45}$$

where T_{DN} = delay time of n stages

T_{RN} = rise time of n stages

Note however that a similar relationship holds for the ith stage (*one* stage only).

$$a_i(p) = 1 - pT_{Di} + \frac{p^2}{2!}\left(\frac{T_{Ri}^2}{2\pi} + T_{Di}^2\right) - \cdots \qquad (4\text{-}46)$$

Now, since $a_n(p)$ is the continued product of the $a_i(p)$, we take a continued product of series like Eq. (4-46) and gather terms in like powers of p, giving

$$a_n(p) = \prod_1^n a_i(p)$$

$$= 1 - p\sum T_{Di} + \frac{p^2}{2!}\left(\sum \frac{T_{Ri}^2}{2\pi} + \sum T_{Di}^2 + \sum_{i\neq j} T_{Di}T_{Dj}\right) - \cdots$$

$$(4\text{-}47)$$

Finally, we compare Eqs. (4-45) and (4-47) and equate the coefficients of the p and p^2 terms, giving the following results:

$$T_{DN} = \sum_1^n T_{Di} \qquad (4\text{-}48)$$

$$T_{RN}^2 = \sum_1^n T_{Ri}^2 \qquad (4\text{-}49)$$

The important result of Eq. (4-49) verifies the previous equation (4-37) in showing that the over-all rise time is the root mean square of the individual rise times of the stages. And Eq. (4-48) shows that the over-all delay time is the simple sum of the individual stage delay times.

Some of the by-products of Elmore's analysis are of interest. For instance, if one has a cascade of similar stages, should all stages be designed for the same gain and the same rise time, or might something be gained by having high gain and slow speed in one stage but counterbalanced by low gain and high speed in another stage? The answer apparently is that the *minimum over-all rise time for a given gain is achieved by making all stages the same* (see Sec. 4-10). The proof involves the use of Lagrange's method of undetermined multipliers, operating upon Eq. (4-49) and another equation which results from the fact that the rise time of each stage is proportional to the gain, since both are proportional to R_L.

$$\prod_1^n A_i \sim \prod_1^n T_{Ri} = \text{const} \qquad \text{for specified gain}$$

For the condition of all stages identical, a simpler form of Eq. (4-49) is

$$T_{RN} = T_{Ri} \sqrt{n} \tag{4-50}$$

Another interesting result is that there can be found an optimum gain per stage in order to achieve a given over-all gain with the minimum over-all rise time. This gain per stage is $\sqrt{\epsilon} = 1.65$ and is derived as follows:

$$
\begin{aligned}
T_{RN} &= \sqrt{n}\, T_{Ri} \\
&= \sqrt{n}\, \frac{A_i}{A/T_R} \\
&= \sqrt{n}\, \frac{A_i}{\eta g_m / \sqrt{2\pi}\, C} \\
&= \sqrt{n}\, \frac{(A_n)^{1/n}}{\eta g_m / \sqrt{2\pi}\, C}
\end{aligned} \tag{4-51}
$$

where η = efficiency compared with RC stage (as in Fig. 4-3).

Solve for $\partial T_{RN}/\partial n$, and equate to zero to find $T_{RN,\,min}$. The minimum value of T_{RN} for a specified A_n occurs when

$$n = 2 \ln A_n \tag{4-52}$$

$$A_i = A_n{}^{1/n} = A_n{}^{1/(2 \ln A_n)} = \epsilon^{\frac{1}{2}} = 1.65 \tag{4-53}$$

$$T_{RN,\,min} = \frac{\sqrt{2\epsilon \ln A_n}}{\eta g_m / \sqrt{2\pi}\, C} \tag{4-54}$$

Note that Eq. (4-54) gives the rise time according to Elmore's definition; for the 10 to 90 per cent value, replace $\sqrt{2\pi}$ by 2.2.

The relationships given above have some practical limitations and require some judicious interpretation and application. First note that n has the same value, regardless of tube type or circuit type; for instance, if A_n is 10^5 (100 db), Eq. (4-52) says that 23 stages are called for, whether one uses 6AU6 or 6AH6 tubes. But it does *not* say that the same minimum rise time results in either case; from Table 4-1 and Eq. (4-54) it will be found that the 6AU6 would give $256/163$, or 1.57, times as great a rise time as the 6AH6.

Also, the minimum of the rise-time function in Eq. (4-51) is a broad one, and one can violate the minimum conditions by quite a margin without serious detriment to the over-all rise time. Elmore provides an example of a 6AC7 amplifier ($g_m = 0.009$ mho, $C = 22$ pf, $\eta = 1.5$) in which the 23 stages required for the 100-db gain give a rise time of 0.032 μsec; yet, with only 9 stages to give the same gain, the rise time is 0.044 μsec. Thus, the rise time is impaired only 37.5 per cent for a 48 per cent reduction in the number of stages. Moreover, the larger load resistor in the latter case (400 ohms instead of 180 ohms) permits a larger output voltage to be realized from the amplifier (222 per cent greater).

4-10 Amplifiers with Nonidentical Stages. The conclusion that an amplifier should be made up of identical stages for minimum rise time is obtained by an analysis which is invalid if *any* of the stages has overshoot. Hence there would seem to be the possibility of designing an amplifier in which some stages have large overshoots (large *m*) coupled to other stages which tend to reduce the overshoot. Attempts to design such an amplifier on the basis of the steady-state amplitude response have usually resulted in amplifiers with excessive over-all overshoot.[1,2] Designing the amplifier on the basis of a linear steady-state phase response gives an excellent transient response;[3] however, the most straightforward data to interpret for the transient case are given by F. A. Muller.[4] Either the linear-phase response or the Muller data give a faster amplifier than is obtained by using identical stages. Since the Muller data have the advantage of giving a specified overshoot, they will be presented here. Each stage of the amplifier is a shunt-peaked stage (Fig. 4-2*a*, where *L* may be zero in some stages). The data for each stage are given in terms of $m = L/R_L^2 C$, as before, and a normalized *r* which is defined as

$$r_i \stackrel{\Delta}{=} \frac{\tau_i}{(\tau_1 \tau_2 \tau_3 \cdots \tau_n)^{1/n}} \qquad (4\text{-}55)$$

where
$$\tau_i \stackrel{\Delta}{=} R_i(C_{1.i} + C_{2.i}) \stackrel{\Delta}{=} R_i C_i \qquad (4\text{-}56)$$

In Eq. (4-56) C_i is the total capacitance, and R_i is the load resistor of the *i*th interstage. Table 4-3 gives the value of r_i and m_i (where $m_i = L_i/R_i^2 C_i$) for each stage. The rise time of the whole amplifier is given in normalized form as T_1. The actual rise time T_{RN} of the whole amplifier is given by

$$T_{RN} = (\tau_1 \tau_2 \tau_3 \cdots \tau_n)^{1/n} T_1 \qquad (4\text{-}57)$$

If the amplifier is to be designed for a given over-all rise time and the g_m and C_i of each stage are known, the load resistor for an individual stage is

$$R_i = \frac{r_i T_{RN}}{C_i T_1} \qquad (4\text{-}58)$$

[1] A. Easton, Stagger Peaked Video Amplifier, *Electronics*, vol. 22, p. 118, February, 1949.

[2] J. H. Mulligan, Jr., and L. Mautner, Steady-state and Transient Response of Feedback Video Amplifiers, *Proc. IRE*, vol. 36, pp. 545-610, May, 1948.

[3] G. A. Caryotakis, Iterative Methods in Amplifier Interstage Synthesis, Stanford Electronics Lab., TR-86.

[4] Muller, *op. cit.*

Table 4-3

	Over-shoot, %	r_1	m_1	r_2	m_2	r_3	m_3	$r_4 = r_5$	T_1
2 stages......	1	0.482	1.400	2.073	0.325	. . .	. . .	. . . .	1.58
3 stages......	1	0.252	4.434	1.058	0.953	3.754	0.318	. . . .	1.78
2 stages, 1 coil......	1	0.838	0.753	1.190	0	. . .	. . .	. . . .	1.91
3 stages, 2 coils......	1	0.335	3.087	1.580	0.565	1.889	0	. . . .	1.92
5 stages, 3 coils......	1	0.135	16.5	0.63	1.90	2.23	0.51	2.30	2.28

The resulting over-all gain is

$$A_n = \prod_1^n g_{m,i} R_i = \prod_1^n \frac{g_{m,i} r_i T_{RN}}{C T_{i1}}$$

$$= \left(\frac{T_{RN}}{T_1}\right)^n \prod_1^n \frac{g_{m,i}}{C_i} \tag{4-59}$$

since
$$\prod_1^n r_i = 1$$

If the stages have identical capacitances and g_m's, the above equation reduces to

$$A_n = \left(\frac{T_{RN}}{T_1} \frac{g_m}{C}\right)^n \tag{4-60}$$

On the other hand, if the over-all gain is specified, the individual load resistors are given by

$$R_i = \frac{r_i}{C_i}\left[\frac{A_n}{\prod_1^n (g_{m,i}/C_i)}\right]^{1/n} \tag{4-61}$$

The resultant over-all rise time is

$$T_{RN} = T_1\left[\frac{A_n}{\prod_1^n (g_{m,i}/C_i)}\right]^{1/n} \tag{4-62}$$

If the stage g_m's and capacitances are again identical, we obtain

$$R_i = \frac{r_i(A_n)^{1/n}}{g_m} \tag{4-63}$$

$$T_{RN} = \frac{T_1(A_n)^{1/n}}{g_m/C} \tag{4-64}$$

As may be seen from the data in Table 4-3, the rise time of a multiple-stage amplifier using these "staggered-peaking" techniques increases much more slowly than as the $\sqrt{n}$. Muller notes that the increase in rise time is more nearly proportional to $n^{1/4}$.

A specific example will indicate the advantages of the stagger peaking. Comparing the rise time of a five-stage amplifier using shunt peaking ($m = 0.315$ to give approximately 11 per cent over-all overshoot) with the rise time of a five-stage staggered amplifier having the same gain *but with coils in only three stages*, we find that the ratio of over-all rise times is 1.4:1. Thus the staggered amplifier is faster and has fewer elements. If the amplifiers are compared for the same over-all rise times, the ratio of the gains obtained is $(1.4)^5 = 5.4$; thus the staggered amplifier has about 14.6 db more gain. Staggered amplifiers with fewer stages show significant but less dramatic improvement.

Note that the value of m is great enough in some stages to cause enormous overshoot ($m = 16.5$ in stage 1 of the five-stage example); hence the previous analysis by Elmore is here invalid. The order of stages is of little importance if high output level is of no consequence. However, a stage with large r_i may be used for the output stage to provide a larger available output voltage. Stages with large r_i at the output end of the amplifier have a large overshoot in their input waveform which may cause overdriving of

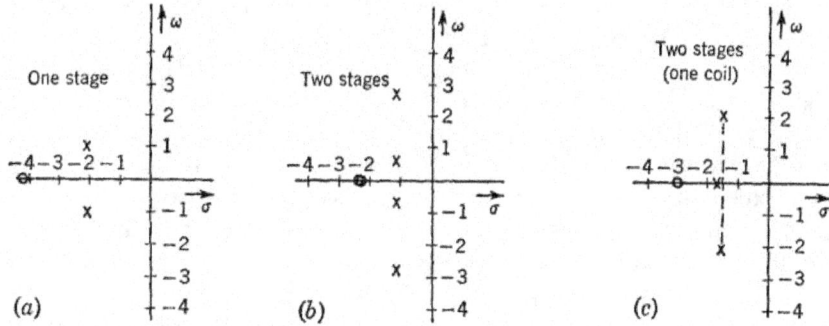

Fig. 4-27 Pole positions for a staggered video amplifier. Each amplifier gives a 1-sec rise time and 1 per cent overshoot. (All coordinates in radians per second.)

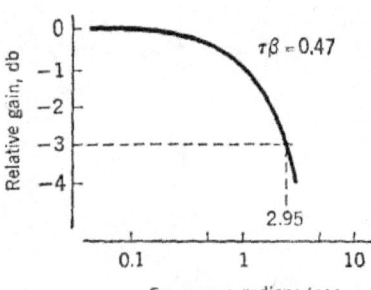

Fig. 4-28 Frequency response corresponding to the pole positions of Fig. 4-27b.

the grid. Thus it is impractical to use the advantage of the large load resistance to the fullest extent.

There is some advantage in using a stage with high r_i for the input stage, too, since this gives large stage gain to help overcome noise in subsequent stages. As a consequence of these two considerations, the stages with large amounts of peaking (the stages with the largest m and lowest r_i) are best placed in the middle of the amplifier.

The pole-zero diagrams for these stagger-peaked amplifiers show that the poles lie very nearly on a line of constant σ. Three examples of the pole-zero diagrams are given in Fig. 4-27. Each of these sets of pole locations yields an amplifier with 1 sec rise time and 1 per cent overshoot. The amplitude response for the two-stage (each stage peaked) example is shown in Fig. 4-28.

Presumably the same sort of improvement in multistage amplifiers could be achieved in a transistor amplifier. However, the situation with transistors is much more complicated because the position of the zero in a shunt-peaked stage does not bear a constant relation to the real part of a complex pole as it does in the vacuum-tube case. In a shunt-peaked vacuum-tube stage $z_1 = 2\,\mathrm{Re}\,p_1$, where z_1 and p_1 are the zero and complex pole positions. Hence the Muller data cannot, in general, be directly applied to the transistor case.

4-11 Output Stages. Most amplifiers have, in addition to the requirement for a certain amount of gain between input and output terminals, a requirement for the amount of voltage (or power) that may be needed at the output. The maximum output voltage is limited by the amount of plate current that can flow through the load resistor of the last stage. The highest value that can be achieved—without regard to the requirements of linearity—is that of a step function which carries the plate current from zero to the maximum rated value for the tube. This rated value differs from tube to tube, and one would tend to choose a tube with a high current rating. But since the rise time of the output stage enters into the total rise time of the amplifier, one must consider this factor as well. Several tubes can therefore be compared on the basis of a new figure of merit suitable for output stages as proposed by Wallman;[1] this is the ratio of maximum voltage output to rise time and depends upon the output capacitance C_o of the tube and the capacitance of the load C_L. As an example, several

[1] Valley and Wallman, *op. cit.*, pp. 103–104.

Table 4-4

Tube	$I_{b,\max}$, ma	C_o	$V_{\max}/T_R$, volts/μsec*
6AU6	10	5	182
6AK5	10	3	200
6CL6	30	5.5	248
6AQ5	45	8.2	725
6V6GT	45	7.5	750
6L6G	75	10	1,130

* Computed for $C_o + 20$ pf.

tubes are compared in Table 4-4 for a 20-pf load, such as might be encountered with the deflection plates of a cathode-ray tube.

$$V_{\max} = I_{b,\max}R_L \qquad (4\text{-}65)$$

$$T_R = 2.2R_L(C_o + C_L) \qquad (4\text{-}66)$$

$$\frac{V_{\max}}{T_R} = \frac{I_{b,\max}}{2.2(C_o + C_L)} \qquad (4\text{-}67)$$

While the data of Table 4-4 give a relative rating to the tubes listed, the actual number of volts per microsecond that is obtainable may vary with the practical circumstances. The values listed assume either that the tube carries rated current with no signal, and that a negative-going step cuts off the current entirely, or the opposite of this, i.e., that the current is cut off in the quiescent state but turned full on by a positive-going step at the grid. If the amplifier must accommodate steps of either polarity, the quiescent operating point would accordingly have to be chosen as approximately $\frac{1}{2}I_{b,\max}$.

A special case exists in which the signals are known to be always pulses of a duration that is short compared with the interval between pulses, i.e., pulses of low "duty cycle." In such a case it may be possible to exceed $I_{b,\max}$ on the positive peaks if the tube rating is based upon heating, i.e., plate dissipation. Sometimes the rating is based on emission limitations or grid current. Each tube needs to be treated as a separate problem.

Occasionally an output stage must operate into a low-resistance load, such as a coaxial cable used to transmit the signal to a distant location. Such a cable is usually terminated in its characteristic resistance R_0, and so the impedance seen from the amplifier is simply a resistance of this value. From consideration of the long-time (or low-frequency) response—to be taken up in Chap. 5—an unreasonably large coupling capacitor would be required with a low value of R_0 (notice that R_0 corresponds to R_g in the analysis associated with Fig. 3-4).   Hence the output stage is usually operated as in either Fig. 4-29 or Fig. 4-30.   The cathode-follower arrangement of Fig. 4-29 has the advantage that the coaxial line is at a low d-c

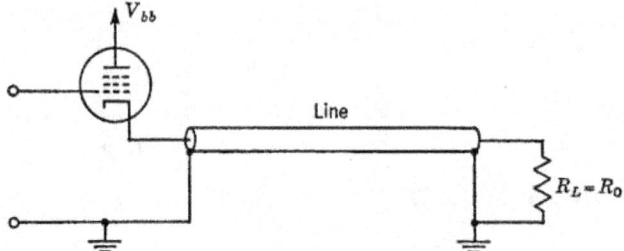

Fig. 4-29 A cathode follower driving a line.

potential.   Otherwise the two circuits are comparable so far as transient response is concerned.[1]

If the output stage must drive a capacitive load, such as an oscilloscope, for instance, the cathode follower will provide a smaller rise time.   (This is at the expense of smaller gain, however, but we are not considering gain

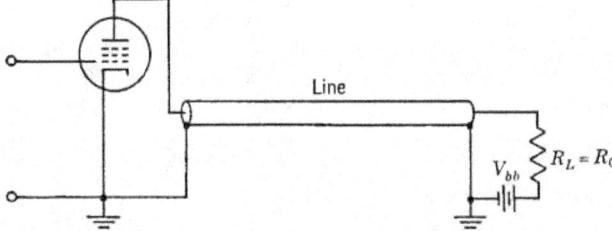

Fig. 4-30 A grounded-cathode amplifier driving a line.

for the output stage.   We *should* if we have alternative ways of providing the needed output voltage.)   In contrast with Eq. (4-66), the rise time for the cathode follower of Fig. 4-31, assuming small-signal operation, is

$$T_R = \frac{2.2 C_L}{G_L + (1 + \mu)/r_p}$$

$$\cong \frac{2.2 R_L C_L}{1 + g_m R_L} \qquad \text{if } \mu + 1 \cong \mu \tag{4-68}$$

[1] For an interesting commentary, see P. I. Richards, Cathode-follower Fallacies, *Rev. Sci. Instr.*, vol. 21, p. 1026, December, 1950.

It should be pointed out that, with a large capacitive load, it is sometimes feasible to use several output tubes in parallel with improved performance. The current is doubled by adding the second tube, but the total capacitance is increased by a smaller percentage.

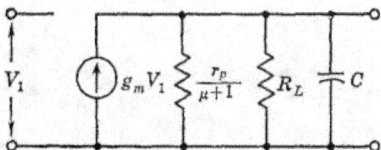

Fig. 4-31 An equivalent circuit for small-signal cathode-follower operation.

Also, it may be that for a given specification of output voltage and rise time there will be several tube possibilities that would be satisfactory. In such a case it would be reasonable to introduce other factors into the comparison, such as gain and input capacitance, the latter influencing the rise time and gain of the preceding stage.

4-12 Transient vs. Steady-state Response. Here we have a topic of long-standing theoretical interest and one of considerable practical importance. The state of our knowledge is substantial but unfortunately is not reducible to a few simple axioms. Although a full-scale recounting of the published papers is not practicable here, the results can be summarized and a few common misconceptions pointed out.

Rise Time vs. Bandwidth. For many years the term "wideband" has been used to describe the type of amplifier one builds in order to obtain a fast response to a step transient, especially in the television art. The various empirical studies have shown a general relationship between rise time and bandwidth to exist as follows:

$$T_R B = 0.35 \text{ to } 0.45 \qquad (4\text{-}69)$$

where T_R = rise time, 10 to 90 per cent

  B = bandwidth, from 0 to upper 3 db frequency

In Eq. (4-69) the value of 0.35 matches best those circuits where the overshoot is small or zero, while 0.45 corresponds to overshoots of, say, 5 per cent or greater.

Theoretical analysis of two idealized situations yields values of $T_R B$ that compare favorably with Eq. (4-69). The first of these is the so-called "ideal filter," having a characteristic as shown in Fig. 4-32. Associated with the amplitude characteristic as shown is a phase shift increasing linearly with frequency, i.e., constant time delay for all frequencies transmitted. Such a response is not physically realizable; this is proved

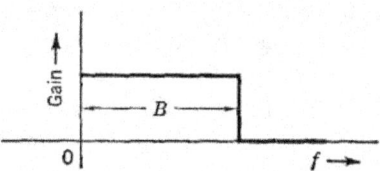

Fig. 4-32 "Ideal" filter-amplifier amplitude response.

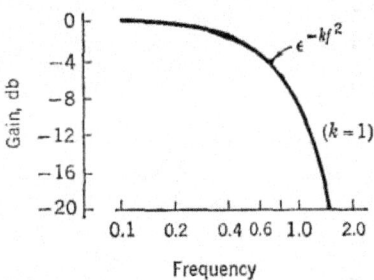

Fig. 4-33 Gaussian amplitude response.

by Wallman [1] but is also apparent from the fact that when a step function is applied at $t = 0$ the computed response shows finite output prior to this time. The nature of the step response is a sine-integral function, yielding always 9 per cent overshoot and $T_R B = 0.51$.

Another amplitude response of theoretical interest is the so-called gaussian function shown in Fig. 4-33. Once again we associate a linear phase characteristic with this amplitude response, and once again the combination is not physically realizable (although the amplitude characteristic is achieved in the limit by an infinite number of resistance-coupled stages). The response of a system of this kind to step function is also a gaussian function, possessing zero overshoot, and

$$T_R B = 0.41 \text{ (0.33 from Elmore's definitions)} \qquad (4\text{-}70)$$

The conclusions to be drawn from these results are that, for a given kind of circuit, and for the same amount of overshoot, a faster rise is obtained with a greater bandwidth. However, merely increasing the bandwidth without regard to overshoot does not necessarily lead to "better" response. Thus, taking a given amplifier whose amplitude response may be a gradually decreasing function of frequency and attempting to speed it up by adding compensating elements in order to make its response wider and squarer (like that of Fig. 4-32) will indeed speed up the amplifier because of the higher value of the $T_R B$ product, but the resulting overshoot may render the amplifier worthless for the intended application.

Transient Distortion vs. Steady-state Distortion. An ideal amplifier from the transient view would have zero rise time and no overshoot. An ideal amplifier from the steady-state view would have an amplitude response that would be constant to infinite frequency and phase shift proportional to frequency. The condition in which these ideals fail to be achieved is termed "distortion." The transient distortion is described in terms such as rise time and overshoot, whereas the steady-state distortion can be described as "amplitude distortion" [2] and "phase distortion." A distortion in either the amplitude or the phase characteristic will lead to transient distortion. This fact is not obvious, and several analyses will be found in

[1] Valley and Wallman, *op. cit.*, pp. 721–723.

[2] Sometimes called "frequency distortion" by authors who use "amplitude distortion" for nonlinear effects.

the literature that examine the relative importance of the types of distortion; we shall shortly consider two of these analyses.

Actually, of course, in most simple amplifier interstage networks of the type already presented in this section the amplitude and phase responses are interrelated in a manner characteristic of the broad class of networks identified by the term "minimum phase." It will not be possible here to go into the details of the definition of this terminology, nor into the details of the amplitude-phase relationship. Suffice it to say that feedback circuits, lattice and bridged-T structures, and distributed-parameter systems are the ones usually falling outside the minimum-phase class.[1] Thus, it is somewhat futile to attempt to place the blame for transient distortion upon either the amplitude distortion or the phase distortion alone. Nonetheless, it is instructive at least to assess that distortion due to amplitude response, for it is possible—and indeed common practice on complicated transmission systems—to exploit a device known as a "phase equalizer." By use of this device, which is an all-pass network of the non-minimum-phase class, it is feasible to make the phase response more nearly linear, without influencing the amplitude response. The use of such equalizers is far beyond the scope of this treatment, although they are of great importance in long-distance television transmission systems. In fact, it might be said that their utility lies in systems that are limited to narrow frequency channels rather than in wideband systems where the gain/rise-time quotient of the amplifier tubes is the limiting factor.[2]

Of the various analyses that have been undertaken, it will be possible to include only two here, namely, those of Wheeler[3] and DiToro.[4] The first of these is suitable for small amplitude or phase distortion, or combinations of both, whereas the latter is more suited to large distortions.

Before plunging into the details of the Wheeler and DiToro papers, it might be well to remind the reader that there are well-established relation-

[1] A basic reference on the subject is H. W. Bode, "Network Analysis and Feedback Amplifier Design," D. Van Nostrand Company, Inc., Princeton, N.J., 1945. The relationship between amplitude and phase stems from a basic property involving the real and imaginary parts of a class of complex variables; see E. A. Guillemin, "Mathematics of Circuit Analysis," pp. 330–349, John Wiley & Sons, Inc., New York, 1950. A brief discussion is also to be found in Terman, *op. cit.*

[2] Two references of interest on phase equalizers are T. C. Nuttall, Some Aspects of Television Circuit Technique: Phase Correction and Gamma Correction, *Bull. assoc. suisse électriciens*, vol. 40, pp. 615–622, Autumn, 1949; G. G. Gouriet, V.H.F. Amplifier Couplings, *Wireless Engr.*, vol. 27, pp. 257–265, October–November, 1950. See also Bell System publications. A. E. Brain, The Compensation for Phase Errors in Wideband Video Amplifiers, *Proc. IRE*, vol. 97, pp. 243–251, July, 1950.

[3] H. A. Wheeler, The Interpretation of Amplitude and Phase Distortion in Terms of Paired Echoes, *Proc. IRE*, vol. 27, pp. 359–385, June, 1939.

[4] M. J. DiToro, Phase and Amplitude Distortion in Linear Networks, *Proc. IRE*, vol. 36, pp. 24–36, January, 1948, includes an extensive bibliography.

ships between the transient and the steady-state responses. These are the Fourier series and integral, and the Laplace transformation. Choice among them depends upon the nature of the driving function and the initial conditions of the system. Not only are these relationships fundamental to the analyses about to be examined, but they are useful in circumstances where it is possible to measure in the laboratory one kind of response only but where one wishes to find the opposite response by computation. For instance, if one can measure the amplitude and phase response and wishes to compute the step response, the Fourier series or integral can be used.[1]

In the paired-echo analysis the amplifier is driven by a transient signal $v_1(t)$ which is chosen to have no singularities on the $j\omega$ axis in the $V_1(p)$ domain. Thus, the inverse Laplace transform reduces to the Fourier integral as follows:

$$v_1(t) = \frac{1}{2\pi j} \int_{c-j\infty}^{c+j\infty} V_1(p)\epsilon^{pt}\,dp \tag{4-71}$$

$$= \frac{1}{2\pi} \int_{-\infty}^{\infty} V_1(j\omega)\epsilon^{j\omega t}\,d\omega \tag{4-72}$$

$$= \frac{1}{2\pi} \int_{-\infty}^{\infty} |V_1(\omega)|\,\epsilon^{j[\omega t+\phi(\omega)]}\,d\omega \tag{4-73}$$

where
$$V_1(j\omega) = |V_1(\omega)|\,\epsilon^{j\phi(\omega)} \tag{4-74}$$

Now let the signal be applied to the amplifier, whose steady-state response $A(j\omega)$ is as shown in Fig. 4-34. The cosine variation of the amplitude response is rather artificial because of its repetitive nature. The linear phase shift assumed will not cause any distortion but only a time delay, as will be seen. The assumption is that the input signal $V_1(j\omega)$ has a spectrum which virtually vanishes after the first minimum of the cosine; hence the rest of the curve is shown dashed. A trapezoidal pulse of suitable slope and duration would be a satisfactory signal.

To obtain the output signal $v_2(t)$, we first find $V_2(j\omega)$ by multiplying $V_1(j\omega)$ and $A(j\omega)$ and then taking the Fourier integral as in Eq. (4-73). In the process it will be convenient to expand $\cos c_0\omega$ in exponentials.

[1] A. V. Bedford and G. L. Fredendall, Analysis, Synthesis, and Evaluation of the Transient Response of Television Apparatus, *Proc. IRE*, vol. 30, pp. 440–458, October, 1942. Some short cuts suitable for quick appraisal are given by W. J. Cunningham, Simple Relations for Calculating Certain Transient Responses, *J. Appl. Phys.*, vol. 19, pp. 251–256, March, 1948; G. S. Brown and D. P. Campbell, "Principles of Servomechanisms," chap. 11, John Wiley & Sons, Inc., New York, 1948.

If it should happen that only the amplitude characteristic can be measured, the corresponding phase characteristic can be computed if the networks are of the minimum-phase variety. See Bode, *op. cit.*; D. E. Thomas, Phase of a Semi-infinite Attenuation Slope, *Bell System Tech. J.*, vol. 26, pp. 870–899, October, 1947.

$$\cos c_0(\omega) = \tfrac{1}{2}(\epsilon^{jc_0\omega} + \epsilon^{-jc_0\omega})$$

$$v_2(t) = \frac{1}{2\pi} \int_{-\infty}^{\infty} V_2(j\omega) \epsilon^{j\omega t} \, d\omega$$

$$= \frac{1}{2\pi} \int_{-\infty}^{\infty} V_1(j\omega) A(j\omega) \epsilon^{j\omega t} \, d\omega$$

$$= \frac{1}{2\pi} \int_{-\infty}^{\infty} |V_1(\omega)| \epsilon^{j\phi(\omega)} \left(\frac{a_0}{2} + a_1 \cos c_0\omega\right) \epsilon^{-jb_0\omega} \epsilon^{j\omega t} \, d\omega$$

$$= \frac{1}{2\pi} \int_{-\infty}^{\infty} |V_1(\omega)| \left(\frac{a_0}{2} + \frac{a_1}{2} \epsilon^{jc_0\omega} + \frac{a_1}{2} \epsilon^{-jc_0\omega}\right) \epsilon^{j[\omega t + \phi(\omega) - b_0\omega]} \, d\omega$$

$$= \frac{1}{2\pi} \frac{a_0}{2} \int_{-\infty}^{\infty} |V_1(\omega)| \epsilon^{j[\omega(t-b_0) + \phi(\omega)]} \, d\omega$$

$$+ \frac{1}{2\pi} \frac{a_1}{2} \int_{-\infty}^{\infty} |V_1(\omega)| \epsilon^{j[\omega(t-b_0+c_0) + \phi(\omega)]} \, d\omega$$

$$+ \frac{1}{2\pi} \frac{a_1}{2} \int_{-\infty}^{\infty} |V_1(\omega)| \epsilon^{j[\omega(t-b_0-c_0) + \phi(\omega)]} \, d\omega \qquad (4\text{-}75)$$

We now compare Eqs. (4-73) and (4-75), and note that the latter contains three terms of the same form, differing only in the constant multiplier $a_0/2$ or $a_1/2$ and the shift in the time scale from t to $t - b_0$, etc. Thus Eq. (4-75) describes three signals (time functions) which have the same waveform as the input signal $v_1(t)$ but which are changed in amplitude and shifted in time. The result is depicted in Fig. 4-35; it consists of a delayed main signal and two "echoes."

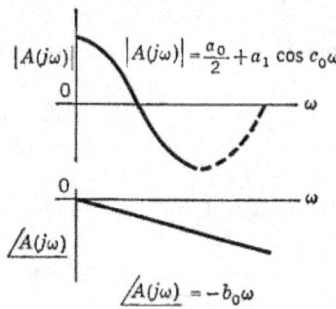

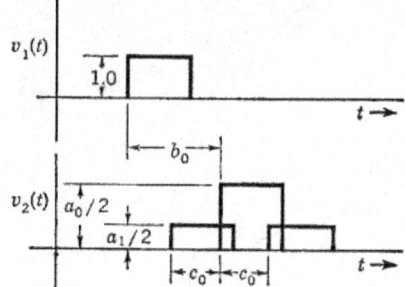

Fig. 4-34 Amplifier response assumed for the Wheeler "paired-echo" analysis (amplitude distortion).

Fig. 4-35 "Paired echoes" resulting from amplitude distortion.

A similar result is obtained if phase distortion is introduced, as in Fig. 4-36. The amplitude characteristic is now assumed to be ideal, i.e., constant amplification from zero to infinite frequency.

$$v_2(t) = \frac{1}{2\pi} \int_{-\infty}^{\infty} V_1(j\omega) A(j\omega) \epsilon^{j\omega t} \, d\omega$$

$$= \frac{1}{2\pi} \int_{-\infty}^{\infty} |V_1(\omega)| \, \epsilon^{j\phi(\omega)} \frac{a_0}{2} \epsilon^{j(-b_0\omega + b_1 \sin c_1\omega)} \epsilon^{j\omega t} \, d\omega$$

$$= \frac{1}{2\pi} \frac{a_0}{2} \int_{-\infty}^{\infty} |V_1(\omega)| \, \epsilon^{j[\omega(t-b_0)+\phi(\omega)]} \epsilon^{jb_1 \sin c_1\omega} \, d\omega$$

But

$$\epsilon^{jb_1 \sin c_1\omega} = \sum_{k=-\infty}^{\infty} J_k(b_1) \epsilon^{jkc_1\omega}$$

For b_1 very small

$$\epsilon^{jb_1 \sin c_1\omega} \cong J_0(b_1) + J_1(b_1)\epsilon^{jc_1\omega} + J_{-1}(b_1)\epsilon^{-jc_1\omega}$$

$$\cong J_0(b_1) + J_1(b_1)\epsilon^{jc_1\omega} - J_1(b_1)\epsilon^{-jc_1\omega}$$

$$v_2(t) \cong \frac{1}{2\pi} \frac{a_0}{2} \int_{-\infty}^{\infty} |V_1(\omega)| \, \epsilon^{j[\omega(t-b_0)+\phi(\omega)]} [J_0(b_1) + J_1(b_1)\epsilon^{jc_1\omega}$$

$$- J_1(b_1)\epsilon^{-jc_1\omega}] \, d\omega$$

$$\cong \frac{a_0}{2} J_0(b_1) \left[\frac{1}{2\pi} \int_{-\infty}^{\infty} |V_1(\omega)| \, \epsilon^{j[\omega(t-b_0)+\phi(\omega)]} \, d\omega \right]$$

$$+ \frac{a_0}{2} J_1(b_1) \left[\frac{1}{2\pi} \int_{-\infty}^{\infty} |V_1(\omega)| \, \epsilon^{j[\omega(t-b_0+c_1)+\phi(\omega)]} \, d\omega \right]$$

$$- \frac{a_0}{2} J_1(b_1) \left[\frac{1}{2\pi} \int_{-\infty}^{\infty} |V_1(\omega)| \, \epsilon^{j[\omega(t-b_0-c_1)+\phi(\omega)]} \, d\omega \right] \quad (4\text{-}76)$$

As before, we can now compare the expression (4-76) with the expression (4-73) for the input signal $v_1(t)$. Each of the three terms of (4-76) corresponds to a replica of the input function, but modified in amplitude and shifted in time. These three output signals are depicted in Fig. 4-37 and can be thought of as a principal signal and two echoes. Notice that one of the echoes is negative, in contrast with the case of amplitude distortion. Thus, the composite output signal

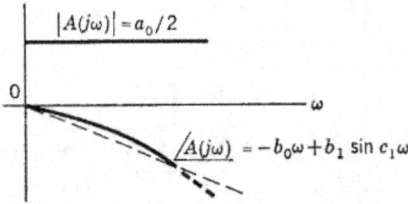

Fig. 4-36 Amplifier response assumed for the Wheeler "paired-echo" analysis (phase distortion).

will be unsymmetrical for phase distortion and symmetrical for amplitude distortion.   A comparative case is illustrated in Fig. 4-38; the dashed line represents the summation that would result if higher-order terms were included in the sinusoidal amplitude and phase expressions, i.e., if the actual amplitude and phase characteristics were expanded in series of cosine and sine terms, respectively.

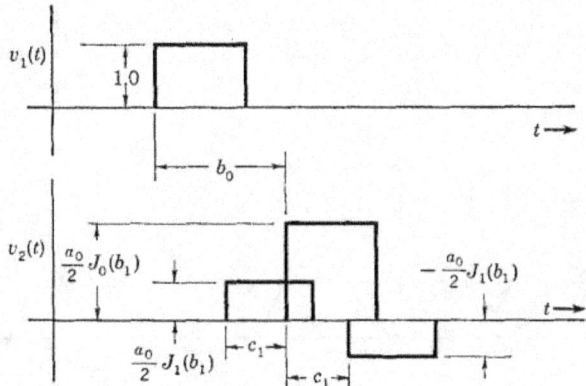

Fig. 4-37 "Paired echoes" resulting from phase distortion.

The paired-echoes analysis is particularly convenient for small distortions (small values of b_1, c_1, etc.) but becomes cumbersome if the distortion is large.   It does, however, provide ample evidence that distortion of the time response can result from either steady-state amplitude or phase distortion and that the effects of the two kinds of steady-state distortion are different.

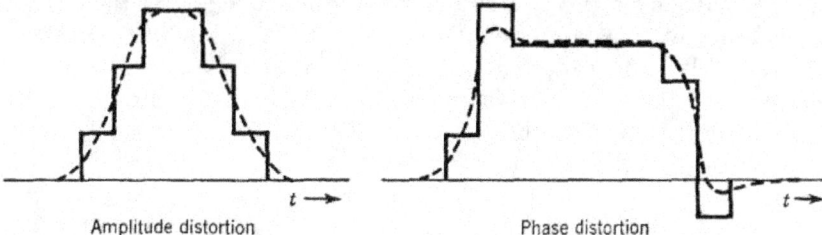

Amplitude distortion   Phase distortion

Fig. 4-38 Response of an amplifier to a pulse input when only amplitude distortion or only phase distortion is present.

One analytical study of interest for cases of larger degrees of amplitude or phase distortion than are normally treated with the paired-echoes technique is that of DiToro.[1] He treats the case of an amplifier (or any other linear network, not necessarily lumped) having a steady-state re-

[1] *Op. cit.*

sponse expressible by the exponential function

$$\frac{A(j\omega)}{A(0)} = \epsilon^{-(a^m\omega^m + jb^n\omega^n)} \tag{4-77}$$

This is not the kind of function that emerges from the analysis of lumped-element amplifier networks; instead rational fractions should be expected. But it has certain advantages and is a reasonably good approximation to the response of actual circuits. Indeed, a large number of resistance-coupled stages approaches this kind of response, with exponents $m = 2$ and $n = 3$.

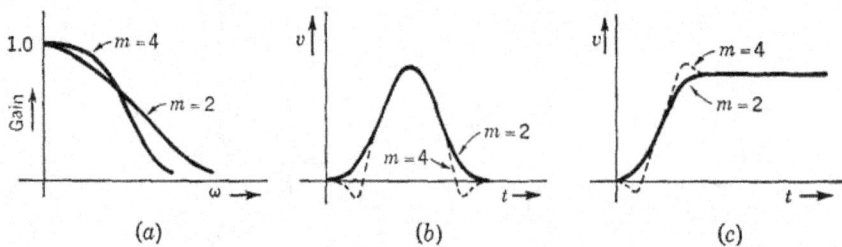

(a)   (b)   (c)

Fig. 4-39 Assumed amplitude response and the resulting transient response from the DiToro analyses. (a) Amplitude response. (b) Impulse response. (c) Step response.

The expression (4-77) is particularly convenient in treating experimental data. The exponents m and n can readily be determined from a logarithmic plot of amplitude and phase data. And from DiToro's paper the transient response can be forecast. This is also true of the paired-echoes analysis, but for smaller distortions; also, as in the case of paired echoes, the DiToro analysis permits separate evaluation of the effects of amplitude or phase distortion taken singly and together. In fact, DiToro has found it necessary in general to find the time response (to an impulse, say) of Eq. (4-77) by first finding it for $\epsilon^{-a^m\omega^m}$, then for $\epsilon^{-jb^n\omega^n}$, and finally for the combination by using the convolution integral.[1]

Some examples of the extensive results obtained by DiToro can be shown. For instance, in Fig. 4-39a is depicted an amplitude response corresponding to $m = 2$, and also one for $m = 4$. On the assumption that the phase distortion is nil, i.e., that $n = 1$ and $b \neq 0$, the corresponding impulse and step responses are as shown in Fig. 4-39b and c.

Similar results for phase distortion—in the absence of amplitude distortion—are depicted in Fig. 4-40.

Notice that amplitude distortion leads to symmetrical responses for the impulse and step, whereas phase distortion leads to unsymmetrical ones.

[1] M. F. Gardner and J. L. Barnes, "Transients in Linear Systems," pp. 228–236, John Wiley & Sons, Inc., New York, 1942.

This is comparable to the paired-echo analysis. Notice also that phase distortion gives overshoot, even with $n = 3$, which is the smallest exponent possible (except for $n = 1$, of course, which would give no distortion but only a time delay). It should be pointed out that the nature of amplitude and phase responses in general requires that m be even and n odd.

The case of combined amplitude and phase distortion has been analyzed by DiToro, as already mentioned. It is necessary to specify, in addition to m and n, the relative magnitudes of a and b [see (4-77)]; the relative rates of amplitude and phase cutoff influence considerably the shape of the time

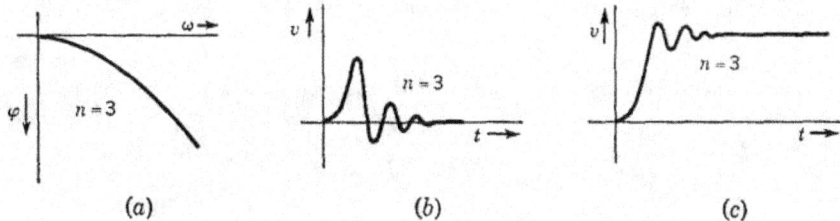

(a)   (b)   (c)

Fig. 4-40 Assumed phase response and the resulting transient response from the DiToro analyses. (a) Phase response. (b) Impulse response. (c) Step response.

response. An extensive assortment of results will be found plotted in DiToro's paper.

4-13 Special Amplifiers for High Speed. Although the details will be reserved for Chap. 6, it is appropriate to mention here that when fast amplifiers are required (amplifiers with small rise time T_R)—faster than can be provided by the networks of Figs. 4-2a through 4-12—there is the possibility of an altogether different kind of amplifier configuration. The circuits discussed in the sections above have been of the *product* variety, or *cascade*, the former term describing the fact that the over-all gain function (of frequency p) is the product of the individual-stage gain functions, while "cascade" implies that one stage is connected after another. There is another class of amplifiers, which can be called *additive* because of the additive nature of their gain functions, and which permit of far greater speeds with conventional tubes than the cascade amplifier provides. This class includes the *distributed* amplifier and the *split-band* amplifier. Research is still being conducted in both these categories, but the present-day status will be discussed in Chap. 6.

PROBLEMS

4-1. Apply Laplace-transform analysis to the complete circuit of Fig. 3-3 to determine $v_2(t)$ if $v_1(t)$ is a step of voltage of amplitude V_0 applied at $t = 0$. (Assume that $r_p \gg R_L$, $C_{cc} \gg C_1$ or C_2, and $R_g \gg R_L$.)

a. Show that for small values of t the result reduces to the response of the simple equivalent circuit of Fig. 3-12a.

b. Show that for large values of t the result reduces to the response of the equivalent circuit of Fig. 3-4.

4-2. Shown in Fig. P4-2 is the so-called "series-peaking" circuit, or "series-compensated" circuit. This circuit is somewhat more difficult to analyze than the corresponding transistor circuit because the extra reactive element makes the circuit equation a cubic. However, for a certain case a relatively simple result may be obtained which illustrates the kind of results obtainable.

a. Show that the equation for the transfer impedance may be written in the form

$$\frac{V_2(p)}{I_1(p)} = \frac{-R^3(C_1 + C_2)^3}{LC_1C_2} \frac{1}{p^3 + p^2\dfrac{C_1 + C_2}{C_2} + p\dfrac{(C_1 + C_2)^2}{mC_1C_2} + \dfrac{(C_1 + C_2)^2}{mC_1C_2}}$$

This equation is in terms of the normalized variables: $p = sR(C_1 + C_2)$, and $m = L/R^2(C_1 + C_2)$.

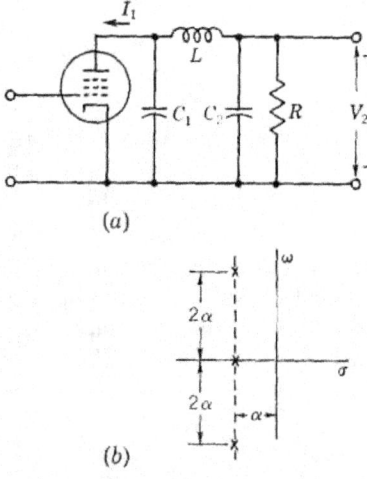

(a)

(b)

Fig. P4-2

b. Arrange the poles according to the pole-zero diagram shown in Fig. P4-2 by finding the capacitance ratio and value of m required.

HINT: Assume that $C_1 = nC_2$, and rewrite the above denominator polynomial in terms of p, n, and m. Express each pole in Fig. P4-2 as a factor; i.e., the top pole would be $p + \alpha - j2\alpha$. Then multiply the factors together to give a cubic. This cubic must be identical with the previously obtained polynomial; hence, the coefficient of each term may be equated to the corresponding term in the other cubic. In this way the value of n and m may be found.

c. Find the rise time and overshoot (both in terms of α and in real time) produced by the circuit when driven by a unit step. (The rise time will be about $1.5/\alpha$.) This computation will have to be done by plotting the response.

d. Show that the improvement factor over an uncompensated circuit, η, is slightly greater than 2.

4-3. The over-all normalized gain function for a linear amplifier comprising lumped elements will always be of the form

$$A_n(p) = \frac{1 + \alpha_1 p + \alpha_2 p^2 + \alpha_3 p^3 + \cdots + \alpha_m p^m}{1 + \beta_1 p + \beta_2 p^2 + \beta_3 p^3 + \cdots + \beta_j p^j}$$

Show by comparison with Eq. (4-45) that

$$T_{DN} = \beta_1 - \alpha_1$$
$$T_{RN}^2 = 2\pi[\beta_1{}^2 - \alpha_1{}^2 + 2(\alpha_2 - \beta_2)]$$

Elmore definitions

4-4. Compute the rise-time–bandwidth product for a five-stage resistance-coupled amplifier. Obtain the bandwidth from an amplitude response determined graphically on the p plane.

4-5. Show that the equivalent circuit given in Fig. 4-31 is correct. (HINT: Find the Norton equivalent looking into the cathode and ground terminals toward the tube.)

4-6. Refer to the shunt-peaking circuit of Fig. 4-2a.

a. Write the equation for gain in terms of a constant containing g_m, R, L, and C times a polynomial containing only s, constants, and the parameter m. (HINT: This will require normalizing the frequency variable; the most useful normalization is $s = pRC$.)

b. Sketch pole-zero diagrams as in Fig. 4-4, giving the pole and zero positions in terms of m. Do this for $m < 0.25$, $m = 0.25$, and $m > 0.25$. Indicate also the limiting pole and zero positions as $m \to 0$ and $m \to \infty$.

c. For $m > 0.25$ write the equation for $v_{g2}(t)$ in terms of a constant containing R, L, and C times an equation in m. Assume that $v_{g1}(t)$ is a unit step.

4-7. Consider the transfer function of any linear network, $A(p)$, which is normalized so that the gain at zero frequency is unity. The steady-state amplitude response $|A(\omega)|$ and phase shift $\phi(\omega)$ are obtained by replacing p by $j\omega$,

$$A(p)|_{p=j\omega} = |A(\omega)|\epsilon^{j\phi(\omega)}$$

a. Show that the following relationships exist between the amplitude and phase functions and the Elmore definitions of delay time and rise time:

$$\left.\frac{d\phi}{d\omega}\right|_{\omega=0} = -T_D \qquad \left.\frac{d^2|A(\omega)|}{d\omega^2}\right|_{\omega=0} = \frac{-T_R}{2\pi}$$

SUGGESTION: Make use of Eq. (4-45). The above result really indicates that this definition of rise time is not of *general* utility, since it is quite possible and often desirable to design a network or amplifier in which $T_R = 0$ (or even $T_R < 0$) according to the above equation; yet the actual 10 to 90 per cent rise time will certainly be greater than zero.

b. As an example, consider the transfer function

$$A(p) = \frac{1}{p^2 + \sqrt{2}p + 1}$$

Find T_D and T_R. What do you deduce concerning the nature of the time response to a step function in the case of this transfer function?

4-8. The transfer function $A(p) = 1/(p^2 + \sqrt{2}p + 1)$ considered in Prob. 4-7 is also known as the two-pole, maximally flat function. (This is discussed further in Chap. 9.) The transient response of this function to a step input is given as the curve for $n = 1$ in Fig. 4-23. The rise time is about 2.16 sec, with 4.3 per cent overshoot.

a. A rise time of 50 mμsec is desired. Find the pole locations of $A(p)$ corresponding to this rise time.

b. These pole locations may be realized by a shunt-peaked pentode stage (which provides the requisite two complex poles plus an undesired zero) in cascade with an RC stage (whose pole can be chosen to cancel the zero). Find the required element values for each stage if the total shunting capacity per stage is 15 pf.

c. Find the gain of the above amplifier if the tubes used have $g_m = 10,000$ μmhos. Compare this gain with the gain of an alternative amplifier *having equal rise time* but comprised of two identical shunt-peaked stages with $m = 0.35$.

d. The series-peaked transistor stage of Sec. 4-7 can be arranged to have two complex poles and no zero in the finite part of the p plane; therefore these two poles may be arranged to give the pole locations for the maximally flat interstage. Find the value of L in terms of R_1, R_{eq}, C, and R which is required to give the maximally flat transfer function. What is the value of η for this function?

4-9. Consider two amplifiers, each consisting of five identical stages. Similar circuits (the same η) are employed in both, but one amplifier uses 6AH6 tubes and the other 6AU6's. (Use the data in Table 4-1.)

a. If both are designed for the same gain, find the ratio of their over-all rise times.

b. If both are designed for the same over-all rise time, find the ratio of their over-all gains.

4-10. All the transistor high-frequency calculations have been on the basis of an approximate equivalent circuit derived from the hybrid pi. One direct way of evaluating the accuracy of the simplified circuit with respect to its prototype is to compare the

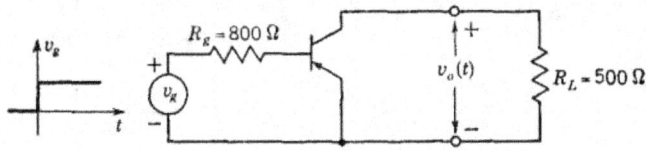

Fig. P4-10

step responses of the two circuits for a given generator and load resistance. For this purpose find the equation for the step response for the three conditions listed below.

a. Neglect C_c (that is, assume that $C_c = 0$).

b. Include C_c in the equivalent circuit as in Fig. 3-38c.

c. Use the hybrid-pi equivalent circuit as shown in Fig. 3-38b. In this case, the equation for the output will not be a single exponential. Show why one of the exponentials is unimportant so that one pole essentially dominates the response function.

d. Calculate the 10 to 90 per cent rise time for the three cases above. (Calculate only an approximate value for case c, based upon the dominant pole.) As a basis for calculation use the circuit in Fig. P4-10 and typical transistor data: $1 - \alpha_0 = 0.02$; $r_e' = 25$ ohms; $r_b' = 200$ ohms; $\omega_t = 3 \times 10^7$ radians/sec; $C_c = 10$ pf.

4-11. Using the high-frequency equivalent circuit of Fig. 2-3, find the Thévenin equivalent looking to the left of the terminals a-a' of the common-collector stage (emitter follower) shown in Fig. P4-11. (The collector capacitance may be neglected here.)

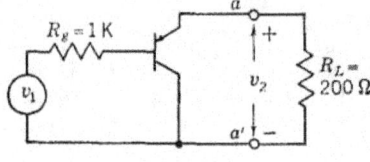

Fig. P4-11

Note the resemblance to Fig. 2-22. Using the Thévenin equivalent you have found, write the equations for $V_2(p)$ and $v_2(t)$. Show that the effect of the capacitance $1/\omega_a r_e'$ would usually be small. Neglecting the effect of this capacitance, compute the rise time for the source and load shown in Fig. P4-11 and the transistor parameters of Prob. 4-10.

5

Step Response of Lowpass Amplifiers for Large Values of Time: Sag, etc.

The failure of practical amplifier circuits to transmit a step function perfectly for large values of time is in one sense a failure of the circuits to transmit direct current or, what is almost the same thing, very low frequencies. The long-time response to a step is generally better if the so-

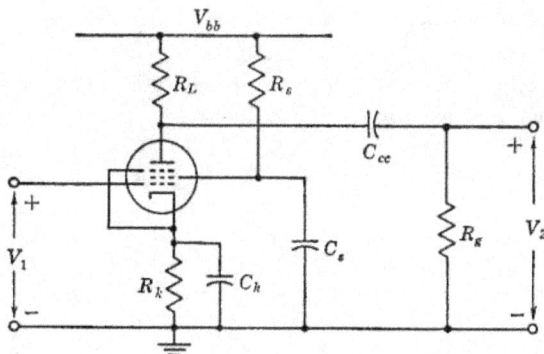

Fig. 5-1 Pentode amplifier stage including all causes of sag.

called low-frequency response of the amplifier is good. Indeed, most discussions of the subject in textbooks are phrased in terms of the low-frequency steady-state behavior, as in Chap. 3. However, the step response is often the desired criterion, and it can be dealt with directly. For an example, in a television system, a picture in which there is a background (sky, for instance) with uniform intensity across the scene would require that the video amplifiers maintain a virtually constant voltage for the

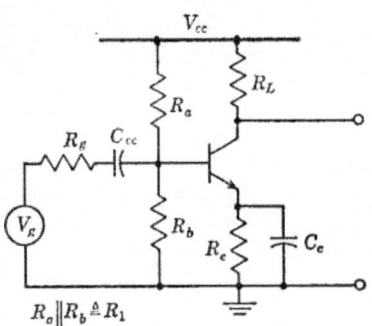

Fig. 5-2 Common-emitter stage including all sources of sag.

duration of each horizontal scan period; thus, for an interval of about 60 μsec the step response must be constant to within some specified sag. The corresponding low-frequency behavior is purely incidental, and specification of the amplitude and phase response at low frequencies is at best indirect and not necessarily unique for a given sag.

It was shown in Chap. 3 that there are only certain portions of the complete circuit of Figs. 5-1 and 5-2 that influence the long-time behavior; thus, the simplified circuits of Figs. 3-4 and 3-33 adequately describe the low-frequency amplitude response or the long-term step response. These circuits contained only the coupling elements C_{cc} and either R_g for the pentode or R_g and the resistances R_g and $r_b + r_e/(1 - \alpha)$ for the transistor, whereas in a practical amplifier circuit it is necessary to consider also the pentode cathode (transistor emitter) bias circuit and the imperfectly bypassed screen-grid voltage supply. These will be taken up individually; the analysis follows closely that of Wallman.[1]

5-1 Coupling Circuit. The effect of the coupling circuit on the amplitude response has already been studied in Chap. 3. In terms of the vacuum-tube circuit first, the gain of a stage, considering only the effect of the coupling elements shown in Fig. 3-4, was found to be

$$A = -g_m R_L \frac{p}{p + 1/R_g C_{cc}} \tag{3-9}$$

The output voltage is found by multiplying the Laplace transform of the input voltage (in this case a step voltage V_1/p) by the gain and taking the inverse transform,

$$v_2(t) = \mathcal{L}^{-1} \left[-g_m R_L \frac{V_1}{p} \frac{p}{p + 1/R_g C_{cc}} \right]$$

$$= -g_m R_L V_1 \epsilon^{-t/R_g C_{cc}} \tag{5-1}$$

The output is therefore an exponential with an initial value of $-g_m R_L V_1$ which decays to zero with a time constant of $R_g C_{cc}$. In considering multiple stages it is convenient to normalize the gain so that the midband gain is

[1] G. E. Valley, Jr., and H. Wallman (eds.), "Vacuum Tube Amplifiers" (vol. 18, M.I.T. Radiation Laboratory Series), pp. 84-92, McGraw-Hill Book Company, Inc., New York, 1948.

unity and then to drive the amplifier with a unit step. Under these conditions the output of an n-stage amplifier is

$$V_n(p) = \frac{1}{p}\left(\frac{p}{p + 1/T_1}\right)^n \tag{5-2}$$

where $$T_1 = R_g C_{cc}$$

The frequency variable may also be normalized—define $s \stackrel{\Delta}{=} pT_1$. Equation (5-2) then becomes

$$V_n(s) = \frac{s^{n-1}}{(s + 1)^n} \tag{5-3}$$

The inverse Laplace transform of Eq. (5-3) is

$$\mathcal{L}^{-1}\left[\frac{s^{n-1}}{(s + 1)^n}\right] = v_n(t) = \frac{d^{n-1}}{dt^{n-1}}\left[\frac{t^{n-1}\epsilon^{-t}}{(n - 1)!}\right] \tag{5-4}$$

For specific values of n this equation gives the following for $v_n(t)$:

n	$v_n(t) = \mathcal{L}^{-1}[V_n(s)]$	Initial slope $v_n'(0^+)$
1	ϵ^{-t}	-1
2	$\epsilon^{-t}(1 - t)$	-2
3	$\epsilon^{-t}\left(1 - 2t + \dfrac{t^2}{2}\right)$	-3
4	$\epsilon^{-t}\left(1 - 3t + \dfrac{3t^2}{2} - \dfrac{t^3}{6}\right)$	-4

The situation regarding initial slope is of particular interest, since this determines the sag. It can be seen from the table above that the slopes (or sags) are additive. That is, the sag will increase directly with the number of stages. This is also true for the case of nonidentical stages and may be easily shown as follows: Assume that the different stages have time constants of $T_1, T_2, \ldots, T_n$. Then Eq. (5-2) becomes

$$V_n(p) = \frac{1}{p} \cdot \frac{p}{p + 1/T_1} \cdot \frac{p}{p + 1/T_2} \cdots \frac{p}{p + 1/T_n} \tag{5-5}$$

The initial slope of the output voltage is easily found using the Laplace transform,

$$V_n'(p) = pV_n(p) - v_n(0^+) \tag{5-6}$$

$[v_n(0^+)$ is the value of v_n immediately after the input step is applied; that is, $v_n(0^+) = 1.]$

$$V'_n(p) = \frac{p^n}{(p + 1/T_1)(p + 1/T_2) \cdots (p + 1/T_n)} - 1$$

$$= \frac{p^n - (p^n + p^{n-1} \sum_1^n \frac{1}{T_i} + \cdots)}{p^n + p^{n-1} \sum_1^n \frac{1}{T_i} + \cdots} \qquad (5\text{-}7)$$

$$v'_n(0^+) = \lim_{p \to \infty} p V'_n(p)$$

$$= - \sum_1^n \frac{1}{T_i} = - \sum_1^n \frac{1}{R_i C_i} \qquad (5\text{-}8)$$

Equation (5-8) shows that the slope of the output is indeed the sum of the individual stage contributions.

In the case of transistor stages, the function describing the behavior of many stages is not exactly the product of the individual gain functions because of the presence of reverse transfer parameters. However, as we have already seen in Chap. 3, the effect of neglecting the transfer parameters is small in common-emitter or common-base stages. Hence, in these cases the sags due to the individual stages obviously may be added. The sag contribution of an individual CE stage may be found by use of Eq. (3-71), which is repeated here,

$$A \approx \frac{p}{p + 1/RC_{cc}} \qquad (5\text{-}9)$$

where
$$R = R_g + \frac{[r_b + r_e/(1 - \alpha)]R_1}{r_b + r_e/(1 - \alpha) + R_1}$$

The initial slope of the output of such a stage in response to a step is $-1/RC_{cc}$, as in the case of the vacuum-tube amplifier. For convenience the initial slope is often given in per cent per unit time equaling -100 per cent$/T$; that is, if $T = 10,000$ μsec, the sag is -0.01 per cent$/\mu$sec, -10 per cent/msec, or $-10,000$ per cent/sec.

5-2 Sag from Other Sources. Additional sources of sag in an amplifier come from the biasing circuits—the cathode (or emitter) bypass capacitor and the screen bypass capacitor. Since we have already written equations describing the effects of these impedances in Chap. 3, let us find a general equation for sag in terms of the gain equation. An equation with sufficient generality to include any source of sag is

$$A(p) \approx \frac{(p + \sigma_1)(p + \sigma_2) \cdots (p + \sigma_n)}{(p + \epsilon_1)(p + \epsilon_2) \cdots (p + \epsilon_m)} \qquad m = n \qquad (5\text{-}10)$$

where σ and ϵ are real and positive.

Treating this equation in a manner similar to that used to find the initial slope of Eq. (5-5) gives

$$v_n'(0^+) = \sum_1^n \sigma_i - \sum_1^m \epsilon_i \qquad (5\text{-}11)$$

Note that this gives the same results as Eq. (5-8), where all the $\sigma_i = 0$. Using Eq. (5-11), we may now find the sag due to the cathode capacitor C_k. From Eq. (3-65) we have

$$A(p) \approx \frac{p + 1/R_k C_k}{p + \dfrac{1 + g_m R_k}{R_k C_k}} \qquad (3\text{-}65a)$$

$$v_2'(0^+) = \sigma_1 - \epsilon_1$$

$$= \frac{1}{R_k C_k} - \frac{1 + g_m R_k}{R_k C_k} = \frac{-g_m}{C_k} \qquad (5\text{-}12)$$

Hence we have the surprising result that the initial slope of the sag due to C_k is independent of R_k. While this is true, the exact effect of the sag is dependent upon the value of R_k, as illustrated in Fig. 5-3, which shows the effect of changing R_k while g_m and C_k are maintained constant.

The equations describing sag from the screen circuit and emitter circuit are tabulated in Table 5-1. As may be seen by inspecting the table, the initial slope of the sag caused by the screen circuit is also independent of the screen dropping resistor R_s. The equations for the transistor appear complicated but may often be simplified when specific values of the parameters are known. The waveforms of the sags due to the cathode (emitter) and screen bypass capacitors are similar, and Fig. 5-3 is representative.

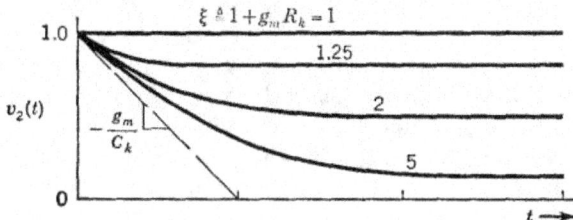

Fig. 5-3 Step response of pentode amplifier showing the effect on the sag due to changing R_k. (C_k and g_m are fixed.)

Table 5-1

Cause	Gain equation	Equation number	% sag/sec
Vacuum tube: Coupling circuit....	$A \approx \dfrac{p}{p + 1/R_g C_{cc}}$	(3-3), (3-9)	$\dfrac{-100\%}{R_g C_{cc}}$
Cathode circuit....	$A \approx \dfrac{p + 1/R_k C_k}{p + \dfrac{1 + g_m R_k}{R_k C_k}}$	(3-65a)	$\dfrac{-g_m\, 100\%}{C_k}$
Screen circuit.....	$A \approx \dfrac{p + 1/R_s C_s}{p + \dfrac{r_{p2} + R_s}{r_{p2} R_s C_s}}$	(3-85)	$\dfrac{-100\%}{r_{p2} C_s}$
Transistor (common emitter): Coupling circuit....	$A \approx \dfrac{p}{p + 1/R C_{cc}}$ $R \overset{\Delta}{=} R_g + \dfrac{[r_b + r_e/(1-\alpha)]R_1}{r_b + r_e/(1-\alpha) + R_1}$	(3-71), (3-72)	$\dfrac{-100\%}{R C_{cc}}$
Emitter circuit....	$A \approx \dfrac{p + 1/R_e C_e}{p + \dfrac{R_s + R_e/(1-\alpha)}{R_s R_e C_e}}$ $R_s \overset{\Delta}{=} \dfrac{R_g R_1}{R_g + R_1} + r_b + \dfrac{r_e}{1-\alpha}$	(3-70)	$\dfrac{-100\%}{R_s C_e(1-\alpha)}$

The parameter ξ in the figure, which determines the ratio of the initial value of $v_2(t)$ to the final value for the different sags, is

Screen circuit
$$\xi = \frac{r_{p2} + R_s}{r_{p2}} \tag{5-13}$$

Emitter circuit
$$\xi = 1 + \frac{R_e}{R_s(1-\alpha)} \tag{5-14}$$

$$R_s \overset{\Delta}{=} \frac{R_g R_1}{R_g + R_1} + r_b + \frac{r_e}{1-\alpha}$$

The sag may be computed from the equations giving per cent per unit time only for times sufficiently short to make the representation of the output $v_2(t)$ by a tangent line valid. Therefore, only for the time interval $t \ll \tau$ is our representation good, where τ is the time constant in the denominator of each of the gain equations in Table 5-1. For instance, in the cathode circuit the important time constant is $R_k C_k/(1 + g_m R_k)$.

5-3 Sag from Multiple Sources. We have already seen that where gain functions (functions of frequency) can be multiplied as in Eq. (5-5) [and proved for the more general case in Eq. (5-10)] the initial slope of the output is the sum of the initial slopes of each sag-producing element. Thus the slopes caused by the cathode circuits and coupling circuits in a chain of pentode amplifier stages can be very accurately added together because the over-all gain function is very nearly the product of the individual gain functions, as discussed in Sec. 3-8. Surprisingly enough, the initial slopes caused by any combination of causes in an amplifier may be added to give exactly the initial slope of the sag in the output of the amplifier. This is true even though the low-frequency cutoff is *not* given exactly in many cases by combining the individual amplitude-response functions.

A simple proof that initial slopes from any source of sag may be added is as follows: Assume a network made up of linear R, C, and active elements (no inductive elements). Assume that the network is discharged initially and is driven by a unit step. Each capacitor initially has zero voltage and $v_{c,i}(0^-) = v_{c,i}(0^+) = 0$. Hence at $t = 0^+$ each capacitor acts like a short circuit, and the currents in each branch of the network may be computed by replacing each capacitor by a short circuit and finding the currents in the resulting resistive network. The derivative of the voltage appearing across each capacitor, the initial current being known, is $dv_{c,i}/dt = i_i(0^+)/C_i$ $(t = 0^+)$. Each voltage $v'_{c,i}$ may be represented by a voltage generator having zero magnitude but the above slope. From the superposition principle, the output voltage is now found by summing the individual outputs caused by each generator acting singly and regarding the others as being short-circuited. In our case each generator v'_c gives a contribution to the initial derivative of the slope of the output voltage. Since in computing the slopes from each cause we have neglected the other causes of sag, we have in essence regarded the other capacitances as short circuits. Hence to find the total sag we need only sum the individual slopes,

$$\text{Total slope (or sag)} = \sum \text{individual slopes (or sags)} \qquad (5\text{-}15)$$

From the evaluation of the separate causes of slope in the long-time response, it becomes evident which of them is the most severe. In the vacuum-tube case the cathode circuit is the worst offender for a given size of capacitor—as an example, consider a 6AU6 tube in which typical values are $g_m = 5{,}000$ μmhos, $r_{p2} = 1/y_{22} = 23$ kilohms, and $R_g = 1$ megohm.

$$\text{Screen slope} = \frac{-100\%}{23 \text{ kilohms} \times C_s} \cong -0.004\%/\mu\text{sec-}\mu\text{f}$$

$$\text{Cathode slope} = \frac{-100 g_m}{C_k} = -0.5\%/\mu\text{sec-}\mu\text{f}$$

$$\text{Coupling circuit} = \frac{-100}{R_g C_{cc}} = -0.0001\%/\mu\text{sec-}\mu\text{f}$$

In the transistor case the emitter circuit requires the largest capacitor for a given slope. Consider the following typical values: $R_g = 500$ ohms, $R_1 = 10$ kilohms, $r_b + r_e/(1 - \alpha) = 1{,}750$ ohms, $1 - \alpha = 0.02$. (Therefore $R = 1{,}990$ ohms, and $R_s = 2{,}226$ ohms.)

$$\text{Coupling circuit} = \frac{-100}{RC_{cc}} \cong -0.05\%/\mu\text{sec-}\mu\text{f}$$

$$\text{Emitter slope} = \frac{-100}{R_s C_{cc}(1 - \alpha)} \cong -2\%/\mu\text{sec-}\mu\text{f}$$

The transistor because of its inherently lower impedance levels requires large capacitors. Fortunately the d-c voltage levels which these capacitors must withstand are also small, so that physically small, high-capacitance electrolytic capacitors may be used. In many cases even the coupling capacitor must be of the electrolytic type. If only small amounts of sag can be tolerated, better performance may be obtained by using no resistance in the emitter lead and resorting to feedback biasing, as shown in Fig. 3-26. [A resistor from the base to a positive supply (assuming a PNP transistor) may be necessary so that the resistor shown as R_f may be sufficiently reduced to give the necessary bias stability.]

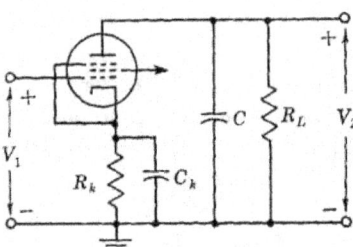

Fig. 5-4   Circuit for cathode peaking.

In the vacuum-tube case the cathode bypass capacitor may be removed [1] to eliminate the greatest cause of sag, but the gain of the stage is also reduced by the factor $1/(1 + g_m R_k)$. This reduction of gain is not so great as that caused by leaving the transistor emitter unbypassed for typical circuit conditions.

5-4 Cathode Peaking. Where strict requirements on slope make it desirable to leave the cathode bias resistor unbypassed, it becomes necessary to reevaluate the short-time transient response. In so doing it has been found beneficial to add a small capacitor across the bias resistor, instead of having none at all. This capacitance is so small, however, that its effect on the long-time response is negligible. In the short-time response, though, the rise time can be improved, with beneficial effects similar to those of shunt peaking, and hence the name "cathode peaking" is usually ascribed to this technique. The analysis proceeds from the circuit of Fig. 5-4.

[1] Removing C_K may have one undesirable side effect: the heater-to-cathode leakage current may cause a heater-frequency noise to appear across R_k and thus in the output.

$$V_2(p) = \frac{-g_m V_1(p)}{1 + g_m Z_k(p)} Z_L(p) \tag{5-16}$$

where $\qquad Z_L(p) = \dfrac{1}{1/R_L + pC} \qquad Z_k(p) = \dfrac{1}{1/R_k + pC_k}$

Let $V_1(p) = 1/p$; then $V_2(p)$ becomes

$$V_2(p) = \frac{-g_m}{C} \left[\frac{p + 1/R_k C_k}{p \left(p + \dfrac{1 + g_m R_k}{R_k C_k} \right) \left(p + \dfrac{1}{R_L C} \right)} \right] \tag{5-17}$$

It will be convenient to define $\rho = R_k C_k / R_L C$ and $K = 1 + g_m R_k$; then $v_2(t)$ becomes

$$\frac{v_2(t)}{-g_m R_L} = \frac{1}{K} \left[1 + \frac{\rho(K-1)}{\rho - K} \epsilon^{-Kt/\rho R_L C} - \frac{K(\rho - 1)}{\rho - K} \epsilon^{-t/R_L C} \right] \quad \text{for } \rho > K \tag{5-18a}$$

$$= \frac{1}{K} \left[1 - \frac{1 - K}{1 - K/\rho} \epsilon^{-Kt/\rho R_L C} - \frac{K(\rho - 1)}{\rho(1 - K/\rho)} \epsilon^{-t/R_L C} \right] \quad \text{for } \rho < K \tag{5-18b}$$

We are interested in trying different values of C_k for the circuit; the other elements are already determined. Thus, we can let ρ take values ranging from zero (cathode resistor unbypassed) to infinity (resistor perfectly bypassed). Three special values of ρ are of interest.

For $\rho = 0$ $(C_k = 0)$

$$\frac{v_2(t)}{-g_m R_L} = \frac{1}{K} (1 - \epsilon^{-t/R_L C}) \tag{5-19}$$

For $\rho = \infty$ $(C_k = \infty)$

$$\frac{v_2(t)}{-g_m R_L} = (1 - \epsilon^{-t/R_L C}) \tag{5-20}$$

For $\rho = 1$ $(R_k C_k = R_L C)$

$$\frac{v_2(t)}{-g_m R_L} = \frac{1}{K} (1 - \epsilon^{-Kt/R_L C}) \tag{5-21}$$

The time response for various values of ρ and for a particular value of K (approximately 2) is shown in Fig. 5-5. Notice that $\rho = 1$ gives a faster rise than does $\rho = 0$ and is the largest value of ρ that gives a monotonic rise (no overshoot); hence, this is the value usually chosen in the cathode-peaking technique. Notice also that $\rho = 100$ is a value in the usual range of cathode bypassing; there is a fast rise—essentially that of the ideal case ($\rho = \infty$)—followed by a long-time sag as previously analyzed for the cathode bias circuit.

It would appear from a comparison of Eqs. (5-20) and (5-21) that cathode peaking with $\rho = 1$ gives a faster amplifier than does the perfectly

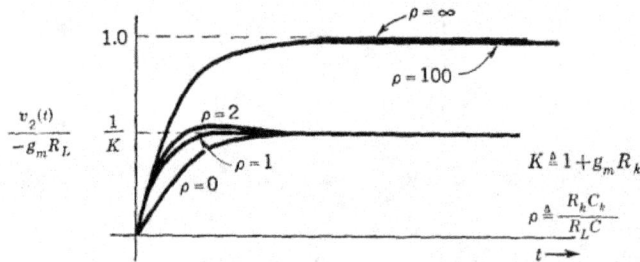

Fig. 5-5 Step response of the cathode-peaked circuit for different values of cathode capacitance. (Drawn for $K = 2$.)

bypassed case. Both responses are simple exponentials with a 10 to 90 per cent rise time already given in Eq. (4-3).

For $\rho = \infty$ $\qquad\qquad\qquad\qquad T_R = 2.2 R_L C$

For $\rho = 1$ $\qquad\qquad\qquad\qquad T_R = \dfrac{2.2 R_L C}{K}$

Thus, so long as K is greater than 1, the rise time is indeed reduced by the factor K. Unfortunately, it is not usually sufficient to consider rise time alone. The peaking circuits discussed in Chap. 4 were compared on their ability to improve the gain–rise-time ratio. On this basis, the cathode-peaking circuit offers nothing, because the gain is reduced by the same factor by which the rise time is reduced, namely, K; hence, the ratio is the same as without peaking.

This perhaps conveys a needlessly unfavorable impression about cathode peaking. The reduction in gain can be restored by increasing R_L by a factor K. Then the gain is the same as in the perfectly bypassed case, the rise time is the same, but a much smaller bypass capacitor can be used, and there is *no sag*.

A similar peaking method may be employed in transistor stages, although the value of R_e (the emitter resistor), which must be employed to give proper peaking, is usually too small to give sufficient bias stability. An excellent treatment will be found in an article by G. Bruun.[1]

[1] Georg Bruun, Common-emitter Transistor Video Amplifiers, *Proc. IRE*, vol. 44, pp. 1561–1572, November, 1956.

5-5 Slope Compensation. This is sometimes called "low-frequency" compensation, and rightly so, inasmuch as the slope or sag in the long-time response to a step is governed by those circuit elements which govern the low-frequency steady-state response. However, a steady-state analysis is perhaps at its worst when it comes to the question of how much compensation should be used for a good step response.

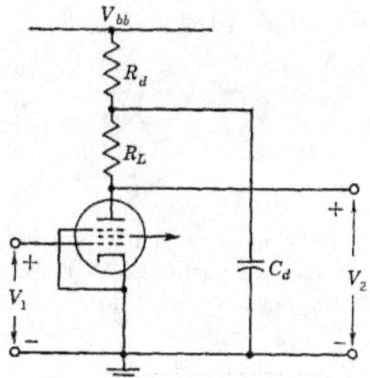

Fig. 5-6 A sag-compensating circuit.

In transient terms, we have already evaluated the long-time step response of the conventional circuit and have found that sag results from several causes: coupling circuit, screen circuit, and cathode bias (or emitter) circuit. Each of these is the source of a negative initial slope, and their effects tend to add directly. What we need is a circuit which will provide a *positive* initial slope, and one which will add to the negative slopes in the right amount to make the sum equal to zero.

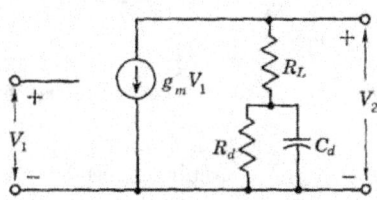

Fig. 5-7 Equivalent circuit for the sag-compensating circuit.

Fortuitously the so-called "decoupling circuit," which is a normal part of a multistage amplifier—introduced to diminish coupling from one stage to the others via the common V_{bb} supply—is a suitable compensating circuit if proportioned properly. It is a long-time-constant circuit, the element values can be varied considerably without hampering its decoupling effectiveness, and it does provide a positive initial slope. The actual circuit is shown in Fig. 5-6 and its equivalent for analysis purposes in Fig. 5-7.

$$V_2(p) = -g_m V_1(p) \left(R_L + \frac{1}{1/R_d + pC_d} \right) \qquad R_L \ll R_g \qquad (5\text{-}22)$$

Let $V_1(p) = 1/p$ (unit step).

$$V(p) \triangleq \frac{V_2(p)}{-g_m R_L}$$

$$= \frac{1}{p} \left(\frac{p + 1/R_L C_d + 1/R_d C_d}{p + 1/R_d C_d} \right) \qquad (5\text{-}23)$$

Using Eq. (5-11), we may immediately find the initial slope,

$$v'(0^+) = \left(\frac{1}{R_L C_d} + \frac{1}{R_d C_d}\right) - \frac{1}{R_d C_d} = \frac{1}{R_L C_d}$$

$$= \frac{100\%}{R_L C_d} \tag{5-24}$$

The slope is evidently positive, and hence, by selecting a suitable value of C_d for the particular R_L being used, it is possible to provide cancellation of the negative slope due to one or more of the causes discussed above.

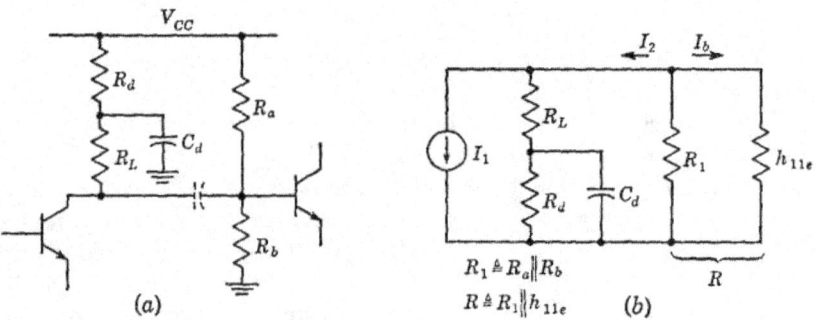

Fig. 5-8 Sag compensation in a transistor amplifier. (a) Actual circuit. (b) Equivalent circuit. [NOTE: $R_a\|R_b \triangleq R_a R_b/(R_a + R_b)$.]

For instance, to cancel the slope due to the coupling circuit, choose C_d so that

$$R_L C_d = R_g C_{cc} \tag{5-25}$$

By way of caution it should be pointed out that, although R_d does not appear in the initial slope, $R_d C_d$ must be large compared with the period of the waveform to be amplified, in order both that the decoupling function be supplied and that the exponential function $v_2(t)$ be adequately approximated by the initial straight-line tangent.

Slope compensation for the transistor video amplifier proceeds from the equivalent circuit of Fig. 5-8. In the circuit R_1 represents the resistance of the biasing network in the base of the transistor. The resistance h_{11e} is the approximate input resistance of the transistor and is equal to $r_b + r_e/(1 - \alpha)$. The base current is

$$-I_b = \frac{R_1 I_2}{R_1 + h_{11e}}$$

$$\frac{I_2}{I_1} = \frac{R_L}{R + R_L} \frac{p + (R_L + R_d)/R_L T}{p + \dfrac{R + R_L + R_d}{(R + R_L)T}} \tag{5-26}$$

where $R = R_1 h_{11e}/(R_1 + h_{11e})$ and $T = R_d C_d$. From Eq. (5-26) the initial slope of I_2 (and hence I_b) may be found by assuming that I_1 is a unit step of current,

$$\frac{I_b'(0^+)}{I_b(0^+)} = \frac{R}{(R + R_L)R_L C_d} \quad \text{or, in \%,} \quad \frac{R \times 100}{(R + R_L)R_L C_d} \% \text{ (5-27)}$$

The positive sag given in Eq. (5-27) may be used in exactly the same manner as Eq. (5-25) to cancel one or more causes of sag in a transistor amplifier. It is not practical in the usual case to compensate for many causes of sag with one compensating circuit because the approximation of the exponential with its initial slope is not good for a long enough time. Therefore in a multistage amplifier the sag-compensating circuits are distributed throughout the amplifier, both to provide the sag reduction and to furnish more decoupling than a single decoupling circuit can afford. The input stages of the amplifier, in particular, should be provided with decoupling since these stages are most prone to signal and noise pickup from the common supply leads.

PROBLEMS

5-1. In a single stage it is possible to compensate exactly for all time the sag introduced by the cathode circuit alone. Find the value of R_d and C_d to compensate exactly a stage with a cathode circuit comprising R_k and C_k. Note that *both* R_d and C_d must be specified for the compensation to be good for all time.

5-2. For an amplifier whose gain function is described by Eq. (5-3), what can be said about the relationship between the sag (expressed as initial slope) and the steady-state low-frequency cutoff frequency? Consider three cases:

a. $n = 1$

b. n arbitrary but known

c. n arbitrary but unknown

5-3. (This problem utilizes the results of several of the previous chapters as well as this one.) An amplifier with a gain of 100 db is required. The amplifier is to be made of identical stages (as far as high frequencies are concerned) using 6AK5's. (Assume that the load connected to the last stage has the same capacity as the 6AK5 input capacity.) The 6AK5 characteristics are

$$E_f = 6.3 \text{ volts}$$

$$I_f = 0.175 \text{ amp}$$

$$C_{gp} = 0.02 \text{ pf}$$

$$C_{in} = 4.0 \text{ pf}$$

$$C_{out} = 2.8 \text{ pf}$$

Assume a total wiring capacitance of 5 pf which is distributed equally between plate and grid circuits.

Typical operation:

$$E_b = 180 \text{ volts}$$
$$E_{c2} = 120 \text{ volts}$$
$$E_{c1} = -1.8 \text{ volts}$$
$$I_b = 7.7 \text{ ma}$$
$$I_{c2} = 2.4 \text{ ma}$$
$$g_m = 5{,}100 \ \mu\text{mhos}$$
$$r_p = 0.5 \text{ megohm}$$
$$r_{p2} = 20 \text{ kilohms}$$

(Note that the parameters change little with moderate changes in E_b.)

a. Assuming two-terminal interstage networks with $m = 0.25$, what is the minimum over-all rise time?

b. How many stages are required to obtain this minimum rise time?

c. The number of stages found in (*b*) is excessive; therefore, it is decided to use only 10 stages to obtain the 100-db gain, but to use the four-terminal networks of Fig. 4-9, 4-10, or 4-11. With the network which will give the fastest rise time, what is the over-all rise time of the amplifier? (Note that capacitance must be added to the interstages to preserve the ratio C_1/C_2 which is required by a given network.)

d. Assume that one sag-compensating circuit will be used for each pair of stages. Let $C_k = 200 \ \mu\text{f}$; $C_s = 1 \ \mu\text{f}$; $R_g = 1$ megohm; and $C_{cc} = 0.1 \ \mu\text{f}$. Compute the necessary values for the sag-compensating circuit. For what length of time will the sag be compensated (make only a rough estimate)?

e. Carefully draw the complete schematic diagram for two stages of the amplifier, including the values of all components. Assume that $V_{bb} = 200$ volts.

5-4. A complete transistor amplifier is shown in Fig. P5-4. The common-collector stage at the input provides a high input impedance; the common-collector stage at the output provides a low output impedance; and the voltage gain is provided by the

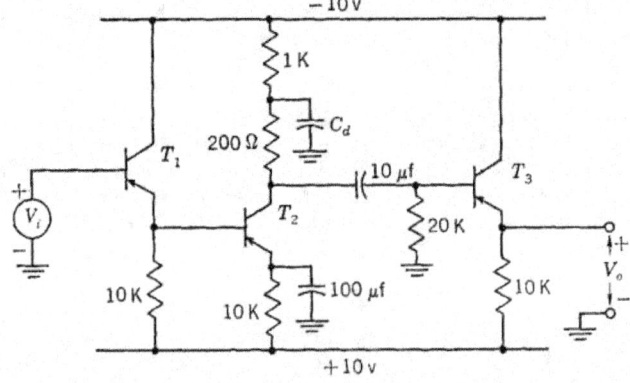

Fig. P5-4

common-emitter stage in the middle. Assume that $r_e = 10$ ohms, $r_b = 1$ kilohm, $r_c = 1$ megohm, and $\alpha = 0.99$ for all transistors.

a. What is the over-all voltage gain?

b. What is the value of C_d required to give zero initial sag in the output voltage $v_o(t)$?

(Be sure to make all reasonable approximations in the representation and calculation of the circuit.)

6

Additive Amplification

Conventional amplifier systems, as discussed in the three chapters preceding this one, could be called "product" amplification. That is to say, when several stages are connected in cascade, the over-all gain function (of the variable p) is the continued *product* of the separate stage gain functions, as expressed for instance in Eq. (4-43). In contrast, the amplifier structures to be described in this chapter have a gain function which is the *sum* of the "gains" provided by the separate elements.

Product, or cascade, amplification is older and more widely used. If the requirements on rise time (or bandwidth, in the steady state) are not too severe, a given amount of gain can be provided with fewer tubes in the cascade connection. But when the requirements are severe, it may be impossible to meet them with the product system, and yet the same tubes may be used in an additive structure to meet the requirements. The criterion that determines whether or not the cascade structure will work is the familiar quotient of gain over rise time, a factor which depends primarily on the tube (Table 4-1) but which may be improved by more complicated circuits, e.g., Figs. 4-9ff.

The difficulty becomes apparent when one is faced with providing a certain over-all gain for n stages (n unknown) and a certain over-all rise time T_{RN}. There are various tube types available, each with a particular g_m/C, and one has at his disposal certain networks, each having an efficiency factor η, expressing the relative speed with respect to the elementary resistance-coupled circuit. These are the "building blocks." Unfortunately, there is a limit to what can be done. As given in Eq. (4-54), the analysis by Elmore has shown that there is a minimum rise time that can be achieved with a given over-all gain which occurs when the stage gain is $\sqrt{\epsilon}$. If the requirements call for a smaller rise time than Eq. (4-54) permits, the job

147

cannot be done with cascade amplification unless better tubes or networks are developed (the networks cannot be expected to provide an efficiency much better than 4, and they become inconveniently complicated at values of 2).

Additive amplification is the solution. In fact, it works even when each tube contributes a "gain" of less than unity! There are two principal types of additive structures. One of these is called the "distributed" amplifier; it was first employed in a British television installation in 1937, and has been in extensive commercial use in the United States since 1948. The other is the "split-band" amplifier and is still under development.

6-1 Basic Theory of Distributed Amplification. This form of amplifier structure was first proposed by Percival in 1935 in a British patent,[1] although the system did not go into active use until after the first published analysis [2] in 1948.

The basic idea of the distributed amplifier is surprisingly simple, although there are, of course, many practical matters that contribute to the difference between actual performance and the first-order theory. The elementary form of the structure is given in Fig. 6-1. The networks $L_1 C_1$ comprise the so-called grid "line," a cascade of filter sections in which the capacitors C_1 are the input capacitance of the tubes. Similarly, the networks $L_2 C_2$ are the plate line,[3] C_2 being the output capacitance of the tubes. The two lines are designed to have the same phase velocity and are terminated in their characteristic resistances R_{01} and R_{02}, respectively, so that no reflections take place. The lines are further assumed to be dissipationless, so that a wave can travel along either of them without attenuation.

Within the limits of these idealized conditions, the following relationships hold:

Characteristic impedance

$$R_{01} = \sqrt{\frac{L_1}{C_1}}$$

$$R_{02} = \sqrt{\frac{L_2}{C_2}}$$

(6-1)

---

[1] W. C. Percival, Thermionic Valve Circuits, British Patent 460562, July 24, 1935–Jan. 25, 1937.

[2] E. L. Ginzton, W. R. Hewlett, J. H. Jasberg, and J. D. Noe, Distributed Amplification, *Proc. IRE*, vol. 36, pp. 956–969, August, 1948.

[3] The conventional filter formulas are open to question where the plate line is concerned, since the structure is driven by a number of current generators along the structure instead of by a single generator at one end. This situation is analyzed by D. V. Payne, Distributed Amplifier Theory, *Proc. IRE*, vol. 41, pp. 759–762, June, 1953; discussion by R. W. A. Scarr, *Proc. IRE*, vol. 42, pp. 596–598, March, 1954.

Phase velocity (sections per second)

$$v_{p,0} = \frac{1}{\sqrt{L_1 C_1}} = \frac{1}{\sqrt{L_2 C_2}} \qquad (6\text{-}2)$$

$$\text{Plate current of each tube} = g_m |V_1| \qquad (6\text{-}3)$$

Because of the equal velocities in grid and plate lines the plate-current contributions of successive tubes will add directly; i.e., they are all in phase at the load.

$$\text{Load current} = \frac{n g_m V_1}{2} \qquad (6\text{-}4)$$

The factor 2 in Eq. (6-4) comes in because half the current contributed by each tube flows to the left in the plate line and is lost in the terminating resistor R_{02}. In spite of the current thus lost, the resistor R_{02} is usually necessary to prevent reflection of the wave traveling to the left of each plate, which at certain frequencies could cancel the wave traveling to the load.

Output voltage

$$V_2 = \frac{n g_m V_1}{2} R_{02} \qquad (6\text{-}5)$$

Amplification

$$A \triangleq \frac{V_2}{V_1} = \frac{n g_m R_{02}}{2} \qquad (6\text{-}6)$$

Equation (6-6) displays the basic property of the distributed amplifier, namely, that the amplification increases linearly with the number of "stages"; i.e., each tube contributes a gain of $g_m R_{02}/2$, and the total gain is the *sum* of the individual contributions. Indeed, each tube may contribute a gain of less than unity, and yet the total gain can be made as large as

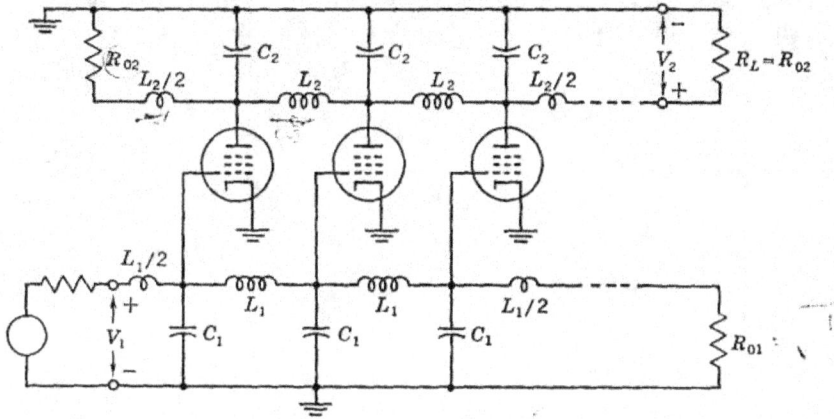

Fig. 6-1 Distributed amplifier.

desired by adding a sufficient number of tubes. This situation is not possible with cascade amplification.

6-2. Cascading Distributed Amplifier Stages. It is feasible—and frequently advantageous—to cascade whole distributed stages of the type of Fig. 6-1. This is done by connecting the grid line of the second stage as the load on the first stage (with a blocking capacitor for d-c isolation, of course). Should R_{01} and R_{02} not be the same, a transformer is in principle required to join the two stages in order to prevent reflection. If very low frequencies as well as the high frequencies must be amplified, a suitable transformer cannot be obtained; if, however, only the band from a few megacycles to the upper limit of the amplifier is required, a suitable coaxial transformer may be constructed. In the former case, it is usual to make R_{01} and R_{02} the same by adding to the smaller of C_1 or C_2 in order to equalize them. The effect is the same, but the gain is reduced by the added capacitance. With the transformer, Eq. (6-6) becomes

$$A = \frac{ng_m \sqrt{R_{01}R_{02}}}{2} \tag{6-7}$$

If, instead of a transformer, capacitance is added to the smaller C,

$$A = ng_m \frac{R_{01}}{2} \qquad \text{if } R_{02} \text{ is made equal to } R_{01} \tag{6-7a}$$

The question to be asked about cascading is: When does one stop adding tubes along the line in one stage and add further tubes in a second stage? This is readily answered for the idealized conditions we have thus far considered, i.e., that the grid and plate networks behave perfectly as lines. Supposing that a total gain A_t is required, how few tubes, N, are necessary in m stages, with n tubes per distributed stage? Let the contribution of each tube to the stage gain A in Eq. (6-7) be A_i, where

$$A_i = \frac{g_m}{2} \sqrt{R_{01}R_{02}} \tag{6-8}$$

Thus

$$A \overset{\Delta}{=} nA_i \tag{6-9}$$

$$A_t = (nA_i)^m = A^m \tag{6-10}$$

$$N = nm \tag{6-11}$$

From Eqs. (6-10) and (6-11)

$$n \ln A_t = N \ln (nA_i) \tag{6-12}$$

Differentiate Eq. (6-12) with respect to n,

$$\ln A_t = \frac{dN}{dn} \ln (nA_i) + \frac{N}{n} \tag{6-13}$$

For a minimum in N, the total number of tubes, set $dN/dn = 0$.   Then

$$\ln A_t = \frac{N}{n} \qquad (6\text{-}14)$$

$$A_t = \epsilon^{N/n} = \epsilon^m = A^m \qquad (6\text{-}15)$$

Thus $A = \epsilon \, (= 8.7 \text{ db})$ to give the most efficient use of the tubes (i.e., minimum N).[1]   Of course, we have learned that cascading stages impairs rise time and reduces bandwidth, but this consideration has been neglected here.   (See Sec. 6-5.)

6-3.   Plate and Grid Line Characteristics.   Before undertaking an analysis which includes the effect of bandwidth shrinkage, we must study the plate and grid lines to determine what effects these will have.   The basic premise in the elementary theory of the distributed amplifier is that these networks simulate smooth lines, terminated in the appropriate R_0 so that no reflections occur.   Apart from the termination problem, the important requirement is that the velocity (or time delay) per section of the grid network equal that of the plate network and also that this velocity be constant with frequency.   This is the same as saying that the phase shift must be proportional to frequency.   Failure to realize this results in phase distortion, which, as was considered in Chap. 4, shows up also in distortion of transient signals.   Naturally, it is true that the amplitude distortion (as a function of frequency) also enters into the transient distortion.   The elementary ladder networks such as those of Fig. 6-1 have a disproportionate amount of phase distortion, and hence an all-out effort to decrease this results in a better balance between amplitude and phase distortion and a better transient response.

The problem of designing the best possible network for use in the distributed amplifier has its counterpart in the design of the best possible delay line from lumped dissipationless elements.   These delay lines are used in oscilloscopes, radar, etc., and provide small time delays with as little distortion as possible, i.e., distortion of rectangular pulses or other transient waveforms.   Interestingly enough, the network most widely used in delay lines is the same as that in the typical distributed amplifier.[2]   Actually, the distributed amplifier presents a somewhat more difficult network problem,

[1] This relationship is given by Ginzton et al., *op. cit.*, and by A. P. Copson, A Distributed Power Amplifier, *Elec. Eng.*, vol. 69, pp. 893–898, October, 1950.

[2] References on the delay line include H. E. Kallman, Equalized Delay Lines, *Proc. IRE*, vol. 34, p. 646, September, 1946; B. Chance et al. (eds.), "Waveforms" (vol. 19, M.I.T. Radiation Laboratory Series), pp. 730–750, McGraw-Hill Book Company, Inc., New York, 1948; J. F. Blackburn (ed.), "Components Handbook" (vol. 17, M.I.T. Radiation Laboratory Series), pp. 191–217, McGraw-Hill Book Company, Inc., New York, 1949; B. Trevor, Jr., Artificial Delay-line Design, *Electronics*, vol. 18, p. 135, June, 1945.

because the input conductance of the tubes connected to the grid line introduces dissipative elements in shunt with the capacitances; this will be discussed later.

Consider now the basic circuit of Fig. 6-1. The grid and plate networks have the appearance of transmission lines, in which the infinitesimal elements of series inductance and shunt capacitance in the continuous line have become finite. It is, indeed, an "artificial line," and the properties of such lines (in the steady state) have been known since the early days of the telephone. Moreover, the whole art of the "wave filter" was first devised by G. A. Campbell from the concept of the artificial line, and we

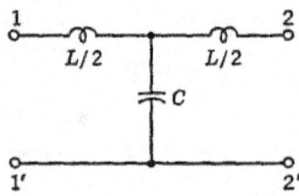

Fig. 6-2 T section.

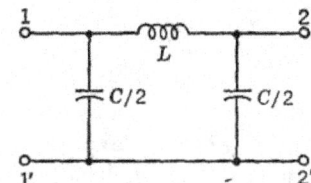

Fig. 6-3 Pi section.

can profitably use the results of much wave-filter analysis in the problem here.

The artificial line fails us because it has one characteristic not possessed by the smooth line, namely, a *cutoff frequency* f_c, where

$$f_c = \frac{1}{\pi\sqrt{LC}} \qquad (6\text{-}16)$$

The significance of the cutoff frequency for the network structure in Fig. 6-1 (which is shown again in both its T and pi equivalents, in Figs. 6-2 and 6-3, respectively) is that a chain or ladder of sections of this type will transmit frequencies below—but not above—this frequency. Moreover, there is an abrupt discontinuity in the phase velocity and the characteristic impedance [1] at the cutoff frequency, and indeed these parameters begin to vary significantly with frequency long before the cutoff is approached. Equations (6-1) and (6-2) for the characteristic impedance and velocity are those which would be appropriate for a smooth line, if L and C represented the values per unit length; they also hold for the lumped network at very low frequencies (far from cutoff). The general expressions for the

---

[1] In filter analysis this is called the *image impedance*, which in the case of the input impedance to an infinite ladder of sections is also the characteristic impedance. The latter is defined in terms of reflections, whereas the former is defined as the input impedance to a section when terminated at the output by an impedance equal to the impedance looking back into the output terminals, i.e., the "image," to use a mirror analogy.

impedance and delay τ per section (defined as $1/v_p$) at any frequency in the region below cutoff are

$$Z_{I,\pi} = \sqrt{\frac{L}{C}} \frac{1}{\sqrt{1-(f/f_c)^2}} \qquad \text{for the pi section} \qquad (6\text{-}17)$$

$$Z_{I,T} = \sqrt{\frac{L}{C}} \sqrt{1-\left(\frac{f}{f_c}\right)^2} \qquad \text{for the T section} \qquad (6\text{-}17a)$$

$$\tau = \frac{\sqrt{LC}}{\sqrt{1-(f/f_c)^2}} \qquad (6\text{-}18)$$

A plot of τ and Z_I is given in Fig. 6-4. It can be seen from the figure that there are two problems which were not considered in the ideal-line case. One is the termination problem, and the other the time-delay, or velocity, problem. It is important that the grid and plate networks be terminated so that there are no reflections; otherwise, as the frequency is varied, the reflected wave could alternately add to and subtract from the forward wave, thus producing variations in the amplifier gain as a function of frequency. From Fig. 6-4 it is apparent that the terminating impedance must vary with frequency in the manner indicated there, instead of being a simple resistor as depicted in Fig. 6-1.

The other problem, time delay, is serious from the standpoint of the transient response. The curve of Fig. 6-4 displays phase distortion; i.e., the time delay is not constant as would be the case with phase proportional to frequency. The discussion in Chap. 4 brought out the undesirable transient performance resulting from phase distortion.

Both of the problems which have been described above, termination and time delay, can be solved sufficiently well for practical purposes, although

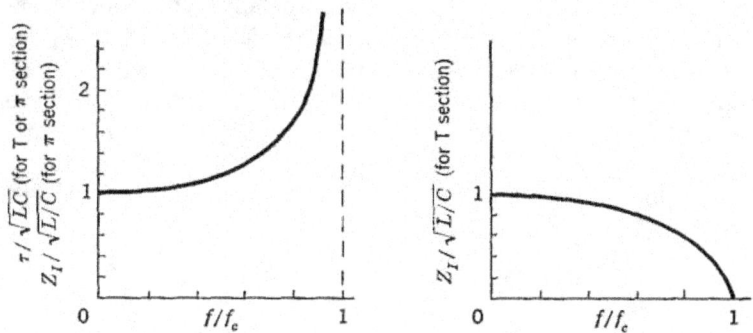

Fig. 6-4 Image impedance and time delay for sections shown in Figs. 6-2 and 6-3 (constant-k sections).

the solutions are by no means the ultimate. Consider first the termination problem. The proper terminating impedance, with an approximation to the characteristic as in Fig. 6-4, can be provided by what the filter experts call an "m-derived half section." A few words might be in order concerning the background of terminology, as well as the device itself.

The network sections of Figs. 6-2 and 6-3 are "lowpass" forms of general filter sections shown in Figs. 6-5 and 6-6. Proper choice of the elements

Fig. 6-5 Constant-k prototype T section.   **Fig. 6-6** Constant-k prototype pi section.

making up Z_1 and Z_2 permits not only lowpass structures but also highpass, bandpass, and band-elimination filters. One basic type of all these is the so-called "constant-k" structure, in which, independent of frequency, the following relationship holds:

$$k^2 = Z_1 Z_2 \qquad k \text{ a real number} \qquad (6\text{-}19)$$

The image impedance of the constant-k section is

$$Z_{I.T} = \sqrt{Z_1 Z_2 + \frac{Z_1^2}{4}} \qquad \text{Fig. 6-5} \qquad (6\text{-}20)$$

$$Z_{I.\pi} = \frac{Z_1 Z_2}{\sqrt{Z_1 Z_2 + Z_1^2/4}} \qquad \text{Fig. 6-6} \qquad (6\text{-}21)$$

The m-derived section evolves from the constant-k one in the following manner: If a network section is assembled as in Fig. 6-7, in which Z_1 and Z_2 are the same as in the constant-k (sometimes called "prototype") section of Fig. 6-5, such a section is said to be m-derived. The coefficient m can be

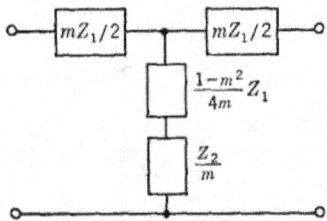

Fig. 6-7 m-derived section.

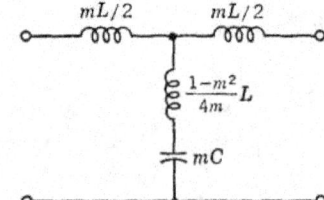

Fig. 6-8 m-derived lowpass section.

any real constant, not necessarily an integer.   The m-derived counterpart of Fig. 6-2 is shown in Fig. 6-8.

If the total shunt impedance is defined as Z_2' and the total series imped- ance is defined as Z_1', that is,

$$Z_1' \triangleq mZ_1 \tag{6-22}$$

and

$$Z_2' \triangleq \frac{Z_2}{m} + \frac{1-m^2}{4m}Z_1 \tag{6-23}$$

then substitution into Eq. (6-20) for $Z_{I,\text{T}}$ gives the new image impedance, which turns out to be exactly the same as that for the constant-k section.

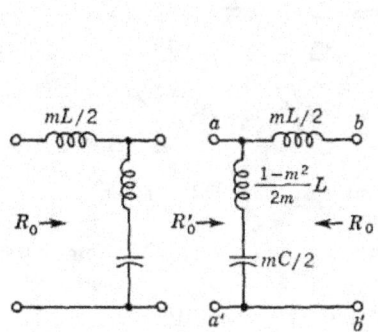

Fig. 6-9 m-derived half section.

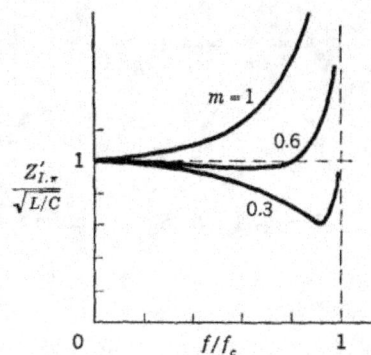

Fig. 6-10   Image impedance of m-derived half section.

Hence, such a section could be inserted in a ladder of prototype sections without producing reflections at any frequency.   There are positive advan- tages to the m-derived section, beyond this permissive attribute of being able to use the section in combination with the basic constant-k structure. The first advantage comes in connection with the termination problem. Suppose that we split the m-derived section of Fig. 6-8 into two half sections as in Fig. 6-9 and then examine the image impedance looking into the terminals a-a'; this is called the mid-shunt impedance $Z_{I,\pi}'$.

The impedance may be found by substituting Z_1' and Z_2' into Eq. (6-21) for $Z_{I,\pi}$,

$$Z_{I,\pi}' = \sqrt{\frac{L}{C}}\,\frac{1-(1-m^2)\,(f/f_c)^2}{\sqrt{1-(f/f_c)^2}} \tag{6-24}$$

For various values of m, $Z_{I,\pi}'$ varies with frequency, as indicated in Fig. 6-10.   This is quite remarkable, because for *all* values of m the image impedance of the main section (terminals b-b') remains as in Fig. 6-4.   The value $m = 0.6$ is particularly useful. Notice that $Z'_{I,\pi}/\sqrt{L/C}$ is real and very close to 1.0 up to a frequency nearly equal to f_c. Our terminating technique is now in hand. We connect a resistance $R = \sqrt{L/C}$ to the terminals a-a'; this matches the image impedance $Z'_{I,\pi}$ (within the limits of the approximation of the $m = 0.6$ curve), and hence looking to the left at b-b' the impedance Z_I is the same as though there were an infinite number of either m-derived or constant-k sections extending to the right, which

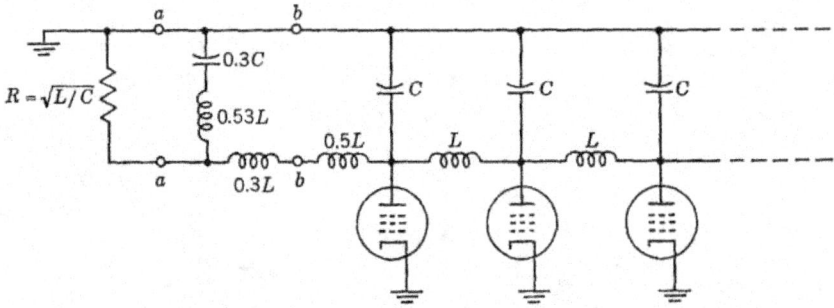

Fig. 6-11 Distributed amplifier showing the terminating half section.

ladder is then terminated. In Fig. 6-11 this is done for the plate network of the distributed amplifier.

Our first problem is thus solved, with the aid of the "terminating half section." Four of these terminations are usually required, one at each end of both grid and plate networks.

The next problem, that of the frequency-variable time delay, is also solved with the employment of an m-derived structure. If one explores the time delay per section for various values of m, curves such as those of Fig. 6-12 are obtained. Notice that $m = 1.27$ has a particularly favorable characteristic, in that the delay time remains approximately constant to a high frequency; this is the value usually chosen for delay lines and distributed amplifiers. Remember that choosing any particular value of m, such as 1.27, has no effect on the image (characteristic) impedance $Z_{I,T}$; so all that has been said about terminations, etc., is unaffected by whether we use constant-k or m-derived sections associated with each tube in our amplifier.

There is one slightly embarrassing feature in choosing a value of m

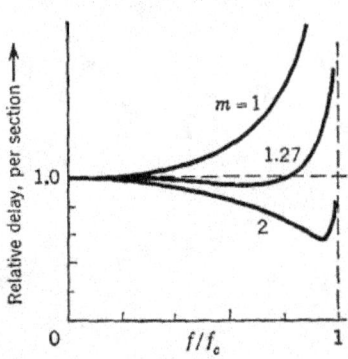

Fig. 6-12 Time delay in m-derived section.

greater than unity: a negative inductance is required in the shunt branch (see Fig. 6-8). Such a requirement can be met, however, by providing mutual inductance of the proper amount between the two series elements. Thus, in Fig. 6-8 we interpret the three inductances to represent the T equivalent of a transformer and then replace the equivalent by an actual transformer. It is possible for the actual transformer to be physically

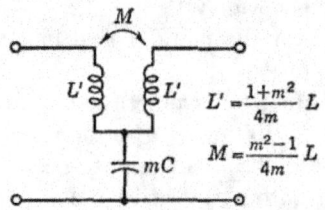

$$L' = \frac{1+m^2}{4m} L$$

$$M = \frac{m^2-1}{4m} L$$

Fig. 6-13 Section using mutual inductance to realize $m > 1$.

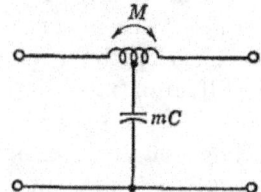

Fig. 6-14 Physical form of Fig. 6-13.

realizable, even though the T equivalent is not. Thus, the circuit of Fig. 6-8 becomes that of Fig. 6-13. In practice the transformer is constructed by tapping onto a single-layer coil of suitable proportions,[1] as indicated in Fig. 6-14.

Thus, each section of the distributed amplifier is arranged as in Fig. 6-14, where the capacitance is either the input capacitance or the output

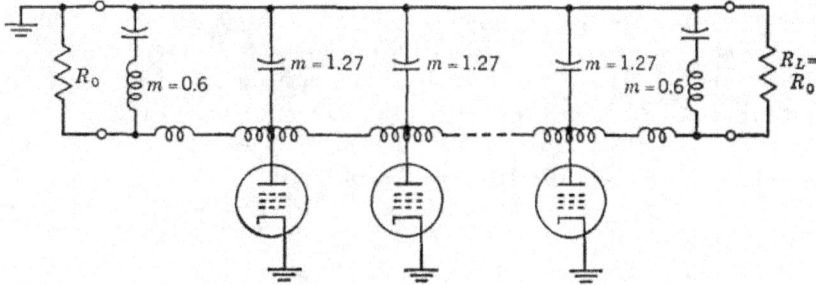

Fig. 6-15 Complete plate-line network showing both phase and impedance equalizing sections.

capacitance of the tube. The complete form of the plate network, for instance, would then be as in Fig. 6-15. This, then, is the basic design philosophy of the distributed amplifier, as currently being used. The steady-state response is quite acceptable to very high frequencies, in spite of the fact that the approximation involved in the terminating half sections

[1] See Kallman, *op. cit.*; also see B. Murphy, Distributed Amplifiers, *Wireless Engr.*, vol. 30, pp. 39–47, February, 1953.

gets worse as the cutoff frequency is approached. From the standpoint of transient response, the approximation does not seem to work out as well, and in practice there is a certain amount of cut-and-try manipulation of the termination on an experimental basis, while the operator observes the shape of the transient response.

Other forms of networks have been employed, such as the bridged T, but the arrangement of Fig. 6-15 is the most widely used. It is possible, for instance, to use a continuous solenoid for the plate or grid line, with taps along its length for the tube connections.[1]

6-4 Effect of Tube Input Conductance. In a practical case, where the amplifier is to operate at frequencies of 100 or 200 Mc, the analysis should be extended to include the effects of the input admittance of the pentode tubes. This input admittance has both a capacitive and a conductive component (see Fig. 2-15b). The latter component is the more serious in the distributed amplifier, because it produces attenuation of the signal traveling down the grid network. The conductance of the tubes increases with the square of frequency, and so as the frequency is increased, the attenuation ultimately reaches a level such that the attenuation per section is greater than the gain provided by the tube; beyond this frequency the gain diminishes—and adding more tubes only makes matters worse. The negative input conductance due to inductance in the screen lead may be used to decrease the total input conductance, but care must be used to prevent regeneration.

The attenuation due to input conductance has been analyzed in the literature.[2] One important conclusion is that, since the magnitude of the input conductance is proportional to the cathode lead inductance and cathode-grid transit time of the tube, the choice of tube must involve these factors as well as the usual ratio of g_m to capacitance.

6-5 Cascade vs. Distributed Amplification. We may now return to the problem of the optimum method of cascading distributed amplifier stages, this time taking into account the bandwidth shrinkage neglected in the preceding analysis. An interesting case to consider would be to regard the *over-all* gain and bandwidth as the fixed parameters and to solve for the arrangement of the amplifier giving the fewest number of tubes. We shall make one assumption to proceed with the analysis,

$$f_{c,t} = \frac{f_{c,i}}{\sqrt{m}}$$

[1] H. G. Rudenberg and F. Kennedy, 200-Mc Traveling-wave Chain Amplifier, *Electronics*, vol. 22, pp. 106–109, December, 1949.

[2] W. H. Horton, J. E. Jasberg, and J. D. Noe, Distributed Amplifiers, Practical Considerations and Experimental Results, *Proc. IRE*, vol. 38, pp. 748–753, July, 1950.

This equation is analogous to the equation giving the combination of rise times [Eq. (4-37)]; i.e., here the over-all bandwidth $f_{c,t}$ of the amplifier is proportional to the individual stage bandwidth $f_{c,i}$ and inversely proportional to the square root of the number of distributed stages, m. This is a reasonable assumption if the individual stages are adjusted to give a good transient response such as an amplitude response resembling a gaussian curve.

The characteristic resistance of the lines may be expressed in terms of the stage cutoff frequency

$$f_{c,i} = \frac{1}{\pi\sqrt{LC}} \tag{6-16}$$

$$R_0 = \frac{1}{\pi f_{c,i}C} \tag{6-25}$$

Equation (6-6) for the distributed stage gain may then be written in terms of the cutoff frequency,

$$A = \frac{ng_m R_0}{2} = n\frac{g_m}{2\pi C}\frac{1}{f_{c,i}} = \frac{nf_0}{f_{c,i}} \tag{6-26}$$

$$f_0 \overset{\Delta}{=} \frac{g_m}{2\pi C} \tag{6-27}$$

The quantity f_0 is a figure of merit for the vacuum tube similar to the gain/rise-time quotient discussed in Chap. 4. Observing that again the total number of tubes $N = nm$, and substituting $\sqrt{m}f_{c,t}$ for $f_{c,i}$, we can write for the over-all gain

$$A_t = \left(\frac{Nf_0}{m^{3/2}f_{c,t}}\right)^m \tag{6-28}$$

Differentiating Eq. (6-28) to find dN/dm and setting it equal to zero gives the proper stage gain, A, to result in the minimum number of tubes, N.

$$A = \epsilon^{3/2} \quad \text{(approx. 13 db)} \tag{6-29}$$

Unlike the cascaded-amplifier example, this particular value of A does not give the maximum over-all bandwidth, but only the most efficient use of the tubes to give the prescribed bandwidth and gain.

A further question which might well be asked is: When does one change from using cascade amplification in favor of distributed amplification? The

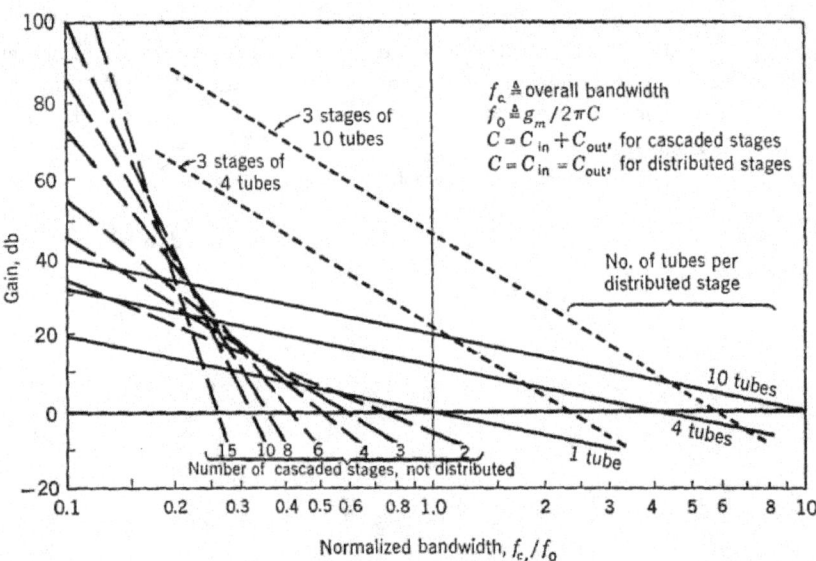

Fig. 6-16 Gain-bandwidth diagram for cascaded and distributed amplifiers.

situation for the two kinds of amplification is depicted in Fig. 6-16.[1] Here
the actual gain of the amplifier is plotted as a function of normalized bandwidth. For a single-stage RC amplifier the gain is unity (0 db) for a band-

---

[1] The equations for the lines in the graph of Fig. 6-16 may be obtained by assuming
that the over-all bandwidth $f_{c,t}$ of a cascaded RC amplifier is related to the stage bandwidth $f_{c,i}$ by $f_{c,t} \cong f_{c,i}/\sqrt{n}$. The gain-bandwidth product of the tube is $f_0 \triangleq g_m/2\pi C$
[Eq. (3-89)]. Then the gain of n stages becomes

$$A_n = \left(\frac{g_m}{2\pi C \sqrt{n f_{c,t}}} \right)^n = \left(\frac{f_0}{\sqrt{n f_{c,t}}} \right)^n$$

and

$$20 \log A_n = -20n \log \frac{\sqrt{n f_{c,t}}}{f_0}$$

This is the equation for the left-hand lines in Fig. 6-16.

For the distributed amplifier the gain is given by Eq. (6-28), with $N = nm$ substituted,

$$A = \left(\frac{n f_0}{\sqrt{m f_{c,t}}} \right)^m \tag{6-28a}$$

$$20 \log A = -20m \log \frac{\sqrt{m f_{c,t}}}{n f_0}$$

This is the equation for the gain of the distributed amplifier. Note the difference in the
definitions for C in the equation for f_0 for the two kinds of amplifier. Also note that
the graph serves only as a guide because the usable bandwidth of a distributed amplifier
is not as great as f_c, the filter cutoff frequency.

width equal to the gain-bandwidth product of the tube. The gain varies inversely with bandwidth as shown for the line marked "1 tube," which could be regarded as the simplest cascaded or distributed amplifier. In the cascaded amplifier increasing the number of stages changes the slope of the line ($A \approx 1/f_c^n$) but moves the intercept of the line to the left since the bandwidth decreases as the number of stages is increased (unity-gain intercept $= 1/\sqrt{n}$). Note that the dashed lines which indicate the gain for an n-stage nondistributed amplifier define a forbidden region to the right of all the lines which it is impossible to enter with a cascaded amplifier. For example, with RC stages a gain of 40 db with a bandwidth of $0.25f_0$ cannot be obtained. The use of more complicated interstages simply has the effect

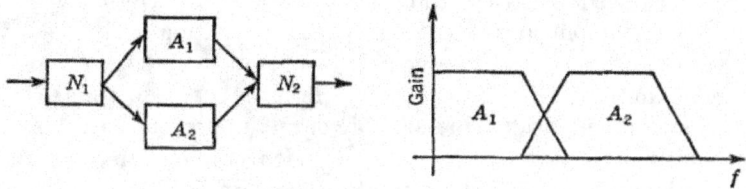

Fig. 6-17 Split-band amplifier.

of multiplying the abscissa by the ratio of the bandwidth of the network to the bandwidth of an RC stage having the same gain. (This ratio is similar to η defined in Chap. 4.) An improvement ratio greater than 3 is unlikely with practical networks. Even with such a complicated network, 40 db of gain could be obtained with only about $0.6f_0$ bandwidth.

The solid curves represent the situation for distributed amplifiers: here the gain (without cascading) is proportional to the number of tubes and is inversely proportional to frequency. Note that gain is obtained in the region which is excluded for cascaded amplifiers. On the assumption that the same laws of bandwidth reduction as used previously apply to cascaded distributed amplifiers, lines giving the gain may be drawn for more complicated combinations. Two such lines are shown, each for three stages, but one for 4 tubes per stage and one for 10 tubes per stage. The bandwidth for a 12-tube amplifier at which the distributed-amplifier approach gives less than a cascaded amplifier is $0.14f_0$. Note, however, that each distributed stage is giving $74/3 = 24.7$ db gain, which is more than the optimum for a minimum number of tubes—hence an example with a smaller number of tubes per stage would give better results.

The decision to select the distributed amplifier is easy when the gain and bandwidth required fall outside those obtainable with cascaded amplifiers, but the choice becomes a matter of economics if either type of amplifier would suffice. One additional consideration may enter into the final selection. In the case of an output stage which must develop a considerable output-voltage swing, the distributed amplifier is advantageous because the total plate-current swing which develops the output voltage is $n/2$ times the plate-current swing of one tube. This property has been utilized to realize wideband power amplifiers where each tube in the distributed chain may be a vhf power tetrode.

6-6 Distributed Amplifiers Using Transistors. Obtaining a distributed amplifier with transistors for active elements is a considerably more difficult task because of two principal problems: (1) the input admittance of the transistor has a considerable conductance component, which loads the input line, and (2) the device is not unilateral, and feedback effects result which are difficult to account for. At least one such successful amplifier has been constructed and gives reasonable performance.[1] The need for such amplifiers is somewhat open to question with the advent of transistors with extremely large gain-bandwidth products. Transistors with gain-bandwidth products of 5,000 Mc have been constructed which enable amplifier bandwidths of 1,000 Mc or more to be obtained. At such a bandwidth lumped-constant circuits are no longer practical; consequently a further increase in bandwidth cannot be obtained by the distributed-amplifier techniques unless some way of utilizing coaxial circuits can be found.

6-7 Split-band Amplifier. The split-band amplifier, sometimes called the parallel-chain amplifier or divided-band amplifier, has been proposed at various times and is the subject of a limited amount of recent research.[2] The basic structure consists of two (or more) amplifiers in parallel, each providing gain over a portion of the entire passband needed, as depicted in Fig. 6-17. The individual amplifiers may be either cascade or distributed. Although the ultimate performance of this split-band structure holds great promise, there is difficulty in designing the branching networks N_1 and N_2 and the characteristics of each amplifier so that the entire assembly has the desired frequency response, particularly in the critical "crossover" region.

[1] P. H. Rogers and L. H. Enloe, "Transistor Distributed Amplifier," U.S. Signal Corps Contract DA-36-039 SC-75021, Final Report, Mar. 15, 1958–Feb. 1, 1959, Applied Research Laboratory, University of Arizona, Tucson, Arizona.

[2] J. C. Linvill, Amplifiers with Prescribed Frequency Characteristics and Arbitrary Bandwidth, *M.I.T. Research Lab. Electronics Tech. Rept.* 163, July 7, 1950; H. A. Wheeler, "Maximum Speed of Amplification in a Wideband Amplifier," Wheeler Monograph 11, Wheeler Labs., Inc., Great Neck, N.Y., July, 1949.

PROBLEMS

6-1. A single-stage distributed amplifier is required to deliver a 100-volt peak-to-peak signal (square wave) and have a cutoff frequency of 50 Mc. The tubes to be used are 6CB6's with $g_m = 6,200$ μmhos, $C_{in} = 9$ pf, $C_{out} = 5$ pf (these capacitances include an allowance for stray wiring capacity); the maximum plate current permitted (per tube) is 25 ma. The networks to be used are m-derived sections designed for constant time delay; that is, $m = 1.27$. In this application there is no necessity to make the grid- and plate-line impedances equal; hence no additional padding capacity is needed.

a. How many tubes must be used to obtain the desired output voltage?

b. What is the gain of the stage?

c. What is the characteristic impedance of the grid line?

d. Draw a terminating half section ($m = 0.6$) for the plate line, and show element values.

6-2. A 5654 pentode has the characteristics given below.

Find the arrangement of tubes in a cascaded, distributed amplifier that will produce 50 db of voltage gain with 100 Mc bandwidth, and use the minimum total number of tubes. Assume that the bandwidth of cascaded stages decreases as $1/\sqrt{m}$. Sketch the arrangement of a single stage including all necessary biasing and coupling impedances, as well as the signal-element values. Only one supply voltage is to be used, but you may specify its value.

Calculate the values which affect the low-frequency behavior so that the *over-all* sag is 0.1 per cent/μsec and each source of sag contributes equally.

$$C_{in} = 4.0 \text{ pf} \qquad\qquad I_b = 7.5 \text{ ma}$$

$$C_{out} = 2.85 \text{ pf} \qquad\qquad I_{c2} = 2.5 \text{ ma}$$

Assume a wiring capacity of 2 pf at grid and plate.

$$E_b = 120 \text{ volts} \qquad\qquad r_p = 0.34 \text{ megohm}$$

$$E_{c2} = 120 \text{ volts} \qquad\qquad r_{p2} = 20 \text{ kilohms}$$

$$E_{c1} = -2 \text{ volts} \qquad\qquad w_p = 1.65 \text{ watts max}$$

$$g_m = 5,000 \text{ μmhos} \qquad\qquad w_{c2} = 0.55 \text{ watt max}$$

(Note that variations of ±20 per cent in E_b make little difference in the tube characteristics.)

7

Introduction to the Filter Amplifier

The previous discussion has centered around the transient response of amplifiers. The only function of the amplifier was to provide gain with as little distortion of the waveform as possible. Suitable test waveforms were the step and square wave, and hence there was developed a technique of evaluating amplifier performance in terms of the response to such waveforms. The steady-state response of the amplifier was purely incidental.

In amplifying portions of many systems, however, there is an additional requirement. Not only is an amplifying unit expected to produce gain over a bandwidth sufficient for the information contained in the signal, but it must also *reject* signals outside the appointed band. This is primarily a steady-state matter; i.e., we are talking in terms of the frequency response of the system. It may of course be true that the desired signal which is to be amplified consists primarily of transient waveforms and that the transient response must therefore not be entirely ignored. Nonetheless, we shall seek to allow for this through use of the interrelationships developed in the preceding chapters between steady-state and transient response.

Our principal concern will be with *bandpass* amplifiers, rather than with lowpass or highpass. In one sense we considered the lowpass amplifier in the preceding chapters, but there were no requirements that the amplifier reject frequencies above any prescribed frequency limit. Bandpass amplifiers are fundamental to all types of systems using the radio-frequency spectrum on a frequency-separation basis (as opposed to time separation, e.g., multichannel pulse-time modulation), including radio, radar, and carrier on wire lines.

There are several aspects of the design requirements, and their relative importance will vary from one application to another. These are the center frequency f_0, the bandwidth B, and the gain magnitude A_0 at band center. It shall be the objective here to present techniques which make possible the achievement of really difficult combinations of requirements, viz., high gain and large bandwidth all in one amplifier, together with good rejection of signals outside the passband.

The bandpass amplifier can, of course, use either vacuum tubes or transistors for active elements; however, the present state of the art of designing transistor bandpass amplifiers is considerably the more rudimentary. Therefore the discussion will center about vacuum-tube examples. The general principles which will be explored are applicable to either transistor or vacuum-tube cases; consequently the discussion should be quite valuable even to one primarily interested in the transistor amplifier.

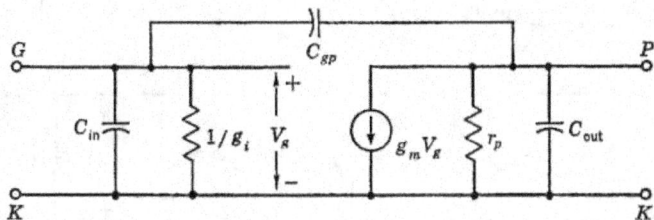

Fig. 7-1 Equivalent circuit for pentode at medium to high frequencies.

The equivalent circuit which is valid for the usual frequency ranges of pentode operation is that of Fig. 2-15, which is repeated here as Fig. 7-1. The tube capacitances C_{in} and C_{out} plus the associated wiring capacitances C_w usually comprise the total interstage capacitance in a wideband amplifier because any additional capacitance, such as a tuning capacitor, reduces the gain which can be obtained for a given bandwidth. The grid-plate capacitance C_{gp} is *usually* small enough in a modern pentode to be neglected as a feedback element because of the relatively low stage gain in a wideband amplifier. If the source and load impedances for the stage are high, however (which implies a narrow-band situation), the feedback due to the grid-to-plate capacitance may be sufficient to alter radically the gain characteristic, or even to cause oscillation. If C_{gp} is neglected, the current generator $g_m V_g$ is the only coupling between input and output; hence the equivalent circuit becomes a *linear unilateral* network. The input conductance, which was given by the empirical equation (2-36), varies predominantly with the square of frequency with a modern tube; however, g_i is usually regarded as a constant across the passband of interest [that is, $g_i \triangleq g_i(f_0)$, where f_0 is the band-center frequency]. The circuit damping conductances, which are added to broaden the interstage, are in parallel with g_i and tend to mask the variation of g_i with frequency.

Our basic problem consists in assembling useful combinations of tubes (or transistors) and interstage networks, usually in the cascade (or "tandem," or "chain") arrangement, in such a way and with such components as to provide a desired gain and amplitude response. (NOTE: *Amplitude response* will be the abbreviated way of saying "amplitude response vs. frequency," or "gain magnitude vs. frequency.")

The problem is similar to that of designing a filter consisting of passive network elements only. Indeed the properties of the over-all transfer function of an amplifier have many points in common with that of the passive filter. It might be said in this regard that we are dealing with applied network theory. Our task is more difficult in two respects: (1) the amplifier must provide gain, and (2) we are restricted to interstage networks that will function properly in combination with the irreducible equivalent network of the active device (e.g., Fig. 7-1). As a consolation, however, it

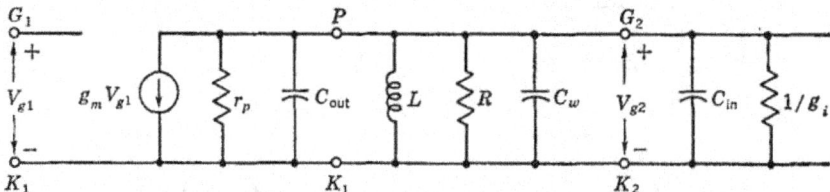

Fig. 7-2 Single-tuned interstage showing all pertinent elements.

will turn out that the unilateral property of the tubes serves to "isolate" the interstage networks, and certain advantages are gained, both in the design and in the adjustment of the amplifier, as contrasted with the completely passive structure. One of the major problems with the transistor case is caused by the nonunilateral nature of the device, which introduces interaction from one stage to the next.

Since the interstage networks can be only those which "fit in" with the tube or transistor, it might well be expected that only certain classes of interstages have been found to have practical value. This is indeed the case, and the *synthesis* of the filter is largely a matter of *selection*, as to both active element and network, based upon suitable *figures of merit* for comparing their value. Therefore we shall commence the study by an examination of several types of interstage network, beginning with the simplest.

We shall not proceed far in the direction of *network* complexity, although the frequency response of a multistage amplifier may be identical in its characteristics to that of a complicated filter. Where many stages are needed to supply gain, it is good strategy to distribute nonidentical, simple, filter-type networks between the stages, each designed in such a way as to yield in the combination the desired over-all response. This comprises the so-called "filter amplifier," a term introduced by Butterworth.[1] When only

---

[1] S. Butterworth, On the Theory of Filter Amplifier, *Exptl. Wireless and Wireless Engr.*, vol. 7, p. 536, October, 1930.

a few stages may be needed to provide the required gain and yet the desired frequency response is that of a filter with many sections, it is possible to employ networks of greater complexity than those commonly used with filter amplifiers. Such networks differ from the usual passive filters in that they must provide for capacitance at input and output terminals and also for finite Q of the inductors.[1] Adjustment of such networks calls for special techniques.[2]

7-1 Properties of a Single Pentode Stage with a Single-tuned Interstage. As a starting point, suppose that our gain requirements are so modest as to permit the use of one stage alone. Or even if they are not,

the single stage is the fundamental "building block" in the multistage amplifier.

Next, let us see what can be done with the simplest possible interstage network, and let us use the performance of such a network as a calibration for comparing other, more complicated networks.

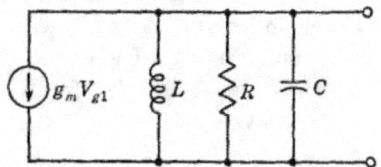

Fig. 7-3 Circuit for a wideband interstage.

An alternative approach might be to study the general properties of two-terminal interstage networks, as the simplest *class* of networks, and without regard to particular configurations. This is more properly in the province of a course in network theory,[3] and time does not permit the full development of these matters here. We shall find it of considerable value, however, to draw upon the results of such development.

As a practical matter, the only form of two-terminal interstage in extensive use today is the single-tuned network. Starting with the basic equivalent network for the tube (Fig. 7-1), the single-tuned interstage is constructed by adding a shunt inductance L, tuned to resonance at the desired center frequency with the total interstage capacitance $C_{in} + C_{out} + C_w$, and a resistance R as illustrated in Fig. 7-2. With most pentodes r_p and g_i may be ignored in comparison with R, in which case the circuit reduces to that of Fig. 7-3. In any event, R can represent the parallel combination which includes r_p and g_i. The capacitor C likewise represents the parallel combination of C_{out}, C_{in}, and stray wiring capacitance C_w.

---

[1] M. Dishal, Design of Dissipative Band-pass Filters Producing Exact Amplitude-frequency Characteristics, *Proc. IRE*, vol. 37, pp. 1050–1069, September, 1949; T. C. Wagner, The General Design of Triple- and Quadruple-tuned Circuits, *Proc. IRE*, vol. 39, pp. 279–285, March, 1951.

[2] M. Dishal, Alignment and Adjustment of Synchronously Tuned Multiple-resonant-circuit Filters, *Proc. IRE*, vol. 39, pp. 1448–1455, November, 1951.

[3] D. F. Tuttle, Jr., "Network Synthesis," vol. 1, John Wiley & Sons, Inc., New York, 1958; see also H. W. Bode, "Network Analysis and Feedback Amplifier Design," D. Van Nostrand Company, Inc., Princeton, N.J., 1945.

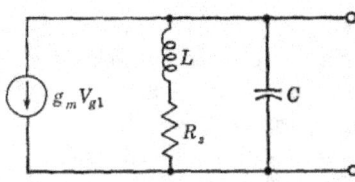

Fig. 7-4 Circuit for a narrow-band interstage.

It must be pointed out here that in adding parallel R we have elected to explore the wideband case. If we wanted a narrow-band amplifier, we would add no shunting resistance and would be careful that the inductor L had appropriate losses (Q) to give the desired bandwidth. The inductor losses, usually considered as a series resistance R_s, are generally the dominant factor in determining the bandwidth, and hence the proper equivalent circuit is that of Fig. 7-4. This case is thoroughly explored in the radio-engineering books,[1] and we leave it now in favor of the more difficult task of maintaining a wide band simultaneously with high gain.

The gain function $A(p)$ for the circuit of Fig. 7-3 may be written in several equivalent forms,

$$A(p) = \frac{V_{g2}}{V_{g1}} = \frac{-g_m V_{g1} Z(p)}{V_{g1}} \qquad \text{where } Z(p) = \frac{1}{\dfrac{1}{pL} + \dfrac{1}{R} + pC} \qquad (7\text{-}1)$$

$$A(p) = -g_m R \frac{1}{1 + R(pC + 1/pL)}$$

$$= -g_m R \frac{1}{1 + \dfrac{R}{\omega_0 L}\left(\dfrac{p}{\omega_0} + \dfrac{\omega_0}{p}\right)} = -g_m R \frac{1}{1 + Q\left(\dfrac{p}{\omega_0} + \dfrac{\omega_0}{p}\right)} \qquad (7\text{-}2)$$

where [2]
$$Q \triangleq \frac{R}{\omega_0 L} \qquad \omega_0 \triangleq \frac{1}{\sqrt{LC}}$$

For $p = j\omega$
$$A(j\omega) = -g_m R \frac{1}{1 + jQ(\omega/\omega_0 - \omega_0/\omega)}$$

$$= -g_m R \frac{1}{1 + jQ(f/f_0 - f_0/f)} \qquad (7\text{-}3)$$

where
$$f = \frac{\omega}{2\pi} \qquad f_0 = \frac{\omega_0}{2\pi}$$

[1] For example, F. E. Terman, "Electronic and Radio Engineering," 4th ed., McGraw-Hill Book Company, Inc., New York, 1955.

[2] The definition of Q here is that suitable for a parallel resonant circuit (Fig. 7-3); for a circuit in which R_s is in series with L, as in Fig. 7-4, $Q \triangleq \omega_0 L/R_s$.

The maximum gain, which is obtained at midband ($f = f_0$), is

$$|A(j\omega_0)| = g_m R \tag{7-4}$$

At the upper and lower band-edge frequencies ω_2 and ω_1 (or f_2 and f_1), respectively, the gain is 3 db less than at midband.

$$A(j\omega_1, j\omega_2) = -g_m R \frac{1}{1 \pm j1}$$

$$|A(j\omega_1, j\omega_2)| = |A(j\omega_0)| \frac{1}{\sqrt{2}}$$

or

$$\frac{f_2}{f_0} - \frac{f_0}{f_2} = \frac{f_0}{f_1} - \frac{f_1}{f_0} = \frac{1}{Q}$$

The bandwidth of the stage is

$$B \stackrel{\Delta}{=} f_2 - f_1 = \frac{f_0}{Q} \tag{7-5a}$$

$$= \frac{1}{2\pi RC} \tag{7-5b}$$

The two frequencies f_1 and f_2, at which the magnitude of the gain function (sometimes called the "amplitude response," or simply "response") is down 3 db from that at band center f_0, are related by

$$\frac{f_1}{f_0} = \frac{f_0}{f_2} \tag{7-6}$$

Thus, f_0 is the geometric mean of f_1 and f_2, and the passband of the amplifier is said to possess "geometric symmetry" about f_0.

Notice that in Eq. (7-5b) the bandwidth is independent of the center frequency f_0. In Eq. (7-5a) f_0 appears, but it is also implicit in Q and cancels out in the quotient.

Notice also that in Eq. (7-4) the gain is independent of the center frequency. Most important of all, however, observe that, in the product of Eqs. (7-5) and (7-4), the gain-bandwidth product GBP is not only independent of frequency but is also independent of the resistance R,

$$\text{GBP} \stackrel{\Delta}{=} \text{gain} \times \text{bandwidth} = \frac{g_m}{2\pi C} \tag{7-7}$$

Since this is the same equation obtained for the lowpass case [Eq. (3-89)], the GBP is independent of whether the amplifier is low- or bandpass. In either case gain is inversely proportional to bandwidth. (This is not necessarily true for the transistor amplifier.)

The product of gain times bandwidth for the single-tuned circuit as given by Eq. (7-7) is determined primarily by the tube. Except for stray wiring capacitance, the quantity C is largely due to the tube, and so also, of course, is g_m. This tells us that, in selecting a tube for large gain and bandwidth, the g_m/C ratio should be as large as possible.[1] Moreover, there is no benefit in a large g_m—such as might be achieved by connecting several tubes in parallel—if the capacitance C is increased in the same proportion.

The expression in Eq. (7-7) is the gain-bandwidth *product* (GBP), a term which we shall distinguish carefully from the *gain-bandwidth factor* (GBF). The latter is a figure of merit that is useful in comparing various interstage *networks* against the single-tuned stage as a yardstick. This is what we did in the transient analysis of fast amplifiers; the relative speed of various circuits such as shunt peaking was compared with the resistance-coupled stage. Thus, the gain-bandwidth *factor* of a circuit is obtained by taking the gain-bandwidth *product* and dividing it by the gain-bandwidth product of the single-tuned stage, namely, $g_m/2\pi C$. We shall find that for two-terminal interstages the maximum gain-bandwidth factor is 2.0, while with four-terminal networks it can theoretically be as great as $\pi^2/2$, or 4.93.

The gain function $A(j\omega)$ given in Eq. (7-3) can be simplified if the circuit bandwidth is, say, 10 per cent or less.

$$\frac{\omega}{\omega_0} - \frac{\omega_0}{\omega} = \frac{\omega^2 - \omega_0{}^2}{\omega_0\omega} = \frac{(\omega + \omega_0)(\omega - \omega_0)}{\omega_0\omega}$$

$$\cong \frac{2(\omega - \omega_0)}{\omega_0} \qquad \text{since } \omega \cong \omega_0$$

Therefore the gain expression may be approximated for this *narrow-band* case,

$$A(j\omega) \cong -g_mR \frac{1}{1 + j(2Q/\omega_0)(\omega - \omega_0)} \qquad \text{for } \omega \cong \omega_0 \qquad (7\text{-}8a)$$

$$\cong -g_mR \frac{1}{1 + j(2Q/f_0)(f - f_0)} \qquad (7\text{-}8b)$$

Notice that, in this narrow-band case, the function has "arithmetic symmetry" about ω_0; that is, the two frequencies ω_1 and ω_2 at which the response is down to 70.7 per cent are equally displaced in frequency increment above and below ω_0, or

$$\omega_1 - \omega_0 = \omega_0 - \omega_2$$
$$f_1 - f_0 = f_0 - f_2 \qquad (7\text{-}9)$$

[1] For a chart showing the g_m, C, and g_m/C for currently available commercial tube types, see J. R. Whyte, Choosing Pentodes for Broad-band Amplifiers, *Electronics*, vol. 25, p. 150, April, 1952; see also Figs. 4-1 and 10-8.

7-2 Multistage Single-tuned Amplifiers (Bandwidth Shrinkage and Gain-Bandwidth Factor). It would be unusual if the gain of one stage were adequate; most applications require several stages in cascade. The intermediate-frequency amplifier in a radar set might have 10 or more stages, while a transcontinental relay link would have hundreds of stages.

In cascade amplification the gain functions combine in a continued product. For a cascade of identical stages the midband gain will be the nth power of the stage gain, namely,

$$|A(j\omega_0)| = (g_m R)^n \tag{7-10}$$

Simultaneously, the bandwidth will decrease as the number of stages is increased. Thus the -3-db points for one stage become the -6-db points for two stages in cascade, with a resultant smaller separation of the -3-db points for the pair. To solve analytically for the manner in which the bandwidth "shrinks," it will be convenient to rewrite Eqs. (7-3) and (7-8) with a normalized frequency variable x (see Fig. 7-5) as follows:

$$\text{Selectivity function} = F(jx) = \left(\frac{1}{1 + jx}\right)^n$$

$$x \triangleq Q\left(\frac{\omega}{\omega_0} - \frac{\omega_0}{\omega}\right) \tag{7-11}$$

In terms of this normalized variable the half bandwidth of a single stage is 1.0, since the response is down 3 db at $x = \pm 1$, and against this we can compare the bandwidths for larger numbers of stages.

$$|F(jx)| = \left(\frac{1}{\sqrt{1 + x^2}}\right)^n \tag{7-12}$$

The particular value x_1, the band-edge frequency, i.e., the value of x for which the over-all response of n stages is $1/\sqrt{2}$ (-3 db) compared with the midband value of 1.0 (0 db), is given by

$$\left(\frac{1}{\sqrt{1 + x_1{}^2}}\right)^n = \frac{1}{\sqrt{2}} \tag{7-13}$$

$$x_1 = \sqrt{2^{1/n} - 1} \tag{7-14a}$$

$$x_1 \cong \frac{0.833}{\sqrt{n}} \qquad n > 1 \tag{7-14b}$$

Now, since the bandwidth for one stage is 1.0, Eqs. (7-14) are also the ratio of the bandwidth of n stages compared with the bandwidth of a single stage,

$$\frac{\text{Bandwidth of } n \text{ stages}}{\text{Bandwidth of one stage}} = \sqrt{2^{1/n} - 1} \qquad (7\text{-}15a)$$

$$\cong \frac{0.833}{\sqrt{n}} \qquad (7\text{-}15b)$$

The shrinkage of bandwidth due to cascading stages is graphically shown in Fig. 7-5. Note that these curves are plotted in terms of the frequency variable x, which is not a linear function of actual frequency ω. Hence the arithmetic symmetry of the gain curves is preserved only for narrow bandwidths on a linear frequency scale. The bandwidth ratio of Eqs. (7-15), however, is valid in terms of ω as well as x.

In a cascade of single-tuned stages the gain increases with the number of stages according to one law, Eq. (7-10), while the bandwidth varies accord-

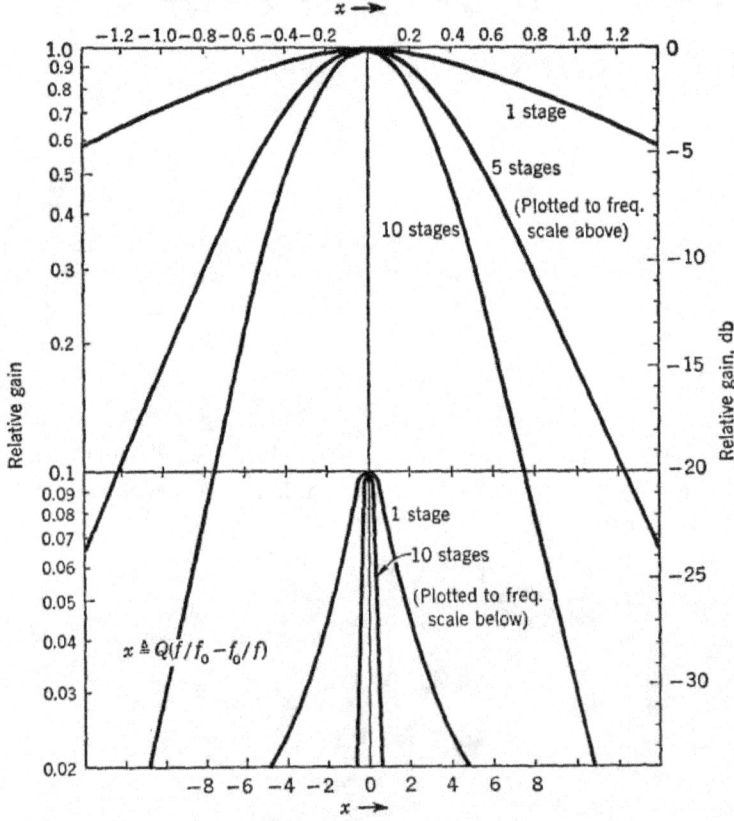

Fig. 7-5 Normalized gain-frequency curves for single-tuned interstages.

ing to another, namely, Eq. (7-15a). Thus the contribution of one stage to the over-all gain-bandwidth product will be different if the stage is used alone or in a cascade with other stages. To allow for this in our figure of merit, the *gain-bandwidth factor* (by means of which we shall compare the effectiveness of various circuits), we do two things: (1) take the nth root of the over-all gain, giving a *mean stage gain*; (2) divide the mean stage gain by $g_m/2\pi C$. The gain-bandwidth factor is then the product of this normalized mean stage gain times the over-all bandwidth and is a function of the circuits used, rather than of the tubes.

$$\text{GBF} = \frac{(\text{over-all gain})^{1/n}}{g_m/2\pi C} \times \text{over-all bandwidth} \qquad (7\text{-}16)$$

For a cascade of n identical stages, the following holds:

$$\frac{\text{GBF for } n \text{ stages}}{\text{GBF for one stage}} = \sqrt{2^{1/n} - 1} \cong \frac{0.833}{\sqrt{n}} \qquad (7\text{-}16a)$$

A phenomenon of both mathematical and practical interest is that for a given bandwidth there is a definite maximum of gain which cannot be exceeded. There is likewise a maximum bandwidth attainable for a specified gain. The latter occurs when each stage contributes a gain A_i of 1.65 to the over-all gain A_N. The number of stages needed and the resulting maximum bandwidth $B_{\max}$ are given below:

$$A_i = \sqrt{\epsilon} = 1.65 \equiv 4.34 \text{ db} \qquad (7\text{-}17)$$

$$n = 2 \ln A \qquad (7\text{-}18)$$

$$B_{\max} = \frac{g_m}{2\pi C} \frac{0.833}{\sqrt{2\epsilon \ln A_N}} \qquad (7\text{-}19)$$

Note that the gain A_i which gives the maximum over-all bandwidth is the same as that obtained in the lowpass case to get minimum over-all rise time [see Eq. (4-53)].

7-3 Selectivity Ratio. It was stated that the filter amplifier has the dual objective of amplifying signals within the desired passband and of rejecting signals lying outside this band. The ability to reject unwanted signals is sometimes called the "selectivity" of the system. Assigning a measure to this selectivity is somewhat arbitrary. The simplest approach would be merely a graphical plot of amplitude vs. frequency, i.e., the amplitude response. The steepness of the portions of the curve *far removed* from the passband is one kind of measure; it is sometimes called the *skirt selectivity*, since it involves the "skirts" of the response curve.

A better quantitative measure can be provided when it is known that in the system where the amplifier will be used the undesired signals will be separated from the desired one by fixed frequency intervals. Thus, in broadcasting, the stations are assigned to "channels" separated by 10 kc, and the severest problem is to reject the signals in adjacent channels. In broadcasting, a term is used, *adjacent-channel selectivity*, which denotes simply the ratio of the gain at midband to the gain 10 kc above or below midband. Similarly, *second-channel selectivity* refers to 20 kc above or below midband. The same philosophy would apply to multichannel carrier systems.

In the general case, however, where frequency assignments are not so ordered, there is another numerical measure of the selectivity that has gained acceptance in recent years. This is the so-called *selectivity ratio*, or *bandwidth ratio*. It is the ratio of the bandwidth at which the amplitude response is down 60 db from that at midband to the bandwidth at which the response is down 6 db. Thus it could be called the "60:6-db bandwidth ratio." Other levels could have been chosen, but these are convenient and serve the purpose of evaluating the response well outside the desired passband. Typical values obtained are on the order of 2 for a good communications receiver and perhaps 12 for a radar system. Clearly, a value of unity would be the limiting case.

For the single-tuned amplifier the selectivity ratio can be easily found by using Eq. (7-12) to find the -6- and -60-db frequencies,

$$\left(\frac{1}{\sqrt{1 + x_a{}^2}}\right)^n = \frac{1}{2} \qquad x_a = -6\text{-db frequency}$$

$$\left(\frac{1}{\sqrt{1 + x_b{}^2}}\right)^n = \frac{1}{10^3} \qquad x_b = -60\text{-db frequency}$$

Solving for x_a and x_b and forming the ratio x_b/x_a, we obtain

$$\frac{x_b}{x_a} = \text{selectivity ratio} = \sqrt{\frac{\sqrt[n]{10^6} - 1}{\sqrt[n]{4} - 1}} \qquad (7\text{-}20)$$

For $n = 1$ Eq. (7-20) has the value of 577, an almost useless selectivity ratio. As the number of stages n is increased, the ratio drops rapidly toward a limiting value of 3.15. For only six stages the ratio is 5.9. Other interstage arrangements will turn out to be superior, however.

PROBLEMS

7-1. A receiver must receive a signal at 20 Mc in the presence of an interfering signal at 21 Mc. How many amplifier stages using single-tuned circuits with a Q of 50 must be used to give at least 50 db desired-signal-to-undesired-signal ratio? What is the 3-db bandwidth of the resulting amplifier?

7-2. An amplifier consisting of identical single-tuned stages is to be built using 6AH6 tubes for which $g_m = 9{,}000$ μmhos, $C_{in} + C_{out} + C_w = 17$ pf. If the over-all bandwidth is to be 20 Mc, what number of stages should be used for maximum gain? Find the resulting individual stage bandwidth, over-all gain, and the circuit constants R and L for a center frequency of 30 Mc. What is the gain-bandwidth factor of the resulting amplifier?

7-3. For an amplifier employing identical stages using 6AK5's, what is the maximum bandwidth obtainable with an over-all gain of 60 db? Assume that $g_m = 5{,}000$ μmhos, $C_{out} = C_{in} = 8$ pf, and $C_w = 5$ pf.

7-4. Prove that the selectivity ratio for n single-tuned stages approaches 3.15 as n approaches infinity.

7-5. Derive the expression below which gives the maximum gain which can be achieved with a cascade of identical single-tuned stages when the over-all bandwidth B and the $g_m/2\pi C$ for the tube to be used are specified.

$$A_{max} \approx 2^{a^2/2e} = (1.136)^{a^2}$$

where

$$a \triangleq \frac{g_m/2\pi C}{B}$$

7-6. A certain pentode tube has a gain-bandwidth product of 75 Mc including an allowance for wiring capacity.

a. Find the interstage element values for a single-stage amplifier to give a bandwidth of 20 Mc at a center frequency of 100 Mc. (Assume that $C = 15$ pf.) What is the stage gain?

b. Find the interstage element values for a two-stage amplifier to give the same *over-all* bandwidth and center frequency. What is the over-all gain?

c. Show that it is impossible to construct a multistage amplifier using this tube and type of interstage to give a 20-Mc bandwidth and an over-all gain of 10.

7-7. Verify the expression for Q following Eq. (7-2). Start with the basic definition of Q as the ratio of the energy stored to the energy lost per cycle.

7-8. The circuit of Fig. P7-8 represents a simple extension of the circuit of Fig. 7-3. An ideal transformer has been added—an addition which can be approximated by tapping down on the inductor L. The modification is said to be "simple" because the gain function varies with frequency in just the way as before; i.e., no new resonances (natural modes) are introduced.

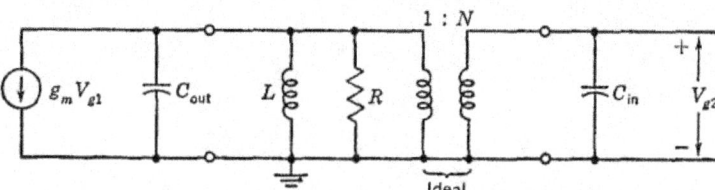

Fig. P7-8

Show that, when the transformation ratio N is chosen for maximum gain-bandwidth product, this product becomes

$$\text{GBP} = \frac{g_m}{2\pi(2\sqrt{C_{\text{out}}C_{\text{in}}})}$$

Show that this gain-bandwidth product is always equal to or greater than that given by Eq. (7-7).

7-9. Assume that an amplifier made up of identical, single-tuned stages employs a tube with a gain-bandwidth product of 75 Mc. For an over-all amplifier bandwidth of 12 Mc compute and plot the over-all gain in decibels vs. the number of stages in the amplifier. Assume that the amplifier is made up of 1, 2, 3, . . . stages up to 15. What is the maximum gain obtained?   Compare the results with those obtained by using the equation in Prob. 7-5.

8

Generalizations and Interpretations

Having had a look at a simple form of bandpass, or filter, amplifier, seeing what it will do in terms of gain, bandwidth, etc., let us now go from the specific to the general. The single-tuned amplifier is only one of many possible interstage networks, all of which will have characteristic properties in common. Most important of all, their gain functions will have similarities, and the proper interpretation of these will aid us in understanding why amplifiers behave the way they do.

Let us see what can be said about the general nature of amplifier gain functions, of which Eq. (7-2) for the single-tuned stage is an example. We can rewrite this equation as follows:

$$A(p) = -g_m R \frac{1}{1 + R(pC + 1/pL)} \qquad (7\text{-}2)$$

$$= -\frac{g_m}{C} \frac{p}{p^2 + p/RC + 1/LC} \qquad (8\text{-}1)$$

$$= -\frac{g_m}{C} \frac{p}{(p - p_1)(p - p_2)} \qquad (8\text{-}2)$$

where

$$p_1, p_2 = \frac{-1}{2RC} \pm j \sqrt{\frac{1}{LC} - \left(\frac{1}{2RC}\right)^2} \qquad (8\text{-}3)$$

$$= \frac{\omega_0}{2Q} \left(-1 \pm j \sqrt{4Q^2 - 1}\right) \qquad (8\text{-}4)$$

$$Q \triangleq R\omega_0 C \qquad \omega_0{}^2 \triangleq \frac{1}{LC}$$

177

The pole-zero diagram of an *underdamped* single-tuned stage as shown in Fig. 8-1a is a pair of conjugate poles and a zero at the origin if the stage Q is greater than $\frac{1}{2}$. The distance from the origin to each pole is always equal to the resonant frequency of the stage, ω_0, which is also the frequency of maximum gain. As the Q of the stage is decreased, the poles move farther from the $j\omega$ axis but maintain the same distance from the origin. Finally for $Q = \frac{1}{2}$ both poles come together on the σ axis, forming a double pole at $\sigma = -\omega_0 = -1/2RC$ (Fig. 8-1b). The stage is then said to be *critically damped* although no special passband shape is produced. Further de-

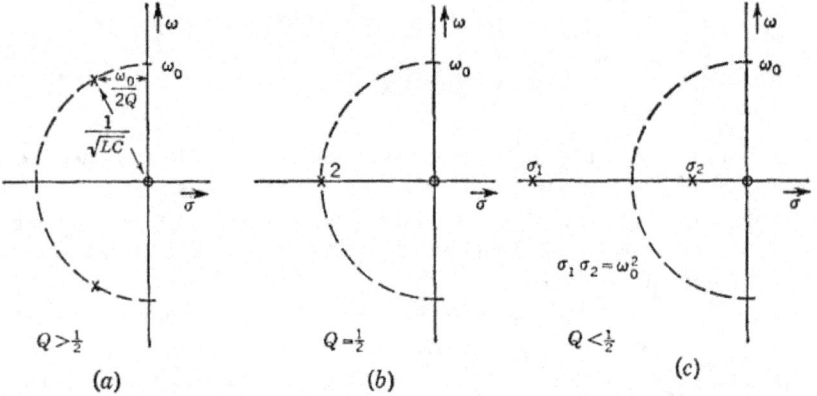

Fig. 8-1 Pole locations for a single-tuned circuit with various amounts of damping (Q).

creases in Q cause the poles to split apart and move along the σ axis in such a way that the geometric mean of their coordinates is the point $\sigma = -\omega_0$ (Fig. 8-1c). Even for Q's so low as to cause the stage to be *overdamped*, the maximum of gain still occurs at the frequency ω_0.

Consider a stage designed to give a narrow passband. Such a stage must have high Q; hence the poles lie very close to the $j\omega$ axis. If we determine the gain of such a stage by the geometric interpretation of the pole-zero diagram (Sec. 3-2), we see, as shown in Fig. 8-2, that the gain is given by

$$A(j\omega) = \frac{-g_m}{C} \frac{j\omega}{(j\omega - p_1)(j\omega - p_2)} \tag{8-5}$$

However, for a high-Q stage, the quotient of the vectors $j\omega$ and $j\omega - p_2$ is almost constant for ω in the region of the passband. Hence $j\omega/(j\omega - p_2) \cong \frac{1}{2}$, and the gain function may be approximated by

$$A(j\omega) \cong \frac{-g_m}{2C} \frac{1}{j\omega - p_1} \tag{8-6}$$

From this we see that the gain is influenced most by the nearby pole, and the narrow-band approximation of Eqs. (7-8) amounts to neglecting the influence of the zero at the origin and the complex pole in the lower half plane. A simple geometrical picture of the narrow-band situation is as shown in Fig. 8-3, where only the pole p_1 is shown with a restricted portion of the $j\omega$ axis. A circle centered at ω_0 with radius $\omega_0/2Q = B_r/2$ goes through the pole p_1 and defines the passband edges (B_r is the bandwidth of the stage in radians per second). Note that for p_1 the radial distance from

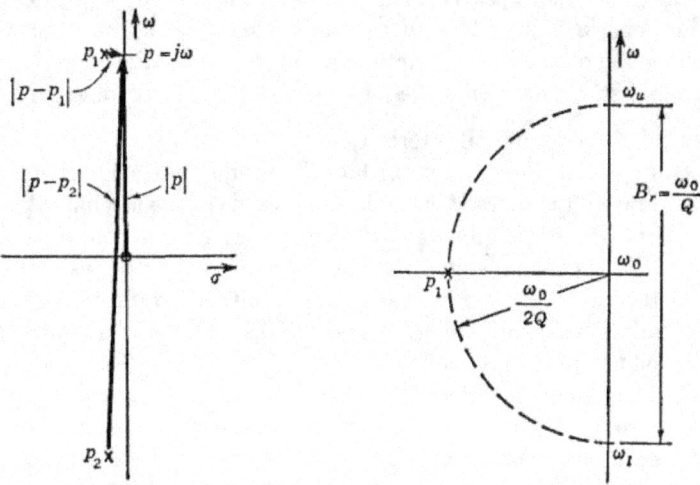

Fig. 8-2 Pole locations for a narrow-band stage.

Fig. 8-3 Approximate pole-zero diagram for a narrow-band stage ($Q \geq 20$).

the origin rather than the imaginary coordinate actually is equal to $j\omega_0$, but in the narrow-band situation the two distances are substantially equal. The semicircular construction also graphically displays the arithmetic symmetry of the narrow-band situation.

8-1 Properties of Amplifier Gain Functions. Before going on to further manipulations in the complex p plane, it might be well to take note of some general properties of the pole and zero locations for typical gain functions of amplifiers composed of lumped linear circuit elements and somewhat idealized linear tubes. The single-tuned circuit which we have just examined is, after all, only one case, and we should like to know whether other amplifiers will be similar, and in what respects.

For one thing, we see that we could design for a given amplitude response vs. frequency right on the p plane, without recourse to the circuit diagram at all. The geometric interpretation shows how the amplitude response depends upon the pole and zero locations, and the potential analogy, which we shall develop shortly, will be even more illuminating and useful. Thus we want to know what are the constraints upon where we might move the poles and zeros in our attempts to produce some particular gain function.

Development of the general properties of gain functions properly belongs in courses on passive network theory. Most of the significant properties are those of the passive structures in the amplifier, changed very little by the addition of the tubes (feedback amplifiers being a possible exception). Hence we shall here present in summary some of the pertinent findings of network theory, but without derivation or proof. The reader is referred to Bode [1] for a more nearly complete treatment.

The remarks will be limited to a single amplifier stage. For a cascade amplifier the over-all gain function is simply the continued product of the individual stage functions (the effect of feedback being disregarded).

I. General four-terminal interstage
   A. Physical realizability, including stability considerations, requires that all the poles of the gain function lie in the left half plane or else on the $j\omega$ axis. In the latter case, they must be simple poles. [2]
   B. Minimum-phase-shift networks (which exclude most all-pass lattice, distributed-parameter, and feedback arrangements) have the additional requirement that the zeros, too, must lie in the left half plane or on the $j\omega$ axis. [3]
   C. The number of poles of the gain function must exceed the number of zeros by at least 1. This results from the stray shunting capacitance requiring that the gain go to zero at infinite frequency.
   D. The poles and zeros must reside either on the σ axis or in conjugate pairs about the σ axis. [4] This is a simple property of the roots of an algebraic equation with real coefficients.
II. Special cases
   A. Two-terminal interstages. This is—or can be considered to be—an elementary form of the general interstage and is subject to greater restrictions on the permissible arrangements of poles and zeros.
      1. For physical realizability alone, both poles and zeros must lie in the left half plane or as simple poles (zeros) on the $j\omega$ axis. [5]
      2. The number of poles must always be 1 greater than the number of zeros. This again comes from the shunt capacitance.
      3. I-D still applies.
   B. All-pass networks, including the lattice and bridged-T as used for phase-correction networks, etc. (equalizers).

[1] H. W. Bode, "Network Analysis and Feedback Amplifier Design," D. Van Nostrand Company, Inc., Princeton, N.J., 1945.
[2] *Ibid.*, pp. 24–27, chap. 7, p. 106 in particular.
[3] *Ibid.*, pp. 121, 242–244.
[4] *Ibid.*, p. 106.
[5] *Ibid.*, p. 121.

1. Each pole in the left half plane must be balanced by a corresponding zero in the right half plane at an equal distance from the $j\omega$ axis. Another way of stating this is that the zeros and poles are negatives of each other; this incorporates I-D.[1]
2. I-A and I-D still apply.
3. I-C is modified to require an equal number of poles and zeros.
4. It will become apparent later that the all-pass phase-correcting network has a gain function whose magnitude is constant and whose phase angle can only decrease with frequency.[2]

8-2 Further Interpretations—Basis of the Physical Analogies. Although it is possible to derive formally all the general properties of amplifier gain functions by use of algebra and calculus of the complex-frequency variable p, it has nonetheless been found instructive to interpret these gain functions in terms of one or more physical analogies. In particular, there are three physical systems which are governed by exactly the same mathematics as our gain functions and which hence are exact analogies. Each of them has certain advantages in making more vivid certain of the gain-function properties, by virtue of better physical intuition in the analogous systems. This better physical intuition is the basis of appeal for this discussion, particularly for engineers.

These analogies supplement the "geometrical interpretation" previously discussed and like it are based upon the fact that the general gain expression can always be written in the factored form

$$A(p) = H \frac{\Pi(p - p_m)}{\Pi(p - p_n)} \tag{8-7}$$

or in the logarithmic form

$$\ln A(p) = \sum \ln (p - p_m) - \sum \ln (p - p_n) + \ln H \tag{8-8}$$

where the $p = p_m$ are the zeros of $A(p)$ (points of infinite loss) and the $p = p_n$ are the poles of $A(p)$ (points of infinite gain). We now rewrite this equation by taking the natural logarithm of the complex number $A(p)$, which has a modulus of magnitude $A(p)$ and argument or phase $\underline{/A(p)}$,

$$\ln A(p) = \underbrace{\ln|A(p)|}_{\mathcal{G}(p)} + j \underbrace{\underline{/A(p)}}_{\phi(p)} \tag{8-9}$$

$$\mathcal{G}(p) = \ln H + \sum \ln|p - p_m| - \sum \ln|p - p_n| \tag{8-10}$$

$$\phi(p) = \sum \underline{/p - p_m} - \sum \underline{/p - p_n} \tag{8-11}$$

[1] *Ibid.*, p. 239.
[2] *Ibid.*, p. 241.

In particular, for $p = j\omega$,

$$\mathcal{G}(\omega) = \ln|A(j\omega)| = \text{logarithmic gain} \tag{8-12}$$

$$\approx \text{gain in db}$$

$$\phi(\omega) = \underline{/A(j\omega)} = \text{amplifier phase shift in radians} \tag{8-13}$$

The function $A(p)$ is mathematically classed as an *algebraic* function, always a *rational fraction*, and therefore is always *analytic* except at the poles. The logarithm [$\ln A(p)$] is also analytic, except at the poles and zeros of $A(p)$. There are useful relationships, known as the *Cauchy-Riemann* conditions, which interrelate the real and imaginary parts of analytic functions with respect to the real and imaginary parts of the variable $p = \sigma + j\omega$,

$$\frac{\partial \mathcal{G}(p)}{\partial \sigma} = \frac{\partial \phi(p)}{\partial \omega} \tag{8-14}$$

$$\frac{\partial \mathcal{G}(p)}{\partial \omega} = -\frac{\partial \phi(p)}{\partial \sigma} \tag{8-15}$$

It also follows that *Laplace's equation* must hold for both real and imaginary parts (except at the poles and zeros),

$$\nabla^2 \mathcal{G} = \frac{\partial^2 \mathcal{G}}{\partial \sigma^2} + \frac{\partial^2 \mathcal{G}}{\partial \omega^2} = 0 \tag{8-16}$$

$$\nabla^2 \phi = \frac{\partial^2 \phi}{\partial \sigma^2} + \frac{\partial^2 \phi}{\partial \omega^2} = 0 \tag{8-17}$$

8-3 Electrostatic Analogy. The appearance of Laplace's equation suggests at once the possibility of using electrostatic potential as the basis of an analogy which is well known to specialists in network theory.[1] The analogy not only provides some degree of physical intuition but also makes available for the solution of network problems a highly developed array of mathematical techniques used in electrostatic-potential problems. In contrast to the other two analogies, which will be described later, it provides no convenient means of experimentation.

The electrostatic interpretation consists in looking at a situation such as that of Fig. 8-2 and imagining the poles to be positive line charges, infinite in length, piercing at right angles the p plane at the locations p_1 and p_2. The zero at the origin is considered a similar line charge, but negative instead of positive. Then the magnitude of the amplifier response $\mathcal{G}(\omega)$ is exactly proportional to the electrostatic potential along the $j\omega$ axis. Intui-

[1] S. Darlington, The Potential Analogue Method of Network Synthesis, *Bell System Tech. J.*, vol. 30, pp. 315–365, April, 1951.

tion becomes available at once, as exemplified by the fact that the gain (potential) is seen to increase when a pole (positive charge) is brought closer to the $j\omega$ axis.

The validity of the analogy can be demonstrated, starting with the application of Gauss's law to the line charge, enclosing a unit length of it with a small cylindrical volume of radius r, as shown in Fig. 8-4. Gauss's law involves integrating the outward flux density over the surface and equating it to the total charge contained within. The geometry of the situation reveals that the flux will be entirely radial; let the flux density then be D_r. It is convenient to let the charge per unit length be q; then

$$\oint \mathbf{D} \cdot \mathbf{ds} = 4\pi q \qquad (8\text{-}18)$$

$$= D_r \cdot 2\pi r \qquad (8\text{-}19)$$

The flux density is related by the dielectric constant to the electric field $\mathbf{E}$, which in turn is equal to the gradient of the potential V,

$$\mathbf{D} = \epsilon \mathbf{E} = \epsilon(-\nabla V) \qquad (8\text{-}20)$$

$$D_r = \left(\frac{2q}{r}\right) = \epsilon\left(-\frac{\partial V}{\partial r}\right) \qquad (8\text{-}21)$$

The potential V is obtained by integrating Eq. (8-21),

$$V = -\frac{2q}{\epsilon}\ln r + K \qquad (8\text{-}22)$$

For convenience let $2q/\epsilon = 1$. Then, if several charges are present, each of the same strength but some positive and some negative, the total potential (a scalar quantity) is the algebraic sum of the individual contributions,

$$V = \underbrace{K + \sum_{}^{m}\ln r_i}_{\substack{m \text{ negative} \\ \text{charges}}} - \underbrace{\sum_{}^{n}\ln r_j}_{\substack{n \text{ positive} \\ \text{charges}}} \qquad (8\text{-}23)$$

In p-plane coordinates, the charges are located at p_i and p_j; hence at any point p

$$r_i = |p - p_i|$$

$$r_j = |p - p_j| \qquad (8\text{-}24)$$

$$V = K + \sum_{}^{m}\ln|p - p_i| - \sum_{}^{n}\ln|p - p_j| \qquad (8\text{-}25)$$

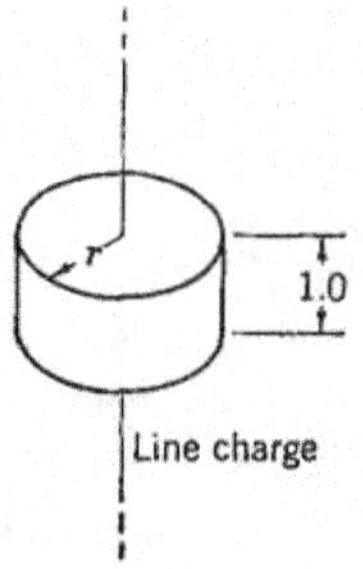

Fig. 8-4 Construction for finding the electric field produced by a line charge.

The analogy between Eqs. (8-10) and (8-25) is now apparent; the potential V is the analogue of the amplifier gain magnitude $\mathcal{G}(p)$. Similarly the scale factor K corresponds to $\ln H$, although this is not an important relationship when we are concerned primarily with how gain functions *vary* as one changes pole coordinates, etc.

We can now generalize the scalar potential into a complex function $V + j\Phi$ by incorporating the orientations as well as the magnitudes of the $p - p_i$ and $p - p_j$. The real part of this complex potential is the same as before, i.e., that of Eq. (8-25), whereas the imaginary part Φ (usually called the "stream function") defines the electrostatic flux distribution,

$$\Phi = \sum^m \underline{/p - p_i} - \sum^n \underline{/p - p_j} \qquad (8\text{-}26)$$

It can be seen that this stream function is the direct analogy of the amplifier phase function ϕ in Eq. (8-11). Laplace's equation and the Cauchy-Riemann conditions hold for both V and Φ,

$$\nabla^2 V = 0$$
$$\nabla^2 \Phi = 0 \qquad (8\text{-}27)$$

$$\frac{\partial V}{\partial \sigma} = \frac{\partial \Phi}{\partial \omega}$$

$$\frac{\partial V}{\partial \omega} = -\frac{\partial \Phi}{\partial \sigma} \qquad (8\text{-}28)$$

Notice now that

$$\frac{\partial V}{\partial \sigma} = -E_\sigma = -\frac{1}{\epsilon} D_\sigma = +\frac{\partial \Phi}{\partial \omega}$$

$$\Phi(\omega) = \int_0^\omega \frac{\partial \Phi}{\partial \omega} d\omega = \int_0^\omega \frac{\partial V}{\partial \sigma} d\omega = -\frac{1}{\epsilon} \int_0^\omega D_\sigma \, d\omega \qquad (8\text{-}29)$$

Thus $\Phi(\omega)$, which is analogous to the amplifier phase shift $\phi(\omega)$ at the frequency ω, is proportional to the total transverse electrostatic flux crossing the $j\omega$ axis between 0 and ω. Other correspondences can be developed and will be found in Table 8-1, page 191.

8-4 Conduction Analogy. There is another analogy which utilizes d-c or low-frequency current flow in a two-dimensional medium corresponding to the p plane. The medium can be an infinitesimally thin conducting sheet, or, what is equivalent, a sheet which is of finite but uniform thickness and in which all electrodes penetrate to full depth so that there is no variation of current flow with depth. The thin sheet can be provided

in the laboratory by a paper with
resistive coating,[1] while the medium
of finite thickness can be a tank of
conducting liquid or electrolyte.[2]
Both arrangements have been used
for experimental work, and by the
exercise of considerable care it is
evidently possible to obtain useful
quantitative results. Another elec-

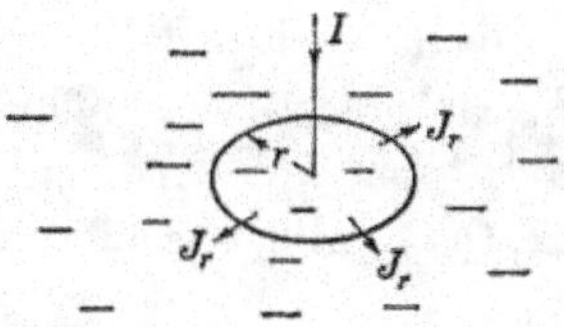

Fig. 8-5 Construction for finding the current density produced by a current source on a conducting sheet.

trical analogy which will be mentioned later is simpler for precise experimental work, while for rough work the elastic membrane to be described
next is very effective.

Consider a two-dimensional conducting sheet, infinite in extent, of resistivity ρ ohms/square, and with a current I entering the sheet at a point.
At a radius r from this point, the current density $\mathbf{J}$ is only in the r direction
because of symmetry and is constant in magnitude around the perimeter
of a circle of radius r (see Fig. 8-5). The relationship between current
density and radius is obtained by summing the total outward current
around the circle,

$$\oint \mathbf{J} \cdot d\mathbf{l} = J_r \times 2\pi r = I \tag{8-30}$$

$$J_r = \frac{I}{2\pi r} \tag{8-31}$$

From the incremental version of Ohm's law, the current density is related to
the electric field $\mathbf{E}$ and hence to the gradient of potential ("voltage
gradient"),

$$\mathbf{J} = \frac{\mathbf{E}}{\rho} = \frac{1}{\rho}(-\nabla V) \tag{8-32}$$

$$J_r = \frac{1}{\rho}\left(-\frac{\partial V}{\partial r}\right) \tag{8-33}$$

$$V = -\frac{\rho I}{2\pi}\ln r + K \tag{8-34}$$

[1] R. E. Scott, Network Synthesis by the Use of Potential Analogs, *Proc. IRE,* vol. 40,
pp. 970–973, August, 1952.

[2] W. W. Hansen and O. O. Lundstrom, Experimental Determination of Impedance
Functions by the Use of an Electrolytic Tank, *Proc. IRE,* vol. 33, pp. 528–534, August,
1945; A. R. Boothroyd, E. C. Cherry, and R. Makar, An Electrolytic Tank for the
Measurement of Steady-state Response, Transient Response, and Allied Properties of
Networks, *Proc. IEE (London),* pt. I, vol. 96, pp. 163–177, May, 1949; A. R. Boothroyd,
Design of Electric Wave Filters with the Aid of the Electrolytic Tank, *Proc. IEE
(London),* pt. IV, vol. 98, mon. 8, 1951; E. C. Cherry, Application of Electrolytic Tank
Techniques to Network Synthesis, *Proc. Symposium on Modern Network Synthesis,*
Polytechnic Institute of Brooklyn, N.Y., April, 1952, pp. 140–160.

This corresponds to Eq. (8-22) for the electrostatic case, and further development can proceed in similar fashion. Thus, if there are several current sources, some positive and some negative, each of unit strength $\rho I/2\pi = 1$, the total potential at any point by use of the principle of superposition is

$$V = K + \sum_{m}^{m} \ln r_i - \sum^{n} \ln r_j \tag{8-35}$$

$$\underbrace{\quad}_{\substack{m \text{ negative} \\ \text{currents}}} \quad \underbrace{\quad}_{\substack{n \text{ positive} \\ \text{currents}}}$$

In p-plane coordinates

$$V = K + \sum^{m} \ln |p - p_i| - \sum^{n} \ln |p - p_j| \tag{8-36}$$

There is also a complex potential in which the stream function corresponds to current flow. Current density $\mathbf{J}$ in the conduction analogy corresponds to flux density $\mathbf{D}$ in the electrostatic analogy.

$$V + j\Phi = K + \sum^{m} \ln (p - p_i) - \sum^{n} \ln (p - p_j) \tag{8-37}$$

As before, both potential V and stream function Φ satisfy the Cauchy-Riemann conditions and Laplace's equation. This can also be shown from the divergence relationship, which expresses the fact that the current entering an elemental area must equal the current leaving (except where there are sources or sinks),

$$\nabla \cdot \mathbf{J} = \underbrace{\frac{\partial J_\sigma}{\partial \sigma} + \frac{\partial J_\omega}{\partial \omega}}_{p\text{-plane coordinates}} = 0 \tag{8-38}$$

As in Eq. (8-33), Ohm's law gives

$$J_\sigma = -\frac{1}{\rho} \frac{\partial V}{\partial \sigma}$$

$$\tag{8-39}$$

$$J_\omega = -\frac{1}{\rho} \frac{\partial V}{\partial \omega}$$

Substituting in Eq. (8-38) gives Laplace's equation in V,

$$\frac{\partial^2 V}{\partial \sigma^2} + \frac{\partial^2 V}{\partial \omega^2} = \nabla^2 V = 0 \tag{8-40}$$

The matter of phase shift in the amplifier is illuminated by the analogy, if we recognize that the stream function Φ is the analogous quantity. Along the $j\omega$ axis,

$$\frac{\partial \Phi}{\partial \omega} = \frac{\partial V}{\partial \sigma} = -\rho J_\sigma \tag{8-41}$$

Thus the amplifier *phase slope*, i.e., the rate of change of phase shift with frequency (also the "envelope delay" or the inverse of the "group velocity"), is proportional to the transverse current density. This suggests that for linear phase shift, i.e., constant phase slope, the poles of an amplifier gain function should be so arranged that in the current analogy a nearly constant transverse current density results along the $j\omega$ axis. An arrangement with uniform pole spacing in the $j\omega$ direction can be expected to be favorable.

The phase shift at a given frequency ω can be determined from the analogy by the total current crossing the $j\omega$ axis between 0 and ω,

$$\phi(\omega) = -\rho \int_0^\omega J_\sigma \, d\omega = -\rho I_\sigma \Big]_0^\omega \qquad (8\text{-}42)$$

In considering the use of the conduction analogy for experimental work, one is confronted by the problem of size. Theoretically the conducting medium should be infinite in extent, but reasonably good results can be obtained if the area is considerably greater than that area near the origin containing all the poles and zeros under study. A conducting boundary is used, usually a circle in shape, at the perimeter of the medium, which provides the return electrode for the current entering or leaving at the poles and zeros. (This current is taken by the poles or zeros at the point of infinity.)

By consideration of property I-*D* in Sec. 8-1, one can reduce the required area of the conducting medium by one-half; since poles or zeros occur in conjugate pairs about the σ axis, there will never be current flow *across* the σ axis and the conducting medium can just as well be terminated in an insulating boundary along this axis.[1] All the information desired about amplitude and phase response can be found in the upper half plane of the analogy. In fact, it will later be shown during the discussion of "image poles" that, by proper attention to boundaries, only the third quadrant need be used, i.e., the left half of the upper half plane.

Experimental work with the conduction analogy is carried out by supplying unit currents to the electrodes—positive for poles and negative for zeros—and then measuring potentials on the resistive paper or in the liquid electrolyte by means of a high-resistance voltmeter, usually of the vacuum-tube variety. Relative potential along the $j\omega$ axis is a direct measure of the logarithmic gain magnitude. Phase shift, on the other hand, involves knowing the total transverse current, which nevertheless can be measured with the same voltmeter. The phase slope is proportional to the transverse current density, which from Ohm's law will be indicated by measuring the

[1] An electrode on the σ axis, representing a pole or zero there, should then introduce only one-half unit current, which is the portion normally entering the upper half plane.

transverse potential gradient. Thus, if the voltage difference is measured between two points close together on a line perpendicular to the $j\omega$ axis, the voltage is proportional to the phase slope at that value of ω. This phase slope can be plotted versus ω, and by graphically or numerically integrating this curve the curve of phase versus ω can be derived.

8-5 Membrane Analogy. Consider a two-dimensional elastic membrane, infinite in extent, with a constant tension T per unit distance; i.e., if the membrane had a cut of unit length, there would be a total force T normal to the cut. For convenience of discussion only, the membrane is assumed to lie in a horizontal plane, although gravitational force is neglected in comparison with the tension forces. We examine the downward

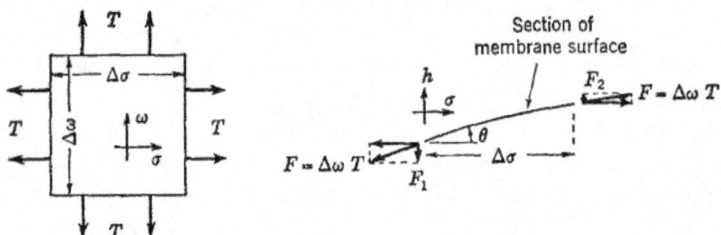

Fig. 8-6 Construction for finding the deformation of an elastic membrane.

vertical forces per unit area occurring when a region of the membrane has been distorted upward from its horizontal position. Let h be the vertical displacement from horizontal. The membrane tension T can be resolved into two components in the σ and the ω directions, to use our familiar p-plane coordinates. Each component contributes to two vertical forces F_1 and F_2 acting on an elemental area, shown in Fig. 8-6 for the contribution due to the σ component of the tension. The expressions for these two forces are given below, together with the net vertical force, as the size of the elemental area is reduced toward zero.

$$F_1 = T \, \Delta\omega \sin\theta \tag{8-43}$$

$$\cong T \, \Delta\omega \tan\theta = T \, \Delta\omega \, \frac{\partial h}{\partial \sigma} \qquad \text{for small } \theta \tag{8-44}$$

$$F_2 = F_1 + \frac{\partial F}{\partial \sigma}\Delta\sigma + \frac{\partial^2 F}{\partial \sigma^2}\frac{(\Delta\sigma)^2}{2!} + \cdots \tag{8-45}$$

$$\cong F_1 + \frac{\partial F}{\partial \sigma}\Delta\sigma \qquad \text{for small } \Delta\sigma \tag{8-46}$$

$$F_2 - F_1 \cong \frac{\partial F}{\partial \sigma}\Delta\sigma \tag{8-47}$$

$$F_2 - F_1 \cong T \, \Delta\omega \, \frac{\partial^2 h}{\partial \sigma^2} \, \Delta\sigma \tag{8-48}$$

A similar difference of vertical forces exists due to the ω component of tension,

$$F_2 - F_1 \cong T \, \Delta\sigma \, \frac{\partial^2 h}{\partial \omega^2} \, \Delta\omega \tag{8-49}$$

The net vertical force on the elemental area is the sum of Eqs. (8-48) and (8-49),

$$\text{Net } F = T \left(\frac{\partial^2 h}{\partial \sigma^2} + \frac{\partial^2 h}{\partial \omega^2} \right) \Delta\sigma \, \Delta\omega \tag{8-50}$$

The net vertical force per unit area, i.e., the pressure, is obtained by dividing by the area $\Delta\sigma \, \Delta\omega$,

$$\text{Net pressure} = T \left(\frac{\partial^2 h}{\partial \sigma^2} + \frac{\partial^2 h}{\partial \omega^2} \right) \tag{8-51}$$

If the region of the membrane is not subjected to external pressure, the net pressure must be zero for equilibrium. Hence, since T is not zero, then

$$\frac{\partial^2 h}{\partial \sigma^2} + \frac{\partial^2 h}{\partial \omega^2} = 0 \tag{8-52}$$

Equation (8-52) is Laplace's equation, and thus there is a direct correspondence or analogy between height of the membrane and the amplifier gain magnitude or the potential in the electrostatic and conduction analogies. Moreover, if on the membrane we draw a family of contour lines, i.e., lines of constant height, or elevation, we can also draw an orthogonal set, the lines of steepest descent. These latter lines correspond to the electrostatic-flux or conduction-current flow lines and contain the phase-shift information concerning our amplifier gain functions.

We are interested primarily in the amplitude variation of the amplifier gain along the $j\omega$ axis. In the membrane analogy we have a sort of "relief map" as shown in Fig. 8-7 for one situation; and the profile of the surface along $j\omega$ is a true plot of the amplitude (logarithmic) response versus ω. Moreover, the "tilt" of the membrane transverse to $j\omega$ is proportional to the phase slope $\partial\phi/\partial\omega$ of the amplifier, implying that for linear phase the transverse tilt of the surface should be closely constant.

In spite of the assumptions in the derivation, namely, a membrane infinite in extent and with very small displacements, simple physical models can be built which vividly portray the amplifier behavior. Remarkably enough, one's physical intuition is sufficiently keen so that with the elastic-membrane analogy most of the insight is acquired from mental images before a model is actually built!

8-6 Summary on Analogies. The principal results are summarized in Table 8-1. While it is evident that any one of the analogies is completely self-sufficient, additional insight is gained from knowledge of several. The electrostatic analogy provides a measure of physical intuition but is quite unsuited to experimental work. It does, however, open up a broad area of mathematical insight and analysis techniques based upon much existing work in electrostatics. The conduction analogy provides a limited amount

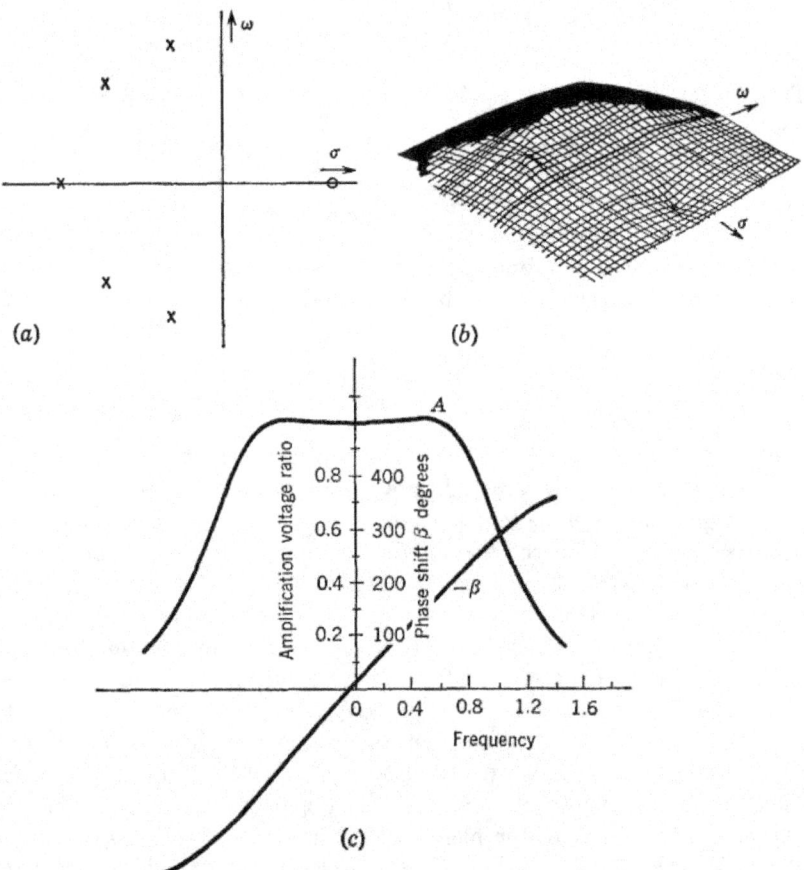

Fig. 8-7 (a) Lowpass pole positions to give an approximation to both constant gain and phase in the passband. (b) View of the membrane analogue. (c) The resulting gain magnitude and phase. (*From W. E. Bradley, Design of a Simple Band-pass Amplifier with Approximate Ideal Frequency Characteristics, Trans. IRE, vol. PGCT-2, pp. 30–38, December, 1953.*)

Table 8-1

Amplifier gain function $A(p)$	Electrostatic analogy	Conduction analogy	Membrane analogy
Gain magnitude (log) on $j\omega$ axis $\mathcal{G}(\omega) = \ln \|A(j\omega)\|$	Scalar electrostatic potential $V(\omega)$	Potential $V(\omega)$	Height (displacement) (horizontal membrane) $h(\omega)$
Phase slope (envelope delay) $T(\omega) = \dfrac{\partial \phi}{\partial \omega} \left(= \dfrac{\partial \mathcal{G}}{\partial \sigma} \right)$	Flux density (transverse) $D_\sigma \left(= -\epsilon \dfrac{\partial V}{\partial \sigma} \right)$	Current density (transverse) $J_\sigma \left(= -\dfrac{1}{\rho} \dfrac{\partial V}{\partial \sigma} \right)$	Slope (transverse) (gradient) $\dfrac{\partial h}{\partial \sigma}$
Phase shift $\phi(\omega)$	Total flux (transverse) $= \displaystyle\int_0^\omega D_\sigma \, d\omega$	Total current (transverse) $= \displaystyle\int_0^\omega J_\sigma \, d\omega$	$\displaystyle\int_0^\omega \dfrac{\partial h}{\partial \sigma} \, d\omega$
Pole (point of infinite gain)	Unit positive line charge	Electrode carrying unit positive current (into plane)	Unit "positive" force (upward)
Zero (point of infinite loss)	Unit negative line charge	Electrode carrying unit negative current (from plane)	Unit "negative" force (downward)
Constant-gain contours (in p plane)	Equipotentials	Equipotentials	Constant-height contours
Constant-phase contours (in p plane)	Flux lines	Current flow lines	Lines of steepest descent

NOTES: Coordinates for all systems: $p = \sigma + j\omega$
ϵ = relative dielectric constant (air = 1)
ρ = resistivity

of physical intuition, but is well suited to accurate experimentation. Finally, the membrane analogy gives the best physical insight and can be readily modeled. Actually, for experimental work, an altogether different approach is perhaps most promising; complex quantities are represented by a-c phasors, and by means of analogue-computer techniques such as logarithmic networks, adders, etc., an all-electronic system can be assembled which provides cathode-ray-tube plots of amplitude and phase after the pole locations have been set.[1]

8-7 Conformal Mapping. Problems such as those presented by the finite size of the conducting medium can be reduced through the use of conformal mappings of the p plane.[2] Moreover, the solution of problems

[1] J. R. Ragazzini and G. Reynolds, The Electronic Complex Plane Scanner, *Rev. Sci. Instr.*, vol. 24, pp. 523–527, July, 1953.

[2] W. H. Huggins, A Note on Frequency Transformations for Use with the Electrolytic Tank, *Proc. IRE*, vol. 36, pp. 421–424, March, 1948.

in terms of the electrostatic analogy can often be expedited through the use of such mappings.[1] The subject cannot be fully developed here, but illustrations can be given. The mapping procedure is mathematically rigorous and is made possible by the fact that we are dealing altogether with

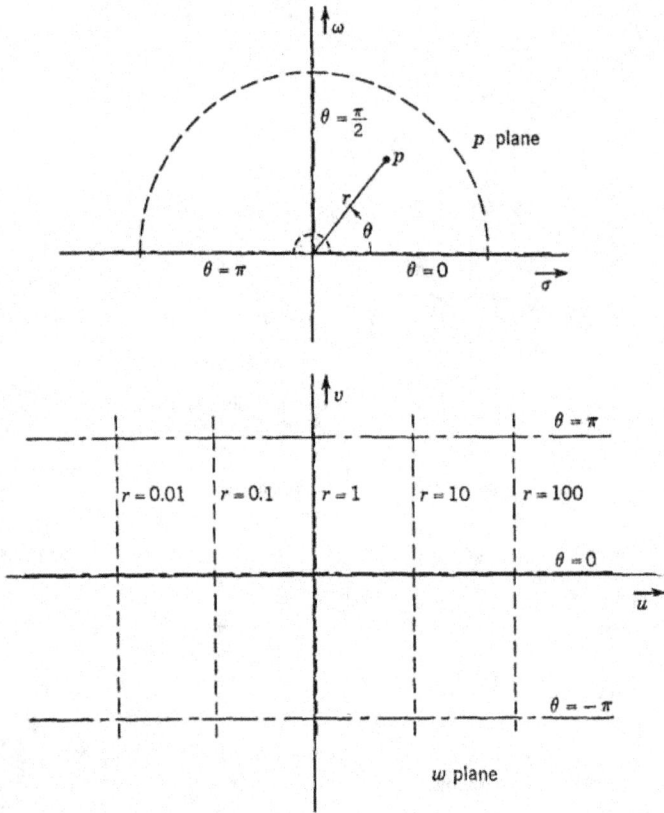

Fig. 8-8 Illustration of the transformation $w = \ln p$.

analytic functions. For our purposes here, the process can be viewed merely as a change in the coordinate system. In this change, quantities like potential in the electrostatic or conduction analogy or height in the membrane analogy remain invariant, but contour lines are warped by the modification of the coordinates. Moreover, large sections of the p plane can be made to "disappear" to infinity or semi-infinite regions made finite by the mapping process.

---

[1] D. L. Trautman, Jr., The Application of Conformal Mapping to the Synthesis of Bandpass Networks, *Proc. Symposium on Modern Network Synthesis*, Polytechnic Institute of Brooklyn, N.Y., April, 1952, pp. 175–191.

A useful mapping function (or transformation) is $w = \ln p$, or more explicitly

$$w = u + jv = \ln p = \ln r + j\theta \tag{8-53}$$

where $\quad\quad\quad p = re^{j\theta}$

This equation maps the p plane into rectangular strips in the w plane, as shown in Fig. 8-8. The entire p plane is contained in an infinitely long strip which is 2π units high in the w plane. As already shown for the conduction analogy, no current crosses the σ axis in the p plane and as a result the plane may be cut along the σ axis. This is the same as putting an insulating boundary along the axis to ensure that no current crosses the axis. The cut σ axis corresponds to cutting along the lines $v = 0$ and $v = \pi$; therefore only that portion of the w plane between these lines need be retained. In the practical use of the analogy, the horizontal dimension of the sheet is actually finite, but the mere addition of each vertical strip corresponds to a tenfold increase in the diameter of the p plane. Hence sufficient accuracy is readily attained. The left and right boundaries are made with vertical conductors (conducting strips) into which flow the currents representing the poles and zeros at the origin and the point of infinity, respectively. This transformation is particularly valuable for band-pass situations where the origin and point of infinity need not be represented. Even though the frequencies of exactly 0 and ∞ are never included, one may go to sufficiently low or high frequencies by using enough decades so that the behavior at 0 and ∞ may be inferred.

Another transformation maps the entire semi-infinite upper half plane into the interior of the unit circle in the w plane.[1]

$$w = \frac{j - p}{j + p} \tag{8-54}$$

The corresponding coordinates are shown in Fig. 8-9, together with several points labeled for comparison. In a conduction analogy, the perimeter of the circle in the w plane would be an insulator, in recognition of the fact that no current crosses the σ axis in the p plane. Likewise, any electrodes on this boundary must carry only one-half unit current, corresponding to the current which enters only the upper half of the p plane.

A third transformation makes possible the design of an amplifier gain function having an "equal-ripple" amplitude characteristic; i.e., within the specified passband the magnitude of the gain function varies about a mean value, equally above and below, within specified limits. This function will be discussed in Chap. 9; it has several practical advantages from

[1] This is one form of the same transformation used in deriving the so-called Smith chart used in transmission-line problems from a semi-infinite rectangular impedance plane.

the standpoints of gain-bandwidth and selectivity. The problem is how to arrange the poles (on the assumption of an all-pole function) in the p plane so that the gain-function magnitude measured along the $j\omega$ axis will have

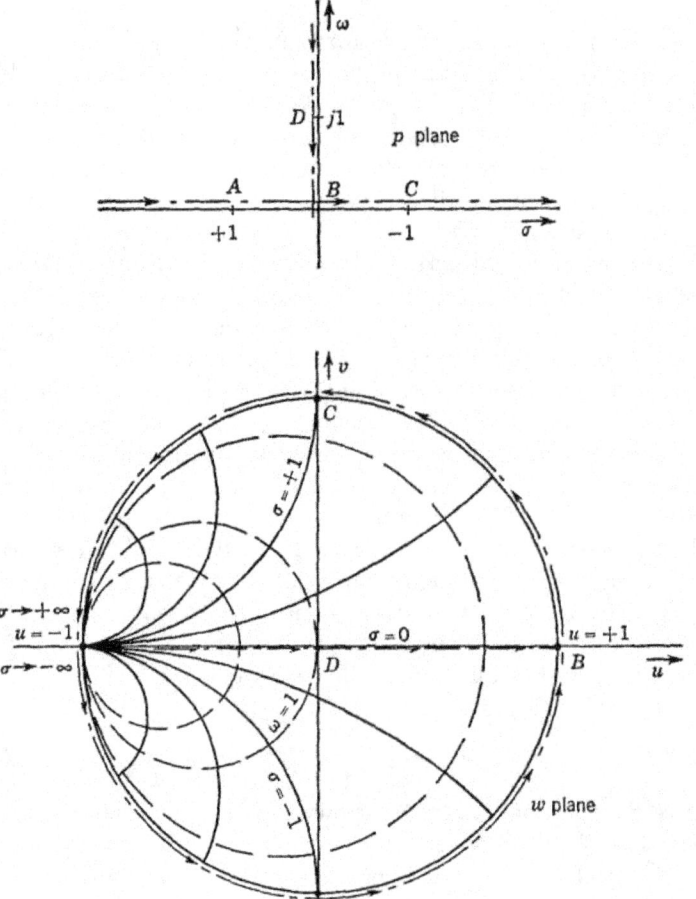

Fig. 8-9 The mapping of the upper half of the p plane into a unit circle by the transformation $w = (j - p)/(j + p)$.

this equal-ripple variation. In terms of the electrostatic analogy, it is not hard to visualize that the potential would have the equal-ripple nature if there were an infinite row of uniformly spaced charges parallel to the $j\omega$ axis and displaced from it by a distance a. Now, if such a situation can result from a suitable mapping of a finite region of the p plane, with a finite number of poles or charges, the problem is solved. A suitable transformation is $p = \sinh w$ and is illustrated in Fig. 8-10.

It will be noted in Fig. 8-10 that the infinite row of poles in the w plane transforms into a finite number of poles in the p plane, distributed around an ellipse and in both left and right half planes. The ones in the right half plane are not permitted in the amplifier gain function for reasons discussed in Sec. 8-1; yet they can readily be discarded. In so doing the electrostatic analogy makes it plausible that there remains an equal-ripple response. From the symmetry of the situation, the poles (charges) in the right half plane will make the same contribution to the potential along $j\omega$ as those in the left half; hence removing them will have the simple consequence of

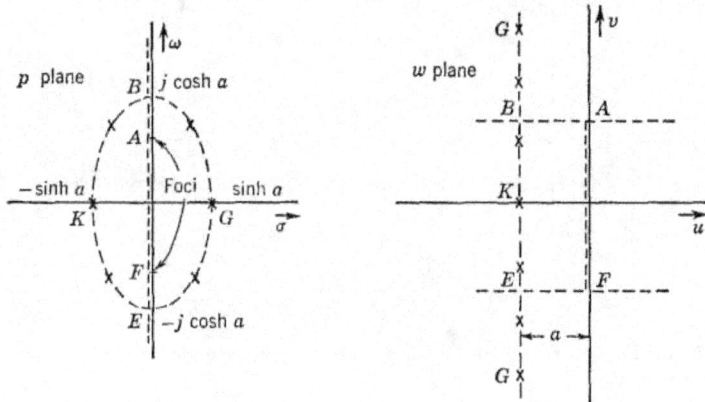

Fig. 8-10 Transformation $w = \sinh p$.

merely reducing the potential at all points along the $j\omega$ axis by $\frac{1}{2}$.[1] The poles in the right half plane can be called "image poles," which brings us to a new topic.

8-8 Image Poles (or Zeros).   When one is concerned primarily with amplitude response along $j\omega$, it is frequently convenient, both in mathematical analysis and in the use of the analogies just described, to employ "image poles." Thus, if Fig. 8-11a represents the singularities of the actual gain function, then the pole-zero arrangement of Fig. 8-11b is the corresponding function with image poles. Analytically, the use of image poles is the same as multiplying the complex function $A(p)$ by $A(-p)$ since the singularities of $A(p)$ occur 180° from the singularities of $A(p)$, as in Fig. 8-11b. The contribution to the potential along the $j\omega$ axis by the poles and zeros of $A(p)$ is exactly equal to the potential contributed by their images

---

[1] In terms of the membrane analogy, removing the right-half-plane poles lowers the height of the membrane by $\frac{1}{2}$, but the membrane still has the same shape along the $j\omega$ axis. However, removing the right-half-plane poles does cause a *tilt* in the membrane transverse to the $j\omega$ axis which was not there before. Hence the phase functions which are represented are different.

since the distance from any point on the $j\omega$ axis to a pole (or zero) is equal to the distance to its image. Hence the potential along the $j\omega$ axis is twice that obtained with no image poles and zeros. Since this corresponds to $2 \ln |A(j\omega)|$, the function we have obtained is

$$A(p)A(-p)|_{p=j\omega} = |A(j\omega)|^2 \qquad (8\text{-}55)$$

That the phase of this function is everywhere zero is shown by the fact that at each point on the $j\omega$ axis the transverse current density due to the singu-

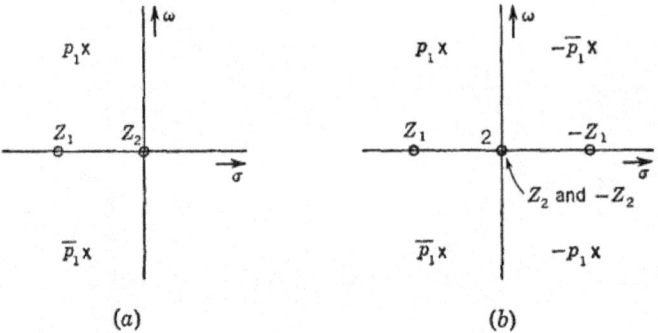

Fig. 8-11 (a) Pole-zero diagram. (b) Pole-zero diagram with image poles to give only amplitude information.

larities of $A(p)$ is exactly canceled by the current density due to the singularities of $A(-p)$. (Note the similarity to the situation along the σ axis, where, also because of symmetry, no current crosses the axis.) The transverse current density is analogous to $d\phi/d\omega$, the phase slope; therefore the phase is everywhere zero (the phase could also be π).

In a similar manner image zeros [for poles of $A(p)$] may be used if only phase information is desired. These added singularities shown in Fig. 8-12

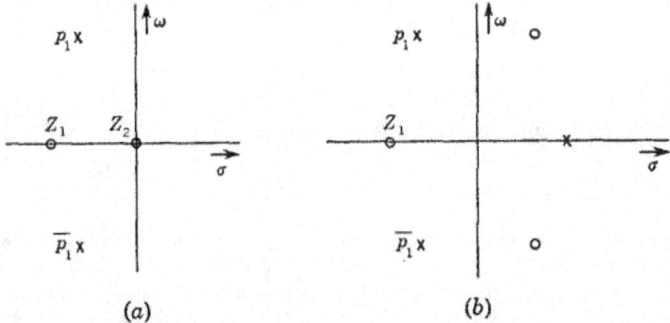

Fig. 8-12 Illustration of the use of image zeros.

correspond to the function

$$\left.\frac{A(p)}{A(-p)}\right|_{p=j\omega} = 2\underline{/A(j\omega)} \tag{8-56}$$

This may be demonstrated by considering the potential along the $j\omega$ axis: the potential at any given point caused by the singularities of $A(p)$ is exactly equal in magnitude, but opposite in sign, to the potential caused by the singularities of $A(-p)$; hence the potential is everywhere zero along the $j\omega$ axis (or constant since the potential is defined only to within an un-

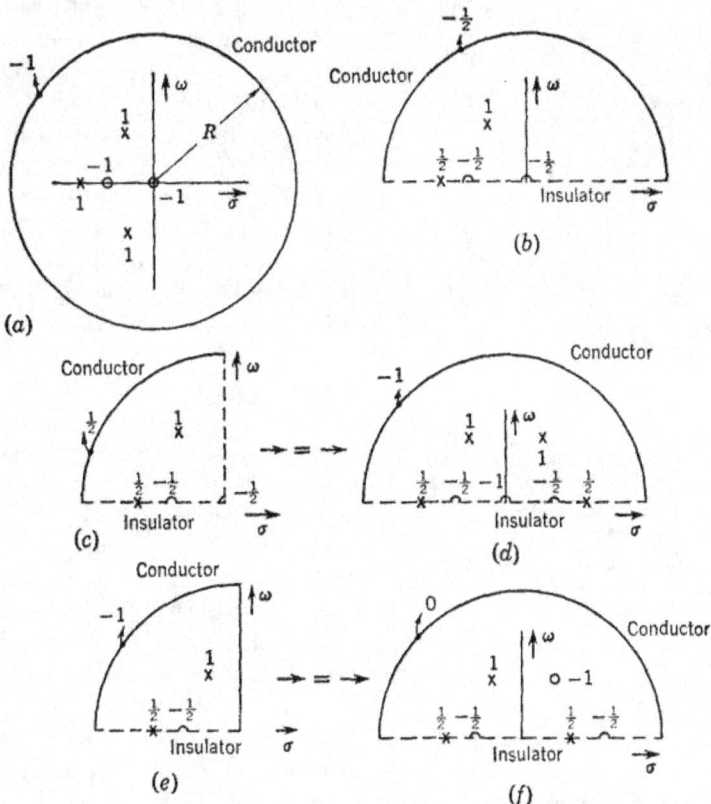

Fig. 8-13 Possible simplifications for specific applications of the conduction analogy. (a) Whole plane approximated by finite circle. All information present—errors increase as poles and ω approach R. (b) Whole plane approximated by semicircle. Same information along ω axis as in (a). (c, d) Planes useful only for *magnitude* information along the ω axis. $V(j\omega_1)$ is twice value obtained in (a) or (b). (e, f) Planes useful only for *phase* information along the ω axis. The $\underline{/A(j\omega_1)}$ is twice that obtained from (a) or (b). (The numbers by the poles and zeros indicate the relative currents which must be applied.)

known additive constant). However, the current density crossing the $j\omega$ axis at each point is doubled, since the current contributions of the poles and zeros of $A(p)$ and $A(-p)$ are equal and additive. Thus the phase function along the $j\omega$ axis is doubled, as shown in Eq. (8-56).

Several examples of the use of image poles and zeros are shown in Fig. 8-13. As mentioned before, only the upper half plane need be used, as shown in Fig. 8-13b. In addition, if only magnitude information is desired, the upper half plane with image poles may be used, as in Fig. 8-13d. With the image poles no current crosses the $j\omega$ axis, and the plane may also be cut, as shown in Fig. 8-13c. On the other hand, if phase information only is desired, image zeros may be used, as in Fig. 8-13f. Use may be made of the fact that the voltage along the $j\omega$ axis is a constant, allowing the plane to be cut along the $j\omega$ axis as well and a conducting boundary added to maintain the required constant potential (Fig. 8-13e). These same concepts may be used in connection with the other analogies, but the maintenance of the necessary boundary conditions may not be so easy as in the case of the conduction analogy.

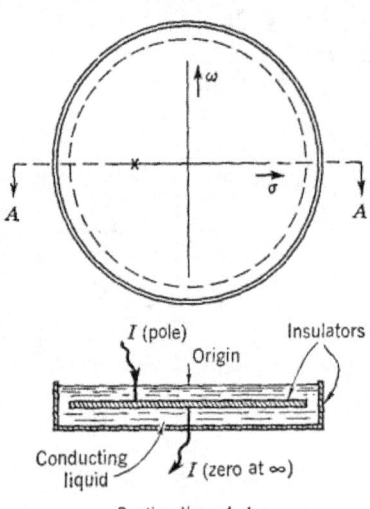

Numerous other techniques may be employed to improve the analogies for use in experimental work. An additional one of interest is a method of reducing the error due to the finite size of the conducting medium, as shown in Fig. 8-14. Here the p plane is inverted outside of the circle used for experiments so that the point of infinity becomes the center of a circle placed under the part of the p plane to be used.[1] The two conducting circles are joined at their rims to form the entire infinite p plane.

Fig. 8-14 A method of eliminating the errors due to finite tank size in the conduction analogy. The conducting medium here is an electrolyte of uniform depth. The conducting path is folded under at the edge, and the point of infinity lies directly beneath the origin.

PROBLEMS

8-1. What is the locus of pole locations in a single-tuned stage as L is varied but R and C remain constant? Make a sketch to show your result, and indicate the limiting positions of the poles as $L \to \infty$ and $L \to 0$.

[1] Boothroyd, Cherry, and Makar, *op. cit.*

8-2. Using graphical means, compute the gain as a function of frequency for the following pair of amplifier stages. Do this for a frequency range slightly greater than the bandwidth of the two stages. What is the gain at band center (midway between the two band edges)? What is the bandwidth of each of the two stages? What would the bandwidth of three pairs of such stages be (i.e., six tubes)?

Stage 1

$$C = 15 \text{ pf}$$

$$f_0 = 20.35 \text{ Mc}$$

$$R = 14{,}530 \text{ ohms}$$

$$g_m = 2{,}000 \ \mu\text{mhos}$$

Stage 2

$$C = 20 \text{ pf}$$

$$f_0 = 19.65 \text{ Mc}$$

$$R = 11{,}280 \text{ ohms}$$

$$g_m = 3{,}000 \ \mu\text{mhos}$$

Use any appropriate simplifications.

8-3. A piece of resistance paper 12 in. wide, as shown in Fig. P8-3, and effectively infinite in length is intended to represent the upper half of the p plane.

a. What distance in the x direction corresponds to 1 octave of frequency (i.e., a frequency ratio of 2:1)?

b. A single-tuned stage with $f_0 = 1$ and $B = 0.5$ is to be represented on the sheet. Show where to place the probe representing the pole, and indicate the relative magnitude

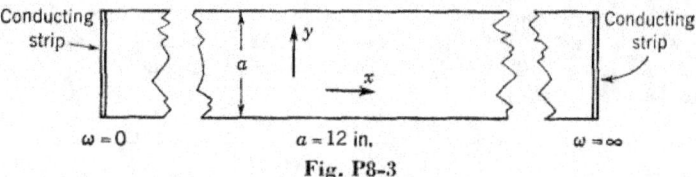

Fig. P8-3

of all the currents entering and leaving the analogy. (HINT: A useful check is Kirchhoff's current law!)

c. A shunt-peaked pentode stage with $m = 0.5$ is to be represented on the sheet. Show the pole and zero locations on the sheet for a bandwidth of 1.0 radian/sec. Again show the relative magnitude of all currents.

8-4. A conduction analogue is set up consisting of a 16-in.-radius circle of resistance paper, with a conducting rim. The analogue is calibrated by supplying 1 ma current to a single probe located at the center of the circle. The potential of a point 8 in. from the center is found to be 0.25 volt relative to the rim.

a. What is the resistivity (ohms per square) of the paper?

b. Is there any error in the above procedure due to the finite size of the medium? Why?

c. It is desired to arrange the analogue so that a 0.1-volt difference in potential corresponds to a 1-db change in gain. What current must be supplied to each pole?

d. If the quarter circle is used with appropriate boundary conditions for finding gain magnitude, what is the required current per pole for 0.1 volt again to correspond to 1 db if (1) the pole is one of a conjugate pair or (2) the pole lies on the σ axis?

8-5. This problem concerns the conduction analogy. A circular electrolytic tank has been set up in the laboratory, as shown in Fig. P8-5, and filled with an electrolyte such as copper sulfate and water. The rim of the tank is a conductor through which flow the currents which go to infinity. The electrode currents are supplied from a transformer through high series resistances to provide constant-current a-c sources. (Alternating current is necessary to prevent errors due to polarization of the electrolyte.) Voltages are measured with a high-impedance vacuum-tube voltmeter. The reference point for the voltmeter can be any constant voltage, since *adding* a constant voltage to all the readings merely amounts to *multiplying* the gain function by a constant. It is convenient to connect the free terminal of the voltmeter to an adjustable tap on the transformer to allow setting the voltmeter reading to zero at some chosen point in the tank. In this way the important voltage differences may be read more accurately. (The data show that the origin was chosen as the point of zero voltage.)

The problem is to convert actual voltmeter readings into relative gain in decibels for the amplifier which is represented by the pole-zero configuration set in the tank. Note that a singularity is represented by a conductor inserted clear to the bottom of the tank so that there is no variation in potential with depth. The conversion can readily be accomplished with the aid of some calibration data obtained by placing a single pole at the center of the tank.

a. Plot the calibration data on semilogarithmic graph paper, and determine the scale factor for the tank, i.e., volts per decibel. How might you explain the relatively large error in the voltage read for the first point? Why is the difference between the readings at 0 and 0.25 in. only 11.2 volts instead of infinity as it should be? Does the fact that this difference is finite influence the accuracy of the calibration of the tank?

b. Plot the measured amplifier characteristic in terms of decibels and inches.

c. The pole-zero diagram shown in Fig. P8-5 could have been a representation of a shunt-peaked pentode stage with $m = 0.414$. Assuming the coordinate of the zero is -3 in., what is the conversion factor between inches measured on the $j\omega$ axis of the analogy and frequency? (Obviously this will involve the R, L, and C of the shunt-peaked stage. Refer to Fig. 4-2.)

Distance from tank center, in.	Calibration data, volts	Amplifier representation, volts
0.00	0.0	0.0
0.25	11.2	0.0
0.50	13.4	0.0
0.75	14.7	0.0
1.00	15.6	0.05
1.25	16.4	0.16
1.50	17.1	0.33
2.00	18.1	0.75
2.50	18.9	1.38
3.00	19.6	1.90
4.00	20.5	3.00

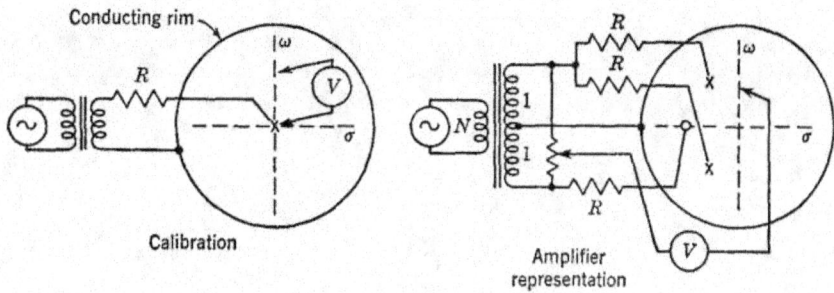

Fig. P8-5

d. Data were also taken on the tank for the determination of phase shift. The two voltmeter probes were positioned parallel to the σ axis and $\frac{1}{16}$ in. on either side of the $j\omega$ axis. From the following data calculate and plot the phase curve versus ω:

ω, in.	V, volts
0.25	0.198
0.75	0.244
1.25	0.261
1.75	0.227
2.25	0.181
2.75	0.133
3.25	0.092
3.75	0.068

9

Commonly Used Functions
for Approximating Constant Gain
or Linear Phase

In the filter amplifier the usual problem is to approximate an "ideal" amplitude response, such as that shown in Fig. 9-1, where the gain is constant in some passband region (ω_1 to ω_2) and zero outside this passband.

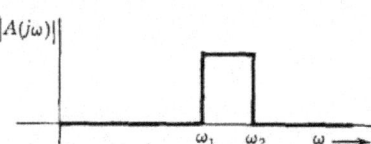

Fig. 9-1 Assumed ideal passband shape for a filter amplifier.

In practice, with only lumped, linear networks neither the constant gain in the passband nor the infinite rejection of signals outside the passband can be obtained. Therefore our problem is to discover suitable gain functions which are physically realizable and which come satisfactorily close to our ideal. In other words we must try to approximate our ideal in a manner suitable for our application. As examples, a radar intermediate-frequency amplifier does not usually require extremely constant gain in the passband, nor does it have particularly stringent requirements on the out-of-band attenuation; conversely an amplifier for telephone repeater service must have extremely constant gain in the passband and possibly very high attenuation out of the passband to remove signals at other carrier frequencies.

Notice that a cascade of identical single-tuned stages provides a crude approximation to constant gain. If we consider the bandpass of Fig. 9-1 moved down to zero frequency, as shown in Fig. 9-2, we can compare the

202

gain with that given in normalized form by Eq. (7-12) and plotted in Fig. 7-5.

$$|A(x)| = \left(\frac{1}{\sqrt{1+x^2}}\right)^n \qquad (7\text{-}12)$$

(Note that such gain functions are always symmetrical about the origin.) Equation (7-12) is not a very good function, either from the standpoint of the shape of the response curve approximating Fig. 9-1 (cf. Fig. 7-5) or from the standpoint of conservation of gain-bandwidth. Two gain functions will be presented which are better in both respects and which can be

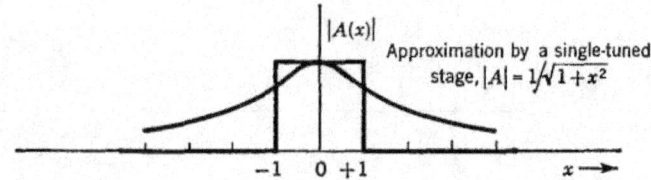

Fig. 9-2 Ideal of Fig. 9-1 and its approximation by one single-tuned circuit. (Passband centered about zero frequency.)

realized almost as readily with cascaded single-tuned or other interstage networks.

Another approximation problem of interest is the attainment of moderately effective filtering, but with the principal interest a linear-phase characteristic in the passband. Such a characteristic can be obtained still using only simple interstage networks.

9-1 Maximally Flat Gain Function.[1] This function, which is also known as the "Butterworth" function [2] and as "approximation in the Taylor sense," has for its normalized magnitude, or amplitude response, the following form:

$$|A(x)| = \frac{1}{\sqrt{1+x^{2n}}} \qquad (9\text{-}1)$$

The shape of the response curve for various values of n (the number of stages) is shown in Fig. 9-3. As n increases, the shape becomes more nearly the rectangle of Fig. 9-2 and the 3-db bandwidth remains constant. The function is always monotonic, i.e., decreases uniformly toward zero on

[1] V. D. Landon, Cascade Amplifiers with Maximal Flatness, *RCA Rev.*, vol. 5, pp. 347–362, January, 1941 (first introduced term "maximal flatness"); W. A. Lynch, The Role Played by Derivative Adjustment in Broadband Amplifier Design, *Proc. Symposium on Modern Network Synthesis*, Polytechnic Institute of Brooklyn, N.Y., 1952, pp. 193–201.

[2] S. Butterworth, On the Theory of Filter Amplifiers, *Wireless Engr.*, vol. 7, pp. 536–541, October, 1930.

either side of band center, and for the n-pole function it represents "maximal flatness" in that the maximum number of derivatives $(2n - 1)$ are zero at band center. This feature can be demonstrated as follows:

$$f(x) = (1 + x^{2n})^{-\frac{1}{2}} \tag{9-2}$$

Expand $f(x)$ in a power series.

$$f(x) = 1 - \tfrac{1}{2}x^{2n} + \frac{3}{4}\frac{x^{4n}}{2!} - \cdots \tag{9-3}$$

Compare Eq. (9-3) with the corresponding Taylor (Maclaurin) series for $f(x)$.

$$f(x) = \underbrace{f(0)}_{1.0} + \underbrace{f'(0)x + f''(0)\frac{x^2}{2!} + \cdots + \frac{f^{2n-1}(0)x^{2n-1}}{(2n-1)!}}_{\substack{\text{These terms missing} \\ \text{in Eq. (9-3); hence} \\ f'(0) = 0 \\ f''(0) = 0 \\ \cdots\cdots\cdots \\ f^{2n-1}(0) = 0}} + \underbrace{\frac{f^{2n}(0)x^{2n}}{(2n)!} + \cdots}_{\substack{\text{This term and} \\ \text{higher even-order} \\ \text{terms are present} \\ \text{in Eq. (9-3)}}}$$

$$\tag{9-4}$$

Thus the first $2n - 1$ derivatives are zero at $x = 0$ for the function $1/\sqrt{1 + x^{2n}}$. Remember that $x = 0$ corresponds to $f = f_0$, the frequency of maximum gain in the bandpass-amplifier case.

The selectivity ratio of the maximally flat response is given in the following equation, which is seen to be similar to Eq. (7-20) for cascaded identical single-tuned stages, but smaller for a given n, hence superior.

$$\text{Selectivity ratio} = \left(\frac{10^6 - 1}{4 - 1}\right)^{1/2n} = (577)^{1/n} \tag{9-5}$$

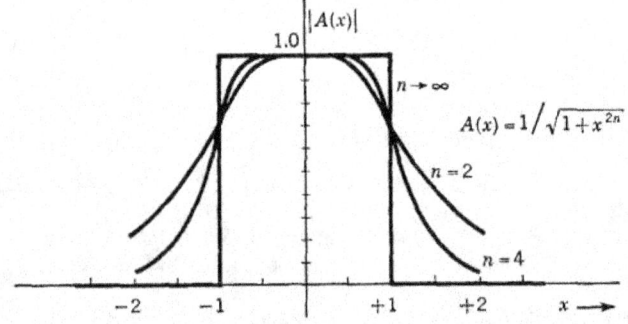

Fig. 9-3 Approximation of ideal by the maximally flat gain function.

Now the question is: Where does one locate the poles in the p plane to give this maximally flat amplitude response for $p = j\omega$? Let us take a specific case and then present the general situation. Suppose that we take a two-pole case, to be designed for a lowpass response; i.e., the frequency of maximum gain and at which the maximum number of derivatives should vanish is $\omega = 0$. Thus $n = 2$, and the desired amplitude response will be

$$|A(\omega)| = \frac{1}{\sqrt{1 + \omega^4}} \tag{9-6}$$

Since we are concerned only with amplitude, it is convenient to work with a function in which phase is absent. As mentioned in the previous chapter (see Fig. 8-11) in the discussion of image poles, a pure amplitude function is produced by taking $A(p)A(-p)$; in terms of the unknown pole locations p_1 and p_2 ($= \bar{p}_1$) this is

$$A(p)A(-p) = \frac{1}{(p - p_1)(p - \bar{p}_1)(-p - p_1)(-p - \bar{p}_1)}$$

$$= |A(p)|^2_{p=j\omega} \tag{9-7}$$

where

$$p_1 = \sigma_1 + j\omega_1$$

$$\bar{p}_1 = \sigma_1 - j\omega_1 = p_2$$

For

$$p = j\omega$$

$$|A(p)|^2 \rightarrow |A(\omega)|^2 = \frac{1}{1 + \omega^4} \qquad \text{from Eq. (9-6)} \tag{9-8}$$

$$|A(p)|^2 \rightarrow |A(\omega)|^2 = \frac{1}{(\sigma_1^2 + \omega_1^2)^2 + 2(\sigma_1^2 - \omega_1^2)\omega^2 + \omega^4}$$

$$\text{from Eq. (9-7)} \tag{9-9}$$

By comparison of Eqs. (9-8) and (9-9), i.e., equating the coefficients of the two equations, it is possible to solve for σ_1 and ω_1.

$$\sigma_1^2 + \omega_1^2 = 1 \tag{9-10}$$

$$\sigma_1^2 - \omega_1^2 = 0 \tag{9-11}$$

$$\sigma_1 = \pm 0.707$$

$$\omega_1 = \pm 0.707 \tag{9-12}$$

Thus the four poles of the amplitude function lie on a unit circle in the p

plane, as shown in Fig. 9-4. The result can be stated in terms of the pole locations having the four values of $\sqrt[4]{-1}$.

In more general terms we may solve the problem for any value of n. Using Eq. (9-1), we may rewrite it in terms of the frequency variable p.

$$|A(j\omega)|^2 = \frac{1}{1 + \omega^{2n}} = \frac{1}{1 + (p/j)^{2n}} \qquad \text{since } p = j\omega$$

$$= A(p)A(-p) = \frac{(-1)^n}{(-1)^n + p^{2n}} \tag{9-13}$$

The roots of Eq. (9-13) are then

$$p_i = [-(-1)^n]^{1/2n} = (-1^{n+1})^{1/2n} = -1^{(n+1)/2n} \tag{9-14}$$

Hence the poles of the gain equation always lie on the unit circle at the $2n$ roots of -1^{n+1}. Specific examples are

$n = 1 \quad p_1, p_2 = \pm 1$

$n = 2 \quad p_1, p_2 = \pm 1\underline{/45^\circ} \qquad p_3, p_4 = \pm 1\underline{/-45^\circ}$

$n = 3 \quad p_1, p_2 = \pm 1\underline{/60^\circ} \qquad p_3, p_4 = \pm 1\underline{/-60^\circ} \qquad p_5, p_6 = \pm 1$

Interpreting our result now in terms of the potential analogy, we would say that positive line charges piercing the page at the pole locations indicated in Fig. 9-4 will yield a maximally flat potential variation along the $j\omega$ axis. We know that the requirements on network functions prohibit the poles in the right half plane, but the potential analogy assures us that we can discard them and still have a maximally flat function. The symmetry about the $j\omega$ axis indicates that, as far as potential along the $j\omega$ axis is concerned, the left- and right-hand sets of poles provide the same potential variation. Thus removing one set simply divides the potential by a factor of 2, or gives the square root of $A(j\omega)A(-j\omega)$, hence $|A(\omega)|$. The 70.7 per cent or 3-db band edges are always $x = \pm 1$, regardless of n [see Eq. (9-1)]. Hence the diameter of the circle is the bandwidth (heavy line).

The shape of the actual amplitude response for various values of n is shown in Fig. 9-5. These would be the actual response curves of a low-pass amplifier using the derived pole locations. We now wish to learn how to change the pole locations to

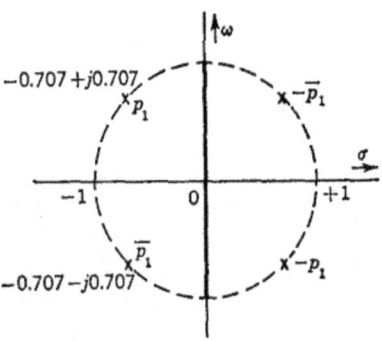

Fig. 9-4 Pole positions for $n = 2$, maximally flat function.

produce a bandpass amplifier. The semicircle could be translated along the $j\omega$ axis until its center was at a point $j\omega_0$, in which case the maximally flat amplitude response would be a bandpass one, centered at ω_0.

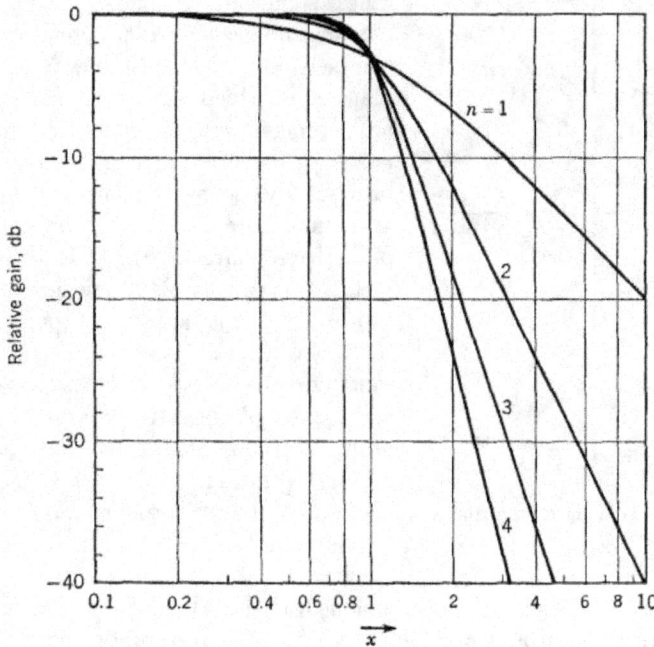

Fig. 9-5 Gain vs. frequency for maximally flat functions and various n's.

The bandwidth can also be varied at will by changing the size of the semicircle; the diameter will always be the 3-db bandwidth, and the response will always be maximally flat. An example is shown in Fig. 9-6. In terms of the single-tuned interstage networks of Chap. 7, the figure would represent a narrow-band approximation, since there are no conjugate poles and no zeros at the origin. Otherwise, the figure corresponds to three single-tuned circuits with different center frequencies and Q. This is the foundation of "stagger tuning," the details of which will be developed in Chap. 10.

9-2 Wideband Transformation. In obtaining a maximally flat response in the bandpass case for large bandwidths, the simple picture of Fig. 9-6 will not suffice. There are zeros at the origin—one for each circuit in the single-tuned case, for example—and a set of conjugate poles (see Fig. 8-1).

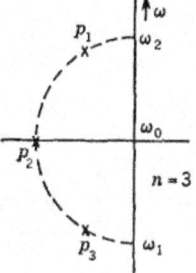

Fig. 9-6 Pole locations for a maximally flat triple in narrow band.

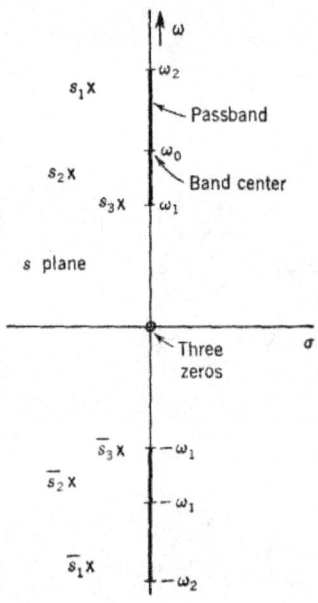

Fig. 9-7   Pole locations for a wideband triple.

In potential terms, these additional charges will influence the potential in the passband and will keep it from being maximally flat unless something is done.

Something can be done, however, and it can be explained in several ways. One explanation, which seems as satisfying as any, proceeds in terms of the potential analogy and the solving of a potential problem by means of conformal mapping. We know how to arrange a single cluster of positive charges (Fig. 9-4) in order to yield a maximally flat variation of potential along the $j\omega$ axis. Yet our problem for the wideband case consists in the arranging of *two* clusters of positive charges and a multiple negative charge at the coordinate origin. These two situations are contrasted in Figs. 9-7 and 9-8; in order to distinguish the two problems, one has been labeled the s plane and the other the p plane.

If somehow we could map the s plane into the p plane, our problem would be solved. This would involve carrying the s-plane origin out to infinity in the p plane, so that the influence of the negative charges would be removed. Second, the two clusters of positive charges in the s plane would have to reduce to one, a feat to be accomplished by causing them to overlie each other in the p plane.

A conformal-mapping function which accomplishes just these desired results is

$$p = \frac{s}{\omega_0} + \frac{\omega_0}{s} \tag{9-15}$$

This transformation, or mapping function, can be derived, but we shall not attempt to do so here. Note that the transformation is a generalization of the change of variable used in Eq. (7-11), where both x and ω are allowed to become complex; i.e.,

$$\frac{x}{Q} = \frac{\omega}{\omega_0} - \frac{\omega_0}{\omega} \tag{7-11}$$

$$p = \frac{s}{\omega_0} + \frac{\omega_0}{s} \tag{9-16}$$

Hence another way of looking at the transformation is that it is a simple change of variable which takes an equation such as Eq. (8-2), which has two poles and one zero, and changes it into an equation with only one pole and no zeros.

The transformation causes the band center in the s plane, namely, ω_0, to go into the origin in the p plane; and the origin in the s plane goes to infinity in the p plane. A more compact expression for the transformation results from normalizing the s plane for any prescribed ω_0 by dividing the plane coordinates by ω_0. Then the transformation becomes simply

$$p = s + \frac{1}{s} \qquad (9\text{-}17)$$

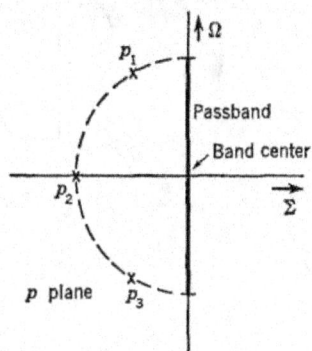

Fig. 9-8 Poles from Fig. 9-7 transformed to lowpass p plane.

The effect of transforming certain lines from the p plane into the s plane is shown in Fig. 9-9. Since we are interested in placing poles on a circle in the p plane, it is interesting to see how circles are transformed to the s plane. Small circles in the p plane are almost circular in the s plane, but centered about $\pm j1$. These approximate the construction used in Fig. 9-6 for narrow bands. As the circles become larger in the p plane, their transformation becomes more distorted in the s plane. Note that the intersection of the circle or its transformation with the $j\omega$ axis defines the edges of the passband and that the geometric mean of the intersections is always unity.

The intersection of the radial lines and a circle in Fig. 9-9 defines the pole positions for a certain bandwidth (determined by the diameter of the circle) and a certain value of n (determined by the angles of the radial lines). In the figure $n = 3$. The angle that these radial lines make with the $j\omega$ axis is the same in the p plane and at the points $\pm j1$ in the s plane because of the conformality of the transformation. This again shows that Fig. 9-6 is correct for sufficiently narrow bandwidths; however, as the bandwidth increases, the transformed radial lines bend and the narrow-band picture no longer holds true.

The s plane in Fig. 9-9 is really the answer to our problem of where to place the singularities of our single-tuned circuits to produce an exact, maximally flat gain function. All we need to do is to choose the transformed circle of proper diameter to give us the desired ratio of bandwidth to center frequency; then we read off the coordinates of the intersections of this transformed circle and radial lines.

A word of caution: The transformation was developed for the single-tuned case as an example and applies only to this case (and others where

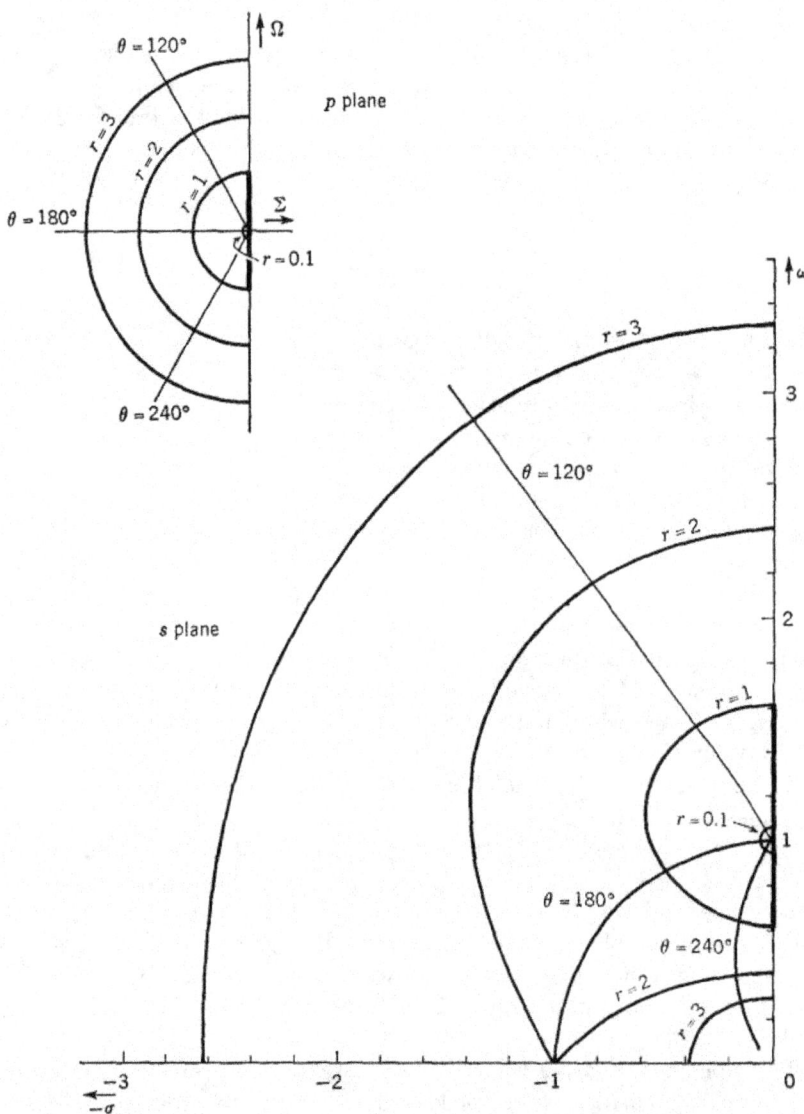

Fig. 9-9 Transformation of pole loci from the p plane to the s plane by the transformation $p = s + 1/s$. The heavy line in the s plane shows the bandwidth which corresponds to the bandwidth (heavy line) in the p plane. Note that the heavy line in the p plane is twice as long as that in the s plane.

the ratio of the number of poles to the number of zeros at the origin is 2). More general forms of transformation exist which will accommodate double-tuned circuits, etc. These slightly modified forms will be introduced in Chap. 11.

The transformation is commonly used in going from the p plane to the s plane since it is in the former plane that we know how to arrange the charges for maximal flatness. The transformation is double-valued in going from p to s, as may be seen by solving Eq. (9-17) for s.

$$s = \frac{p}{2} \pm \sqrt{\left(\frac{p}{2}\right)^2 - 1} \qquad (9\text{-}18)$$

Hence each pole in p goes into two in s. Unless the pole is real in the p plane, conjugate pairs of poles do not result in the s plane, although we know that the end result *must* be conjugate clusters for physical realizability. We are saved by the fact that, when the whole maximally flat cluster in p is transformed, the result is all right. Thus p_1 in Fig. 9-8 taken together with p_3 produces s_1 and s_3, together with their conjugates.

This transformation can also be regarded as a lowpass-to-bandpass one. A realizable network function set up as a pole arrangement in the p plane transforms into a realizable function in the s plane, which gives a bandpass characteristic in the latter when it was lowpass in the former.

A practical matter in using the transformation is that there is a numerical factor of 2.0 between the bandwidths in the two planes. The bandwidth from $-r$ to r in the p plane becomes two passbands in the s plane each of bandwidth r. Thus, if $(\omega_2 - \omega_1)\omega_0$ is desired to be 1.0 in the s plane, then $\Omega_2 - \Omega_1$ must be set up as 2.0 in the p plane.

9-3 Equal-ripple Gain Function. An alternative to the maximally flat approximation to constant gain in the passband is the equal-ripple response illustrated in Fig. 9-10. Such an amplitude response is readily devised. The magnitude of the ripples can be specified, although not their

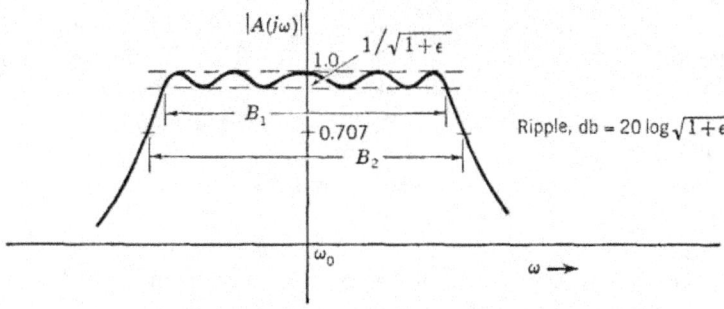

Fig. 9-10 Equal-ripple response.

frequency spacing. This response function turns out to be advantageous from three standpoints: more gain can be achieved for the same bandwidth; the approximation to constant gain is usually better for steady-state applications than is the maximally flat response; and the selectivity ratio is better.

The analytic expression of the amplitude response of Fig. 9-10 is given below. It contains Chebyshev polynomials, and because of this the approximation function is sometimes called the "Chebyshev response."

$$|A(x)| = \frac{1}{\sqrt{1 + \epsilon C_n{}^2(x)}} \tag{9-19}$$

where $\quad \epsilon$ = ripple parameter = $\log^{-1} \dfrac{\text{ripple (db)}}{10} - 1 \tag{9-19a}$

$C_n(x)$ = Chebyshev polynomial

$$C_1(x) = x \qquad\qquad C_4(x) = 8x^4 - 8x^2 + 1$$

$$C_2(x) = 2x^2 - 1 \qquad C_5(x) = 16x^5 - 20x^3 + 5x$$

$$C_3(x) = 4x^3 - 3x \qquad C_6(x) = 32x^6 - 48x^4 + 18x^2 - 1$$

$$C_{n+1}(x) = 2xC_n(x) - C_{n-1}(x) \tag{9-20}$$

There is also a convenient expression for the general polynomial in terms of trigonometric functions,

$$C_n(x) = \begin{cases} \cos(n \cos^{-1} x) & \text{for } -1 < x < +1 \\ \cosh(n \cosh^{-1} x) & \text{for } |x| > 1 \end{cases} \tag{9-21}$$

The property the Chebyshev polynomials have which is of interest to us is shown in Fig. 9-11; for x in the range of ± 1, the maximum and minimum values of $C_n(x)$ are also ± 1, and $C_n{}^2(x)$ oscillates between 0 and 1 in the "approximation band"— $-1 \leq x \leq +1$. Therefore inspection of Eq. (9-19) shows that in this band the maximum value of $A(x)$ is 1 and the minimum value of $A(x)$ is $1/\sqrt{1 + \epsilon}$, as shown graphically in Fig. 9-10 (for $n = 5$). Far outside the approximation band $C_n{}^2(x)$ increases like x^{2n}. Therefore the gain function decreases outside the passband much as it did for the maximally flat case [cf. Eq. (9-1)].

Since Eq. (9-19) seems to provide a very useful gain function, what we now need to know is where to locate the poles of the gain function to realize this gain function along the $j\omega$ axis. From these pole locations we may later determine the necessary element values in the interstages.

The simplest gain function leading to the equal-ripple amplitude response is an all-pole one, with the poles situated around a semiellipse, as opposed to a semicircle for the maximally flat response. This is depicted in Fig. 9-12 and was previously shown in Fig. 8-10. The actual value of the pole locations may be found by the process used to factor Eq. (9-1) and resulting in Eq. (9-14).

The band between $x = 1$ and -1, the foci of the ellipse, corresponds to the bandwidth B_1 in Fig. 9-10 and is the band within which the gain stays

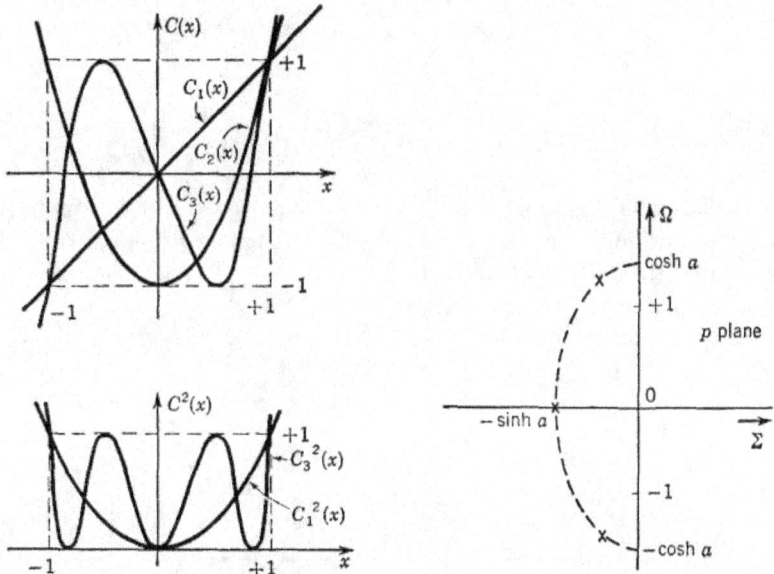

Fig. 9-11 Chebyshev polynomials.   **Fig. 9-12** Pole locations for an equal-ripple gain function.

within the ripple tolerance. On the other hand, the band between $x = \cosh a$ and $-\cosh a$ corresponds very closely to the 3-db bandwidth B_2 in Fig. 9-10, differing from 3 db by the amount of the ripple, which is usually small, say $\frac{1}{2}$ db. We are actually interested in both these bandwidths, although designers of passive networks usually talk only about B_1. The 3-db bandwidth B_2 provides a comparison with the maximally flat case and is also the bandwidth to be used in evaluation of noise in amplifiers.

The parameter a is a function of the desired ripple magnitude. Qualitatively it can be seen, via the potential analogy, that as the minor diameter of the ellipse is made smaller, bringing the poles in closer to the $j\omega$ axis, the potential variation (the "ripple") will be larger. The formal expression is

$$a = \frac{1}{n} \sinh^{-1} \frac{1}{\sqrt{\epsilon}} \qquad (9\text{-}22)$$

The expression for the location of the kth pole is as follows:

For n odd

$$p_k = \sinh\left(-a \pm jk\frac{\pi}{n}\right)$$

where

$$k = 0, \pm 1. \pm 2, \ldots, \pm \frac{n-1}{2}$$

For n even

$$p_k = \sinh\left(-a \pm j\frac{1+2k}{2}\frac{\pi}{n}\right)$$

(9-23)

where

$$k = \ldots, -\frac{n}{2}\ldots, -1, 0, +1, \ldots, +\left(\frac{n}{2}-1\right)$$

It will actually be more expeditious in comparing the equal-ripple case with the maximally flat one—and indeed in designing for equal ripple—if we equalize the 3-db bandwidths in the two cases. Thus, in the maximally

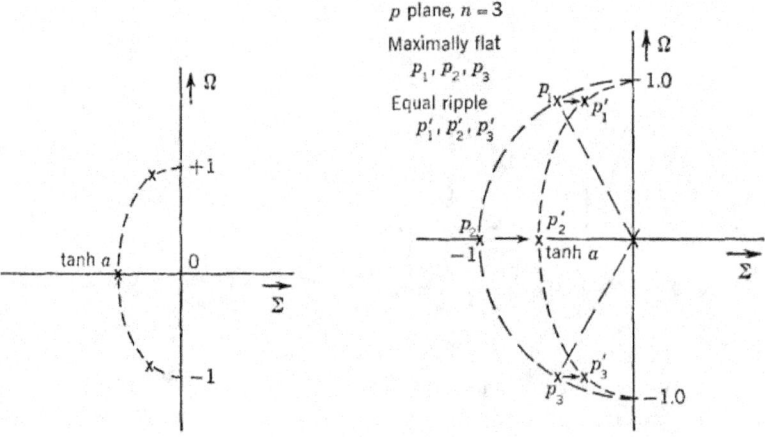

Fig. 9-13 Same as Fig. 9-12, except that the passband (-3 db) is from -1 to $+1$.

Fig. 9-14 Construction used in going from maximally flat to equal-ripple pole locations.

flat case (Fig. 9-3) the 3-db passband extended from $x = +1$ to -1. Hence, if we change the scale of the ellipse in Fig. 9-12 to bring its major diameter between $+1$ and -1, the two cases are closely comparable. This we do by dividing the scale of Fig. 9-12 by $\cosh a$, giving Fig. 9-13.

Remarkably enough, the pole locations are closely related to the maximally flat case. The imaginary components are identical, and the real components for the equal-ripple case are simply tanh a times the real component for maximally flat. This is illustrated in Fig. 9-14.

We can now summarize the two functions, i.e., the two sets of pole locations. The maximally flat response results from placing the n poles on the unit circle (for a band from $+1$ to -1), according to the $2n$ roots of -1^{n+1}. For the corresponding equal-ripple case, i.e., the same 3-db bandwidth (almost), multiply the real components by tanh a.

Rather than write the formulas for the $2n$ roots for arbitrary n, it is really simpler to state (and to remember) the simple geometrical procedure:

> Divide the semicircle (180°) by n, giving the angular separation of the poles. Lay them out on a semicircle of the desired diameter (determined by the bandwidth desired), starting with a *half angle* at the $j\omega$ axis.

Thus in the example above (Fig. 9-14), 180° divided by $n = 3$ gives 60° for the pole separation. The poles are laid out on the semicircle, starting

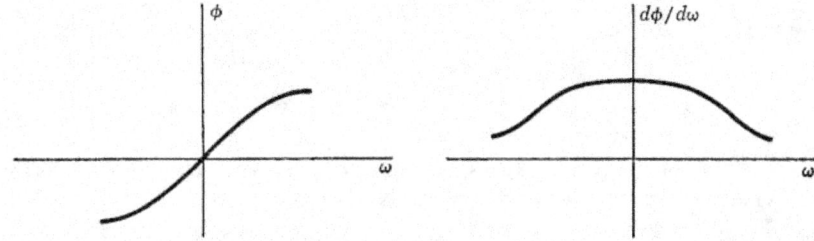

Fig. 9-15 Typical phase function.   **Fig. 9-16** Typical phase-delay function.

with a half angle, namely, 30°, next to the $j\omega$ axis. This gives maximal flatness of amplitude response. For equal ripple, just translate the poles horizontally toward the $j\omega$ axis by multiplying the real component of the pole coordinates by the factor tanh a (tanh a is given in Table 10-6 as a function of the amount of ripple and n).

9-4 Linear-phase Response. In some bandpass or lowpass systems, it may be the objective to make the amplifier have as closely as possible a phase shift ϕ that varies linearly with frequency, or, what is equivalent, a phase delay $d\phi/d\omega$ that is constant with frequency. The amplitude response is taken as it comes.

The phase response is always an odd function of frequency with respect to band center and may look as in Fig. 9-15. The derivative, on the other hand, will be an even function, as in Fig. 9-16. This suggests that an approximation to perfect linear phase (which is not physically realizable over an infinite frequency range) would be to make the phase delay maximally flat.[1] The problem becomes one of deriving a function which expresses the delay curve of Fig. 9-16, in terms of the pole locations, and then solving for the locations necessary to make the delay curve maximally flat. The procedure parallels that for the maximally flat amplitude response.

As an example, let us take once again the two-pole gain function, low-pass, as in Fig. 9-17, for which

$$A(p) = \frac{1}{(p - p_1)(p - \bar{p}_1)}$$

(9-24)

$$p_1 = \sigma_1 + j\omega_1$$

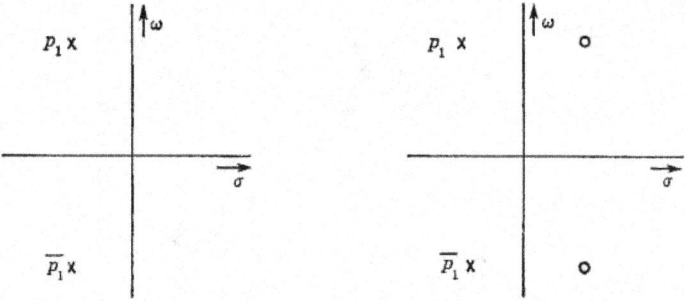

Fig. 9-17 Assumed pole locations of $A(p)$ to give maximally flat delay.

Fig. 9-18 Same as Fig. 9-17 with image zeros, $A(p)/A(-p)$.

The phase response can be derived directly from Eq. (9-24), or it can be obtained by forming a phase function $A(p)/A(-p)$, pictured in Fig. 9-18, which has twice the actual phase shift and no amplitude variation [see Eq. (8-56)]. Taking the latter approach, which possesses no special virtue here except to illustrate a broadly applicable technique,

$$\phi(\omega) = \frac{1}{2} \arg \frac{A(p)}{A(-p)} \bigg|_{p=j\omega}$$

(9-25)

$$\frac{A(p)}{A(-p)} \bigg|_{p=j\omega} = \frac{(-p - p_1)(-p - \bar{p}_1)}{(p - p_1)(p - \bar{p}_1)} \bigg|_{p=j\omega}$$

$$= \frac{[\sigma_1 + j(\omega + \omega_1)][\sigma_1 + j(\omega - \omega_1)]}{[\sigma_1 - j(\omega + \omega_1)][\sigma_1 - j(\omega - \omega_1)]}$$

(9-26)

---

[1] W. E. Thomson, Networks with Maximally Flat Delay, *Wireless Engr.*, vol. 29, pp. 256–263, October, 1952, p. 309, November, 1952; Lynch, *op. cit.*; F. A. Muller, High-frequency Compensation of *RC* Amplifiers, *Proc. IRE*, vol. 42, pp. 1271–1275, August, 1954.

Substituting Eq. (9-26) in Eq. (9-25) gives

$$\phi(\omega) = \tan^{-1}\frac{\omega + \omega_1}{\sigma_1} + \tan^{-1}\frac{\omega - \omega_1}{\sigma_1} \tag{9-27}$$

Differentiating Eq. (9-27) gives the delay function

$$\frac{d\phi}{d\omega} = \frac{1}{1 + \left(\dfrac{\omega + \omega_1}{\sigma_1}\right)^2}\frac{1}{\sigma_1} + \frac{1}{1 + \left(\dfrac{\omega - \omega_1}{\sigma_1}\right)^2}\frac{1}{\sigma_1}$$

$$= \frac{2\sigma_1}{\sigma_1{}^2 + \omega_1{}^2}\frac{1 + \dfrac{1}{\sigma_1{}^2 + \omega_1{}^2}\omega^2}{1 + \dfrac{2(\sigma_1{}^2 - \omega_1{}^2)}{(\sigma_1{}^2 + \omega_1{}^2)^2}\omega^2 + \dfrac{1}{(\sigma_1{}^2 + \omega_1{}^2)^2}\omega^4} \tag{9-28}$$

Equation (9-28) is an even function in ω, as was the amplitude function of Eq. (9-9), and can also be made maximally flat. It is expanded in a power series, in which the coefficients of ω^2, ω^4, etc., correspond to the derivatives at $\omega = 0$. Equating as many of these to zero as is possible yields maximal flatness. Expanding Eq. (9-28) in a power series gives a coefficient of the ω^2 term which is the difference between the numerator and denominator coefficients of ω^2. Equating this difference to zero gives

$$\frac{1}{\sigma_1{}^2 + \omega_1{}^2} = \frac{2(\sigma_1{}^2 - \omega_1{}^2)}{(\sigma_1{}^2 + \omega_1{}^2)^2} \tag{9-29}$$

$$\sigma_1{}^2 = 3\omega_1{}^2$$

$$\sigma_1 = \pm\sqrt{3}\,\omega_1 \tag{9-30}$$

No further coefficients can be made zero, but notice that three are zero—not only ω^2, but also ω and ω^3 because of the function being an even one. Thus, as in the amplitude response, three derivatives are zero for a two-pole function, or $2n - 1$ derivatives for an n-pole function.

Also notice that Eq. (9-30) differs from Eq. (9-12), which dictates that $\sigma_1 = \omega_1$ for maximally flat amplitude. It is of interest to complete the comparison by setting the amplitude bandwidths equal in the two cases. For Eq. (9-12) the bandwidth is 2.0; so we can adjust the flat-delay function to have this amplitude bandwidth. To do this, we arbitrarily let $\omega_1 = 1$ in Eq. (9-24), then $\sigma_1 = \sqrt{3}$, and by substitution we find that Eq. (9-24) becomes

$$A(p) = \frac{1}{p^2 + 2\sqrt{3}p + 4} \tag{9-31}$$

The magnitude of Eq. (9-31) is

$$|A(j\omega)|^2 = \frac{1}{\omega^4 + 4\omega^2 + 16} \tag{9-32}$$

At $\omega = 0$, $|A(j\omega)|^2 = \frac{1}{16}$; therefore the band-edge frequencies are found by solving for ω_b in the equation

$$\omega_b{}^4 + 4\omega_b{}^2 + 16 = 32$$

or
$$\omega_b = \pm 1.57 \tag{9-33}$$

Thus the proper pole locations to give both the linear-phase response and a

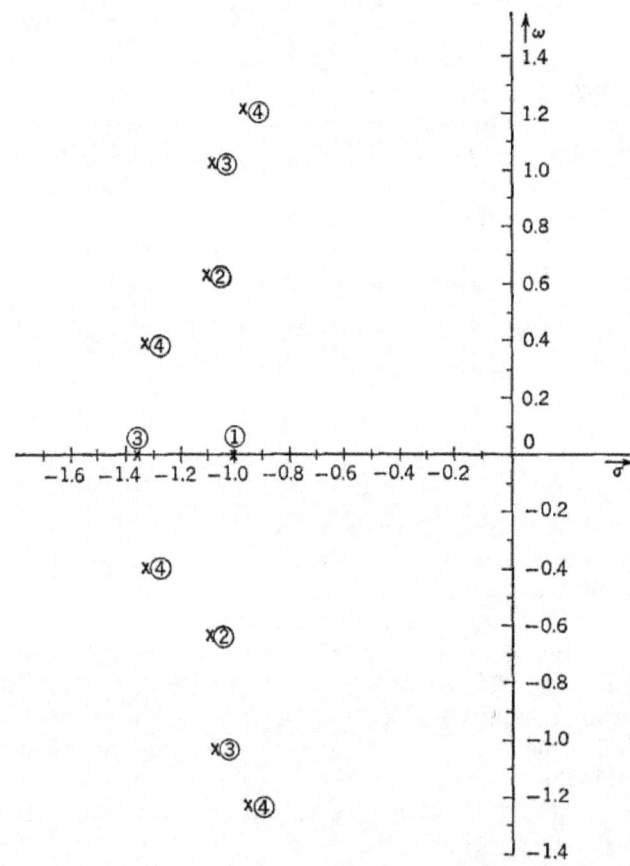

Fig. 9-19 Maximally linear-phase pole locations. (Band normalized to ±1.) Circles indicate order (n) of approximation.

3-db bandwidth of 2.0 (band from -1 to $+1$) are

$$\sigma_1 = \frac{\sqrt{3}}{1.57} = \pm 1.102$$

(9-34)

$$\omega_1 = \frac{\pm 1}{1.57} = \pm 0.636$$

These values contrast with those for maximally flat amplitude response giving the same 3-db bandwidth, which are, from Eq. (9-12), $\sigma_1 = \omega_1 = \pm 0.707$.

For the general case of n poles, things are not so simple as they were for maximally flat amplitude. The poles do not remain on the same contour for all n, nor do they lie on any simple contour for a given n. Moreover, the computational difficulty increases rapidly for large values of n. A few results from Thomson's paper are sketched in Fig. 9-19, the numerals indicating the poles corresponding to a given value of n.

Equal-ripple approximations to linear-phase or constant delay have received study, but no compilation of generally useful results has yet appeared.

9-5 Arbitrary Responses. The formal examples of amplitude or phase responses which have been presented are by no means the only useful ones. In some respects they are too circumscribed by theoretical limitations. For instance, it is not possible to get both maximally flat amplitude and maximally flat delay in a simple amplifier structure; yet Bradley [1] has shown that one can devise a response which, from a practical standpoint, is very flat in both amplitude and delay and which has good selectivity as well.

Other situations call for amplitude or phase response having rather arbitrary shapes. For all such responses other than the idealized ones described in this chapter, it is probable that the most useful approximation procedure is experimentation with one or more of the analogies presented in Chap. 8.

PROBLEMS

9-1a. Find the pole positions in the lowpass plane (p plane) for a three-pole, maximally flat amplifier with a bandwidth extending from $\Omega = -0.3$ to $\Omega = +0.3$.

b. What fractional bandwidth (ratio of bandwidth to center frequency) will result if the pole positions in (*a*) are transformed to the bandpass plane with the transformation $p = s + 1/s$?

c. Find the pole locations in the s plane, using the transformation in (*b*). If the band center were desired at 50 Mc instead of 1.0, what would you do with these pole positions?

[1] W. E. Bradley, Design of a Simple Band-pass Amplifier with Approximate Ideal Frequency Characteristics, *Trans. IRE*, vol. PGCT-2, pp. 30–38, December, 1953.

9-2. Find the locus in the s plane of a point moving from $-\infty$ in toward the origin along the Σ axis in the p plane. What special feature occurs in the s plane at the transformation of the point $p = -2$?

9-3. Assume a gain function of the form

$$A(p) = \frac{1}{p^2 + \alpha p + 1}$$

We wish to make this function maximally flat by properly choosing the constant α. To do this, form the function $|A(j\omega)|^2$. This function may then be differentiated with respect to ω and as many derivatives as possible set equal to zero. From this the value of α may be determined. Show that this value of α gives a pair of poles in the same position as specified by Eq. (9-12).

9-4. The series-peaked transistor stage of Chap. 4 has two complex poles and no zeros in its transfer function. Therefore these poles may be placed in such a manner as to obtain a maximally flat gain function. (Note that this procedure will lead to a transient response with considerable overshoot.)

Using the approximations in Sec. 4-7, find the relations between the stage element values to produce a maximally flat stage with bandwidth ω_a.

9-5. In the p plane, the poles of a maximally flat gain function lie on a circle. Sketch in the s plane the transformed circles and pole locations for $n = 3$ for each of the two conditions

$$\frac{B_r}{\omega_0} = 0.3 \qquad \frac{B_r}{\omega_0} = 1.5$$

B_r is the 3-db bandwidth in the s plane, and ω_0 is the center frequency. Check your result with the curves in Fig. 9-9.

9-6. Find pole locations in the p plane for equal-ripple responses of unit bandwidth having $\frac{1}{2}$ and 1 db ripple. Do this for both $n = 2$ and $n = 3$. Find the ratio of the gain at the origin for each case to the gain at the origin for an equal number of poles arranged in a maximally flat manner. Assume that the gain depends upon the distances from the poles to the origin only.

9-7. Prove that, in general, the intersections of the circles and the $j\Omega$ axis in the p plane when transformed to the s plane display geometric symmetry about the band-center frequency.

9-8. The pentode shunt-peaked stage can also be made maximally flat by the proper choice of the parameter m (see Sec. 4-3 and Fig. 4-2). Form a magnitude function $A(p)A(-p)$, and let $p = j\omega$. Then equate as many derivatives of this function to zero at $\omega = 0$ as possible. This procedure will determine the value of m, which should be $m = 0.414$. What angle does a line connecting the origin to a pole make with the σ axis for this value of m?

10

Stagger Tuning

The term "stagger tuning" refers to an amplifier comprising several stages in cascade, in which the stages are not tuned identically to the same frequency but are "staggered" at frequencies above and below the desired center frequency of the complete amplifier. Not only are the tunings of the individual stages nonidentical, but their bandwidths are also different.

The objectives of stagger tuning are twofold: (1) a greater gain-bandwidth factor is generally achieved than with a cascade of identical stages, and (2) a prescribed amplitude response, such as maximally flat or equal ripple, can be synthesized, either of which is more desirable for filtering than is the response of identical stages.

Historically, the advantages and possibilities of stagger tuning were apparent to a few persons several years before it became a widely used technique. The desirability of synthesizing a complicated gain function from simple networks in a multistage amplifier was first advocated by Butterworth in 1930,[1] although the gain-bandwidth advantage did not become apparent until Schienemann's paper in 1939.[2] The latter paper was apparently not utilized by anyone in this country until about 1943, although Landon in the meantime had published a paper having to do with the maximally flat response function.[3] To Henry Wallman belongs the credit

[1] S. Butterworth, On the Theory of Filter Amplifiers, *Wireless Engr.*, vol. 7, pp. 536–541, October, 1930.

[2] R. Schienemann, Trägerfrequenzverstarker groszer Bandbreite mit gegeneinander verstimmten Einzelkreisen, *Telegraphen Fernsprech Tech.*, 1939, pp. 1–7.

[3] V. D. Landon, Cascade Amplifiers with Maximal Flatness, *RCA Rev.*, vol. 5, pp. 347–362, 481–497, January and April, 1941.

for first exploiting the stagger-tuning technique [1] used in connection with wideband intermediate-frequency amplifiers in the receiver of a radar system.

Wallman's work provided usable data for synthesizing the maximally flat amplitude response with single-tuned amplifier stages. This was extended by Baum [2] to include the equal-ripple function and by Trautman and other workers to include other interstage networks. [3]

The principal elements of the technique have already been described. In Chap. 9 there were presented the pole locations for three kinds of gain functions, yielding maximally flat or equal-ripple amplitude response or maximally flat time delay (linear phase). In Chap. 7 there were developed the equations for the gain function of one single-tuned amplifier stage, and in Chap. 8 the gain function was factored to yield the relationship between the poles and the element values for the single-tuned circuit. Now all that remains is to assign a single-tuned stage for each pole of the desired over-all gain function. Then from the pole locations we shall be able to determine the stage element values, expressed usually in terms of the tuning (center frequency ω_0) and the bandwidth (or Q).

We shall follow Wallman's convention of distinguishing three cases, depending upon the relationship of bandwidth to center frequency. The first case is *narrow-band*, where the bandwidth is less than 5 per cent of the center frequency. At the other extreme is the *wideband* case, where the bandwidth is 30 per cent or more of center frequency. In between is what Wallman calls the "asymptotic," or *intermediate*, case.

10-1 The Narrow-band Case. The narrow-band case gives arithmetic symmetry of the amplitude response and is the case where the zeros at the origin and the conjugate poles are neglected. The single pole has the coordinates shown in Fig. 10-1, which depicts the p plane normalized by 2π to give bandwidths in cycles per second instead of radians per second. Notice that the horizontal coordinate of the pole is $f_0/2Q$, which is also $B/2$, where B is the 3-db bandwidth of the single-tuned circuit as given in

[1] H. Wallman, Stagger Tuned I-F Amplifiers, *M.I.T. Radiation Lab. Rept.* 524, February, 1944; the essential content of this report appears as chap. 4 in G. E. Valley, Jr., and H. Wallman (eds.), "Vacuum Tube Amplifiers" (vol. 18, M.I.T. Radiation Laboratory Series), McGraw-Hill Book Company, Inc., New York, 1948.

[2] R. F. Baum, Design of Broad-band I-F Amplifiers, *J. Appl. Phys.*, vol. 17, pp. 519–529, 921–930, 1946.

[3] D. L. Trautman, Jr., Maximally Flat Amplifiers of Arbitrary Bandwidth and Coupling, *Electronics Research Lab., Stanford, Tech. Rept.* 41, Feb. 1, 1952; J. S. Eddy, Stagger Tuned Amplifiers with Double-tuned Interstages, *Electronics Research Lab., Stanford, Tech. Rept.* 29, January, 1951; D. L. Trautman and J. A. Aseltine, Equal-ripple Bandpass Amplifiers, *Univ. Calif., Los Angeles, Dept. Eng. Rept.* 51-9, August, 1951; M. M. McWhorter, The Design, Physical Realization and Transient Response of Double-tuned Amplifiers of Arbitrary Bandwidth, *Electronics Research Lab., Stanford, Tech. Rept.* 58, February, 1953.

Eq. (7-5). Hence a semicircle drawn through the pole and centered on f_0 gives the band edges for the single-tuned circuit. Do not confuse this geometrical trick with the semicircular construction required for placement of *several* poles in order to yield maximal flatness.

Now, for stagger tuning in the narrow-band case, one merely draws the semicircle, as in Fig. 10-2, of diameter B corresponding to the desired *over-all* bandwidth, and places the desired number of poles on the semicircle, one

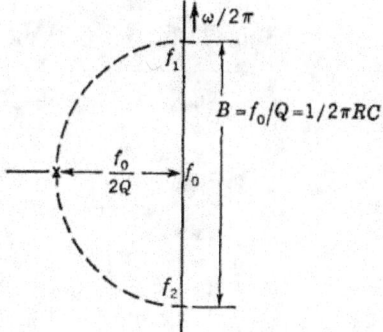

Fig. 10-1 Approximate pole location for a narrow-band single-tuned stage.

for each single-tuned stage. From the pole coordinates one then computes for each stage the individual f_{01} and $B_1 = f_{01}/Q_1$, as in Fig. 10-1. The results can be drawn up in a simple table which can be used without having to repeat the geometrical construction for each new case. Such a table is given in Table 10-1, with values from Valley and Wallman.[1] The table is

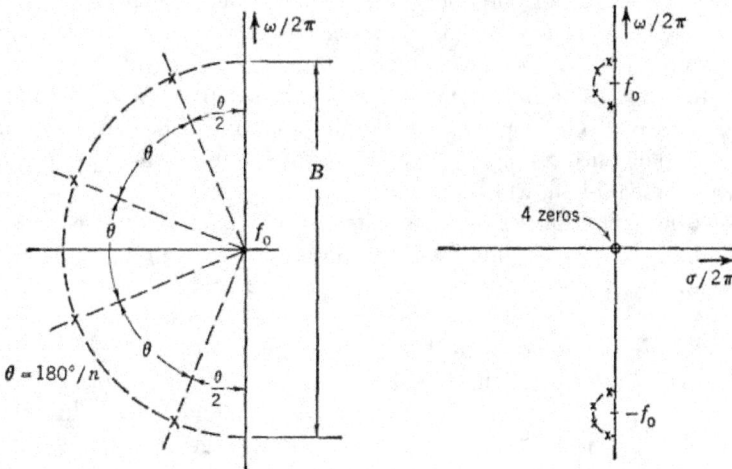

Fig. 10-2 Pole cluster about band center for a maximally flat amplifier and the corresponding situation including all poles and zeros. (Narrow band.)

given only through the staggered quadruple, although obviously it could be extended as far as desired. Practically, however, the pairs and triples are the most widely used. Higher orders require stages with high Q (poles near the $j\omega$ axis), sometimes higher than can be obtained in the presence of

[1] *Op. cit.*, p. 180.

Table 10-1.   Narrow-band Stagger Tuning (Maximally Flat)

1. Staggered pair ($n = 2$)
   Two stages tuned to $f_0 \pm 0.35B$, each having a
   bandwidth $0.707B$
2. Staggered triple ($n = 3$)
   One stage tuned to f_0, with bandwidth B
   Two stages tuned to $f_0 \pm 0.43B$, with bandwidth $0.50B$
3. Staggered quadruple ($n = 4$)
   Two stages tuned to $f_0 \pm 0.46B$, with bandwidth $0.38B$
   Two stages tuned to $f_0 \pm 0.19B$, with bandwidth $0.92B$

NOTE: f_0 is the center frequency of the over-all amplifier, and B is the over-all 3-db bandwidth.

loading due to tube input conductance. Also, the tune-up procedure is lengthened, since additional signal generator frequencies must be used.

The data in the table provide the necessary design data, and that would be the end of the story if electrical components always had the proper values of resistance, capacitance, etc. This is not the case, of course, and particularly in wideband amplifiers, where the principal capacitance is that due to the tube. There is substantial variation from one tube to the next, and hence each stage is usually made tunable over a sufficient range to allow for this. The inductance is readily tuned by means of a brass or powdered-iron "slug," or core, moved into and out of the coil. Although a tuning capacitance could also be added, to do so increases C per stage and reduces the gain-bandwidth product.

The tuning procedure is simplicity itself. A signal generator is connected to the input of the amplifier and a vacuum-tube voltmeter to the output. If the amplifier is, say, a triple, with stages 1, 2, and 3, to be tuned to frequencies f_1, f_2, and f_3, according to Table 10-1, then the signal generator is first set at f_1 and stage 1 tuned for a maximum voltmeter reading. Next, stage 2 is adjusted with the signal generator set at f_2. Finally, stage 3 is adjusted at frequency f_3. The order of adjusting the stages is completely unimportant.

Also, except for certain second-order effects, it is immaterial in which order the stages are connected. To make sure these effects are small, the tunings can be rechecked.

10-2   The Wideband Case. The wideband case requires the use of the transformation developed in Chap. 9.

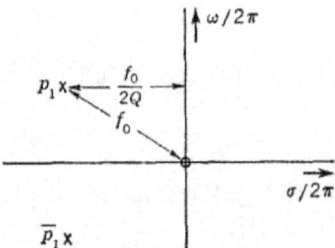

Fig. 10-3   Exact pole-zero locations for a single-tuned stage.

The principal differences are that the poles in the s plane are not on a circle to give maximal flatness and that the radial distance from the origin to the pole of each single-tuned circuit represents the resonant frequency f_0 (see Fig. 10-3). Each new design problem could be worked through by laying out the poles on a circle in the p plane, transforming their coordinates to the s plane, and then determining the stage tunings and bandwidths, as in Fig. 10-3. But for repeated design work it is more convenient to reduce the results of the process to formulas and curves. Two excellent curves are given in Valley and Wallman,[1] for staggered pairs and triples. The corresponding formulas are given in Table 10-2 and the curves in Figs. 10-4 and 10-5.

Table 10-2. Wideband Stagger Tuning (Maximally Flat)

1. Staggered pair ($n = 2$)
   Two stages tuned to $f_0\alpha$ and f_0/α, having the same Q

$$Q^2 = \frac{2}{4 + \delta^2 - \sqrt{16 + \delta^4}}$$

$$\left(\alpha - \frac{1}{\alpha}\right)^2 + \frac{1}{Q^2} = \delta^2$$

2. Staggered triple ($n = 3$)
   One stage tuned to f_0 with bandwidth B
   Two stages tuned to $f_0\alpha$ and f_0/α, same Q

$$Q^2 = \frac{2}{4 + \delta^2 - \sqrt{16 + 4\delta^2 + \delta^4}}$$

$$\left(\alpha - \frac{1}{\alpha}\right)^2 + \frac{1}{Q^2} = \delta^2$$

NOTE: f_0 is the center frequency (geometric center) of the over-all amplifier. B is the over-all 3-db bandwidth, and $\delta \triangleq B/f_0$.

It is of interest to note in passing the locations of the poles in the wideband case for, say, the staggered triple. The center stage has the same f_0 as the triple; hence, its pole will always lie on a circle drawn through the point jf_0 and will be farther from the $j\omega$ axis for greater bandwidths (or on the σ axis for $\delta \geq 2$). The other two stages both have the same Q, and their center frequencies are related by the factor α; this results in the poles for

[1] *Op. cit.*, pp. 188, 190.

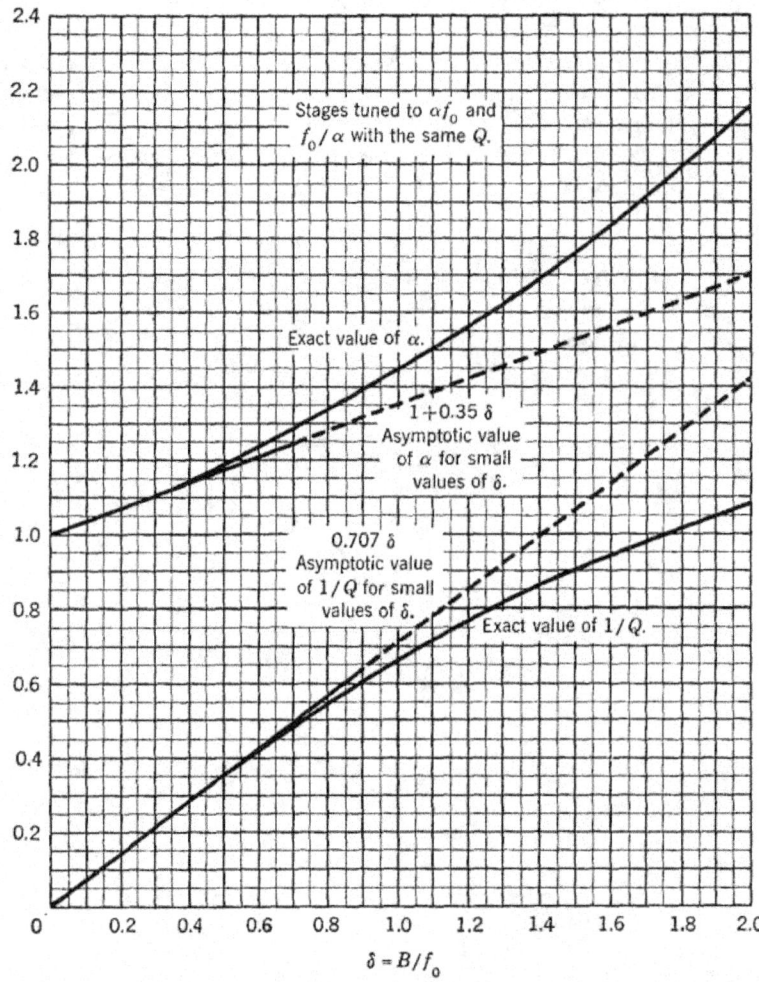

Fig. 10-4 Design curves for an exact flat staggered pair.

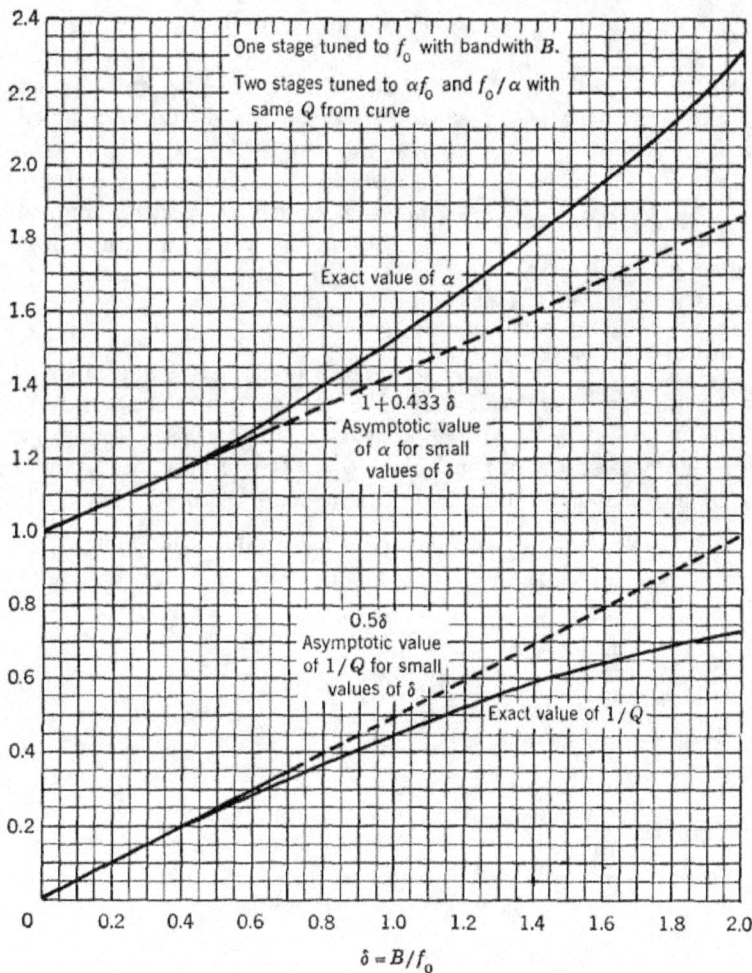

One stage tuned to f_0 with bandwith B.

Two stages tuned to αf_0 and f_0/α with same Q from curve

Exact value of α

$1 + 0.433\,\delta$
Asymptotic value of α for small values of δ

0.5δ
Asymptotic value of $1/Q$ for small values of δ

Exact value of $1/Q$

$\delta = B/f_0$

Fig. 10-5 Design curves for an exact flat staggered triple.

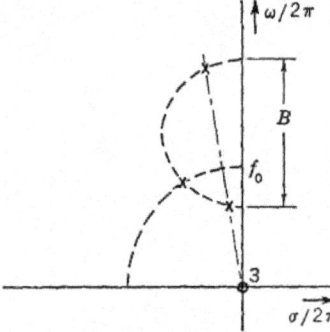

Fig. 10-6 Pole positions for maximally flat triple in the wideband case (lower half plane omitted). Compare with Fig. 9-9.

these two stages lying on the same radial line through the origin. The complete picture (omitting the lower half plane) is shown in Fig. 10-6.

10-3 The Asymptotic Case. The formulas of Table 10-2 are cumbersome to use and hence should be avoided unless the bandwidth is really large. An *intermediate* region of B/f_0 can be defined in which the calculation can be simplified and yet an accuracy of better than 1 per cent can be retained. If B/f_0 is less than 0.3, the values of α and Q in Table 10-2 approach very closely an *asymptotic* limit; this is indeed called the "asymptotic case" by Wallman. For example, in the staggered pair, α approaches $1 + 0.35\delta$, and Q approaches 1.414δ. These formulas for the intermediate case are given in Table 10-3. It will be noted from a comparison of the narrow-band and intermediate cases (Tables 10-1 and

Table 10-3.   Intermediate Bandwidth (Maximally Flat)

$$\delta = B/f_0 \cong 0.05 \text{ to } 0.3 \text{ (or 0 to 0.3)}$$

1. Staggered pair ($n = 2$)
   Two stages tuned to $f_0\alpha$ and f_0/α, same Q

$$Q = \frac{1.414}{\delta} \qquad \alpha = 1 + 0.35\delta$$

2. Staggered triple ($n = 3$)
   One stage tuned to f_0 with bandwidth B
   Two stages tuned to $f_0\alpha$ and f_0/α, same Q

$$Q = \frac{2.0}{\delta} \qquad \alpha = 1 + 0.433\delta$$

3. Staggered quadruple ($n = 4$)
   Two stages tuned to $f_0\alpha_1$ and f_0/α_1, same Q_1

$$Q_1 = \frac{2.63}{\delta} \qquad \alpha_1 = 1 + 0.46\delta$$

   Two stages tuned to $f_0\alpha_2$ and f_0/α_2, same Q_2

$$Q_2 = \frac{1.088}{\delta} \qquad \alpha_2 = 1 + 0.19\delta$$

NOTE: f_0 is the center frequency of the over-all amplifier, and B is the over-all 3-db bandwidth.

10-3, respectively) that the tuning of the low stages is the same, but not that of the high stages. In the narrow-band case the bandwidths of corresponding high and low stages are the same, whereas in the asymptotic case the Q's are the same. Thus the equations of Table 10-3 display the correct *geometric* symmetry about f_0. The differences, though, are quite small. Because the intermediate-bandwidth formulas are so easy to use, they may well be used for a narrow-band example as well.

All three of the preceding tables give the data for the maximally flat amplitude response. To obtain the equal-ripple response in the narrow-band case, it is necessary only to multiply the bandwidth of each stage by the factor $\tanh a$, as defined in Chap. 9. For values of $\tanh a$ see Table 10-6.

For the wideband and intermediate cases, the equal-ripple response can be derived from the transformation $p = s + 1/s$. The resulting formulas will be found in Table 10-4.

Table 10-4. Wideband Stagger Tuning (Equal Ripple)

Ripple factor $\tanh a$ as in Eq. (9-18), Fig. 9-8, and Table 10-6

$$\delta = \frac{B}{f_0} = \frac{\text{desired over-all 3-db bandwidth}}{\text{desired center frequency}}$$

---

1. Staggered pair ($n = 2$)

   Two stages tuned to $f_0\alpha$ and f_0/α, having same Q

   $$Q^2 = \frac{2}{4 + R^2 - \sqrt{(R^2 - 4)^2 + 8\delta^2}}$$

   $$\left(\alpha - \frac{1}{\alpha}\right)^2 + \frac{1}{Q^2} = R^2$$

   $$R^2 = \frac{\delta^2}{2}(1 + \tanh^2 a)$$

2. Staggered triple ($n = 3$)

   One stage tuned to f_0 with $Q = 1/(\delta \tanh a)$

   Two stages tuned to $f_0\alpha$ and f_0/α, same Q

   $$Q^2 = \frac{2}{4 + R^2 - \sqrt{(R^2 - 4)^2 + 12\delta^2}}$$

   $$\left(\alpha - \frac{1}{\alpha}\right)^2 + \frac{1}{Q^2} = R^2$$

   $$R^2 = \frac{\delta^2}{4}(3 + \tanh^2 a)$$

NOTE: These formulas reduce to those of Table 10-2 (maximally flat) when the ripple is reduced to zero, making $\tanh a = 1$. Thus this table includes not only equal ripple but also maximally flat as a special case.

10-4 Cascades of *n*-uples. A stagger-tuned cascade of *n* stages is called an "*n*-*uple*," e.g., a quadruple for $n = 4$, but an *n*-uple where *n* is arbitrary.

Now, an *n*-uple can be designed according to the principles which have been set forth in the preceding pages for as many stages as desired. For practical reasons, however, the order *n* of the *n*-uples usually is not higher than 3 or 4 (occasionally perhaps as high as 6). Hence, if more than three or four stages are required to obtain the required over-all gain, it is customary to cascade several *n*-uples. It turns out, of course, that, when identical triples are cascaded, the bandwidth shrinks. The situation for cascaded identical single-tuned stages was given in Eq. (7-15). For the stagger-tuned case the gain magnitude in terms of the normalized frequency variable *x* is for *one* *n*-uple,

$$|A_n(x)| = \frac{1}{\sqrt{1 + x^{2n}}} \tag{10-1}$$

The gain for *m* such *n*-uples cascaded is

$$|A_{nm}(x)| = \left(\frac{1}{\sqrt{1 + x^{2n}}}\right)^m \tag{10-2}$$

Setting the over-all gain at $1/\sqrt{2}$ and solving for *x* gives the half bandwidth of the resulting amplifier, and also the bandwidth shrinkage factor since the half bandwidth for one *n*-uple is unity.

$$\frac{\text{Bandwidth of } m \text{ staggered } n\text{-uples}}{\text{Bandwidth of one staggered } n\text{-uple}} = (2^{1/m} - 1)^{1/2n} \tag{10-3}$$

Implicit in this equation is the fact that the higher the order *n* the slower will be the bandwidth shrinkage as *n*-uples are connected in cascade.

10-5 Gain-Bandwidth Factor. This useful figure of merit was defined in Chap. 7; it provides a means of comparing the amount of gain-bandwidth realizable from tubes having the same g_m/C when used with various interstages.

For one single-tuned stage, the gain-bandwidth factor was shown to be, by definition, equal to unity. For a cascade of *m* identical single-tuned stages, the gain-bandwidth factor is

$$\text{GBF} = \sqrt{2^{1/m} - 1} \tag{7-15}$$

This is the same as Eq. (10-3) with $n = 1$. Thus the gain-bandwidth factor diminishes as the number of stages increases.

In marked contrast to this, the gain-bandwidth factor of n single-tuned stages arranged in a staggered n-uple (maximally flat) is always 1.00, regardless of n. Thus, stages can be added indefinitely without loss of gain-bandwidth factor. The proof that the gain-bandwidth factor is always unity in an n-uple is simple for the narrow-band case. Here the gain of each stage *at band center* f_0 is the same regardless of n because the distance from f_0 to the pole characterizing one stage is the same as the distance to the pole of any other stage because of the circular pole locus. By use of Eq. (8-6), the gain of any stage at midband is

$$|A(j\omega_0)| = \frac{g_m}{2C}\frac{1}{|j\omega - p_1|} = \frac{g_m}{2C}\frac{1}{B_r/2} \tag{10-4}$$

In Eq. (10-4) B_r is the *over-all* bandwidth in radians per second. Substituting in the definition of GBF [Eq. (7-16)], we obtain

$$\text{GBF} \triangleq \frac{(A_n)^{1/n}B_r}{g_m/C} = \left[\left(\frac{g_m}{CB_r}\right)^n\right]^{1/n}\frac{B_r}{g_m/C} = 1 \tag{10-5}$$

The corresponding proof that the GBF is unity in the wideband case is complicated by the fact that all stages do not then have the same gain at band center. The easiest way of proceeding is to consider gain in the low-pass plane (the p plane) where the poles are equidistant from the band center ($\omega = 0$). Since the transformation from the p to the s plane is conformal, the gain produced by a set of singularities in the p plane is the same as that produced by the corresponding set in the s plane. We have learned, however, that a complex pole in the p plane does not transform into a conjugate pair in the s plane; consequently we must consider a conjugate pair of poles in the p plane which transform into two conjugate pairs in the s plane (or *two* single-tuned stages). The gain of such a pair of stages will be equal to that of any other pair of stages with poles lying on the same circle in the p plane. Also, the pair of stages will have the square of the gain of a stage whose pole lies on the intersection of the σ axis and the circle. Therefore, in the wideband case the geometric mean of the gain of pairs of stages having the same Q's will equal the gain of a centered stage. Since the geometric mean of the over-all gain is used in the equation for determining the GBF, the latter will again be unity. To illustrate the situation described, the gains (in decibels) of the three stages of a relatively wideband triple ($\delta = 2$) are plotted on a logarithmic frequency scale in Fig. 10-7. (The arithmetic symmetry of the curves on the logarithmic frequency scale indicates geometric symmetry on a linear frequency scale.) Note that the gain (decibels) of the low- and high-frequency stages is equally above and below the mid-frequency stage at band center.

When it becomes necessary to cascade staggered n-uples, the gain-bandwidth factor is given by Eq. (10-3). The numerical values which result from these formulas are both of interest and of practical importance. Table 10-5 shows the gain-bandwidth factors obtained with N tubes ($N = mn$) from $N = 1$ to 9 used in various combinations.

To use the table, refer back to Eq. (7-16), where the gain-bandwidth factor was defined. As an example, suppose that one wishes to compare nine tubes used as identical stages or as three staggered triples. The ratio of over-all bandwidth obtainable for the same over-all gain would be

$$\frac{0.80}{0.28} = 2.86$$

On the other hand, the ratio of gain obtainable for the same over-all bandwidth is

$$\left(\frac{0.80}{0.28}\right)^9 = 1.3 \times 10^4$$

Comparison of the gains for equal bandwidths emphasizes the fact that an amplifier made up of a large number of identical stages is indeed an in-

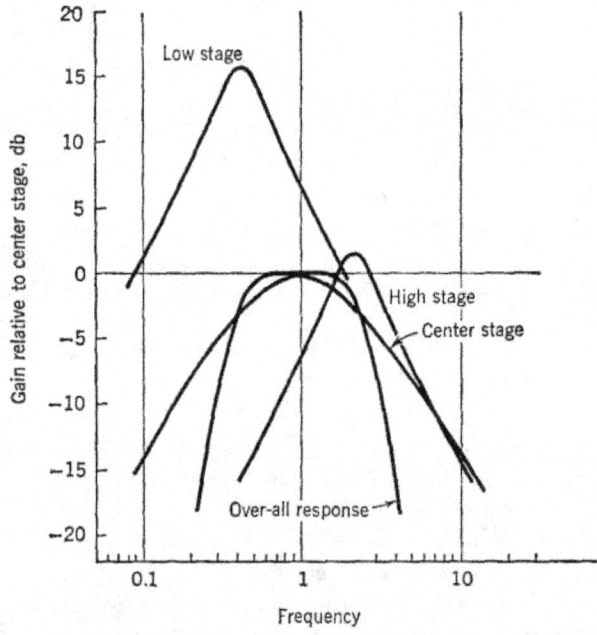

Fig. 10-7 Gain of the three stages in a maximally flat triple, assuming g_m/C identical for each stage. $B/f_0 = 2$.

efficient device. The contrast for the case of equal gain is not so startling, but the same phenomenon is at work.

The gain-bandwidth factor for an n-uple adjusted for an equal-ripple type of response is greater than unity because for the same bandwidth the poles are located closer to the $j\omega$ axis than in the maximally flat case. Since

Table 10-5.   Gain-Bandwidth Factors

No. of tubes, N ($= mn$)	Identical stages $\sqrt{2^{1/n} - 1}$	Cascaded n-uples $(2^{1/m} - 1)^{1/2n}$		
		m pairs $\sqrt[4]{2^{1/m} - 1}$	m triples $\sqrt[6]{2^{1/m} - 1}$	m quadruples $\sqrt[8]{2^{1/m} - 1}$
1	1.00			
2	0.64	1.00 ($m = 1$)		
3	0.51	...	1.00 ($m = 1$)	
4	0.44	0.80 ($m = 2$)	...	1.00 ($m = 1$)
5	0.39			
6	0.35	0.71 ($m = 3$)	0.86 ($m = 2$)	
7	0.32			
8	0.30	0.66 ($m = 4$)	...	0.90 ($m = 2$)
9	0.28	...	0.80 ($m = 3$)	

the poles are closer to the $j\omega$ axis for larger amounts of ripple, we would expect a larger GBF, as is shown in Table 10-6. As an example, the gain of a quadruple with 0.2 db ripple is about 10 db greater than the corresponding maximally flat quadruple.

The GBF for combinations of equal-ripple n-uples can be approximated by taking the GBF for the combination from Table 10-5 and multiplying it by the GBF for the equal-ripple n-uple (Table 10-6). For example, the GBF for three cascaded pairs with 0.1 db ripple per pair is $(1.07)(0.71) = 0.76$; the *over-all* ripple would be $(3)(0.1) = 0.3$ db.

10-6 Practical Design Information. The attempt in this discussion has been to provide the underlying theory and some physical intuition in the matter of how stagger tuning works. Active workers in a field such as

Table 10-6. Equal-ripple Function

Values of tanh a and gain-bandwidth factor as a function
of the ripple in decibels, for pairs, triples, and quadruples

Ripple, db	Number of tubes, n					
	$n = 2$ (pairs)		$n = 3$ (triples)		$n = 4$ (quadruples)	
	tanh a	GBF	tanh a	GBF	tanh a	GBF
0.01	0.953	1.025	0.846	1.080	0.731	1.155
0.03	0.920	1.040	0.786	1.120	0.662	1.202
0.05	0.898	1.050	0.750	1.140	0.623	1.230
0.07	0.880	1.060	0.725	1.160	0.597	1.251
0.10	0.859	1.070	0.696	1.182	0.567	1.275
0.20	0.806	1.100	0.631	1.230	0.505	1.325
0.30	0.767	1.120	0.588	1.265	0.467	1.355
0.40	0.736	1.135	0.556	1.290	0.439	1.380
0.50	0.709	1.150	0.524	1.320	0.416	1.410

this inevitably produce helpful graphs, tables, nomograms, and the like, to shorten the time required for numerical designs.[1]

Design Example. It is instructive to carry through an example, especially to show how certain graphical aids can be devised and put to use.

Suppose that in a given system there is needed a bandpass amplifier to provide a gain (voltage amplification) of 60 db with a bandwidth of 7 Mc. The center frequency is of no consequence in the initial phases of the design procedure and in fact is needed only when one comes to calculate the interstage inductances. Also assume that system considerations, such as reducing the number of tube types, limit the available tubes to the 6AK5 and 6AH6. Measurements on the wiring situation in which the tubes will be

---

[1] R. C. Wittenberg, Broad-banding by Stagger Tuning, *Electronics*, vol. 25, pp. 118–121, February, 1952; E. R. Jenkins, Stagger Gain Calculator, *Tele-Tech*, vol. 9, p. 29, April, 1950; B. A. Wightman, A Graphical Method for Determining the Number and Order (N) of N-uples in Stagger Tuned Amplifier Design, *Natl. Research Council Can. Rept.* ERA-212, December, 1951.

The basic work of Wallman is usually quite adequate for the maximally flat case; in particular, chaps. 4 and 8 in the book "Vacuum Tube Amplifiers" (Valley and Wallman, *op. cit.*) contain many practical details. For the equal-ripple case, the papers by Baum (*op. cit.*) and by Wittenberg may prove useful.

mounted show that a total interstage wiring capacitance of 5 pf can be expected.

The design questions which we shall answer here are: (1) Which tube should be used? (2) Which combinations of single-tuned stages will meet the gain-bandwidth requirements? (3) Which will require the fewest tubes? (4) Which will give the best selectivity ratio?

Choice of Tube. Shown in Fig. 10-8 is a plot of Eq. (7-7) for a number of currently used tubes, including the two allowable types for the example at

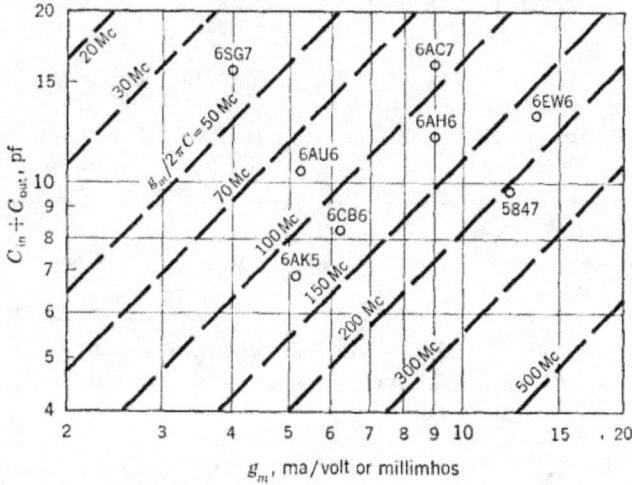

Fig. 10-8 Chart for tube selection. (Each tube should be displaced vertically by the amount of the interstage wiring capacity.)

hand. The points on the graph do not include wiring capacitance, however; so each of the two must be translated upward by 5 pf. Because of the logarithmic capacitance scale, the 6AH6 is displaced the least and hence proves to be the best choice. The actual interstage capacitance is further increased by the change in input capacitance occurring when plate current flows in the tube, as noted in Sec. 2-5. The output capacitance is for all practical purposes unchanged by varying the operating point. Table 10-7 gives the tube-manual value of C_{in}, together with the measured value with the tube cold (heater off) and the measured value at normal plate current. (The values are measured on a small number of tubes.) Note that the tube-manual values are in general 30 to 50 per cent too low. Thus the actual gain-bandwidth for the 6AH6 with this wiring capacitance is 72 Mc.

Choice of Stagger Combinations. The gain-bandwidth performance can be displayed conveniently in the graphical presentation devised by Wight-

Table 10-7.   Comparison of Cold Input Capacitance and the Normal Operating Capacitance for Typical Tubes

Tube	Measured C_{in} (cold)		Measured C_{in} (at normal I_b)	% change
5654 (6AK5)	4.1	(4)	5.3	+29
6AH6V	9.8	(10)	12.8	+31
6AU6	5.8	(5.5)	8.7	+50
6EW6	9.9	(10)	14.2	+43
6BA6	5.8	(5.5)	7.7	+33
6AG5	6.5	(6.6)	8.5	+31
E180F (6688)		(7.5)	(11.1)	+48

NOTES: Values in parentheses are from tube manual.

The measured values shown were obtained at 1 Mc and do not represent the average of a large number of tubes.

man, as shown in Fig. 10-9. Note the similarity of construction to Fig. 6-16. The curves are a plot of the gain-bandwidth factor relationships [see Eq. (7-16)],

$$20 \log A_n = \text{db gain} = -20n \log \frac{B}{g_m/2\pi C} + 20n \log \text{GBF} \qquad (10\text{-}6)$$

The "normalized bandwidth" is obtained by taking the actual over-all bandwidth required of the amplifier (7 Mc, in our example) and dividing it by the $g_m/2\pi C$ of the tube, which including wiring capacitance equals 72 Mc. Thus, for the example at hand, the normalized bandwidth is $7/72$, or about 0.097. All curves in Fig. 10-9 which cross the vertical line through 0.097 at a level of 60 db or greater will meet the requirements. The lowest of these are:

5 × 1 (five identical stages)
1 × 3 (one staggered triple; three tubes)
2 × 2 (two staggered pairs; four tubes)
1 × 4 (one staggered quadruple; four tubes)

Fewest Tubes. It is evident that the staggered triple requires the fewest tubes. This may not be the practical answer, however. System requirements may favor the better selectivity ratio of the 2 × 2 or 1 × 4 in spite of the extra tube (see below). Also, practical consideration of the variability of tube characteristics may call for a margin of safety, thus favoring the other combinations.

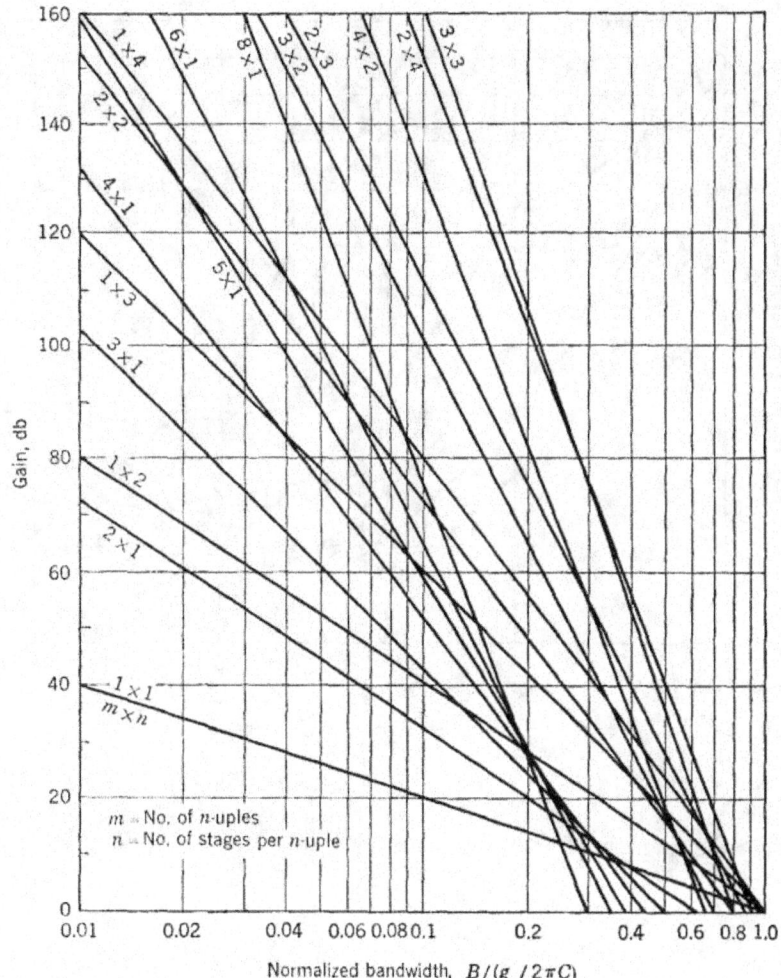

Fig. 10-9 Gain-bandwidth chart for single-tuned stages. (*n*-uples are maximally flat.) [*From B. A. Wightman, A Graphical Method for Determining the Number and Order (n) of n-uples in Stagger Tuned Amplifier Design, Natl. Research Council Can., Rept. ERA-212, December, 1951.*]

Selectivity Ratio. The selectivity ratio for a cascade of identical stages has already been given in Eq. (7-20) and for the maximally flat function corresponding to a single *n*-uple in Eq. (9-5). For a cascade of *m* identical *n*-uples, the selectivity ratio is

$$\text{Selectivity ratio} = \left[\frac{(10^6)^{1/m} - 1}{4^{1/m} - 1} \right]^{1/2n} \qquad (10\text{-}7)$$

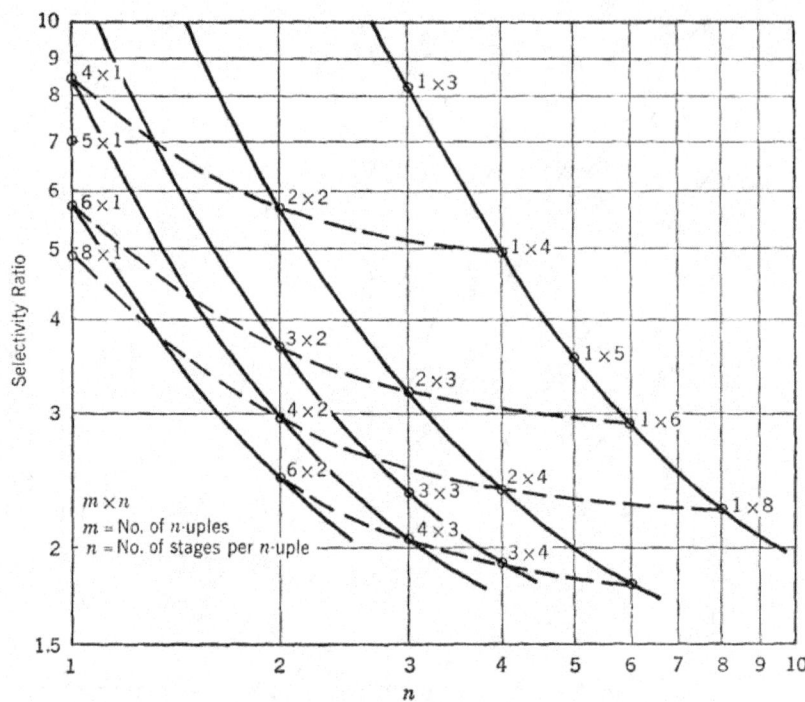

Fig. 10-10 Selectivity ratio of stagger-tuned stages.

It is instructive to display these relationships graphically, as in Fig. 10-10, which permits one to see quickly the relative merits of various combinations.

For the example at hand, the selectivity ratios for the three alternatives are:

$m \times n$	Selectivity ratio
1×3	8.4
5×1	7.0
2×2	5.7
1×4	4.9

System requirements will have to govern the choice here. It would appear that, in going to four tubes, one might as well use the 1×4 combination and achieve the better selectivity ratio and higher gain. Here again, however, the system situation might favor the 2×2, since it has only two different interstage types to manufacture and to align.

PROBLEMS

10-1. The following questions pertain to a staggered triple composed of three cascaded single-tuned circuits (stages) arranged to give a maximally flat amplitude response. Assume that the bandwidth B_r is sufficiently small compared with the center frequency ω_0 so that the *narrow-band approximation* may be used.

a. In terms of g_m/C and B_r, how much gain and phase shift are provided by *each* stage at ω_0? (Note that ω_0 is the *over-all* center frequency.)

b. What is the maximum gain of each stage, and at what frequency does it occur?

c. What are the bandwidth and Q of each stage?

d. Suppose that the g_m of the center stage falls to one-half of its normal value. Why does this *not* result in a dip in the center of the response curve?

e. Show that, for small fractional bandwidths, the wideband transformation becomes approximately $p = 2(s - j\omega_0)$. In terms of p, g_m/C, and B_c, write the gain function $A(p)$ for the over-all triple.

10-2. Derive the relationship which gives the gain-bandwidth *factor* of a stagger-tuned pair and triple for the equal-ripple case. The expression should be a function of $\tanh a$. Check your expression with Table 10-6 by calculating the GBF for a ripple of 0.1 db.

10-3. Engineer Jones (a fictitious character!) hears about stagger tuning and the benefits it produces in giving a flat passband. He decides to try it "on his own" and reasons that one should take several single-tuned stages of equal bandwidth and space them in frequency at equal intervals. He tries the scheme with three stages. Now it turns out that his choice of bandwidth and frequency separation places the poles of the over-all gain function as shown in Fig. P10-3. (Jones is, of course, blissfully unaware of this.) He finds that his amplitude-response curve has a pronounced peak near the frequency of the center stage; so he reasons that the center stage has too much gain. This he corrects by reducing the tube g_m of this stage, but the solution does not work. Why not? By graphical means compute the gain curve that Jones obtained.

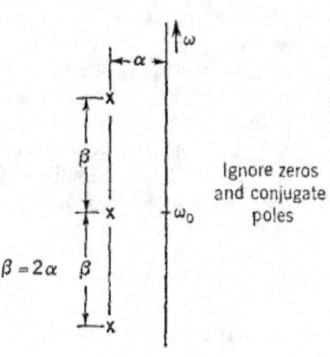

Fig. P10-3

10-4. This is a typical design problem which may be solved in the main by use of the design charts in the chapter.

a. Suppose that the tubes available for a given amplifier are limited to the 6AK5, 6CB6, or 6AH6. Choose the tube which will give the best gain-bandwidth product if the stage wiring capacitance is 5 pf. Use the chosen tube for the following parts of the problem.

b. The tube having been chosen, the next problem is to choose a type and number of interstage networks to meet the following requirements: gain $\geqq$ 80 db; bandwidth = 7 Mc. Which arrangement of single-tuned stages will meet these requirements with the fewest tubes?

c. Which of the "fewest-tube" arrangements gives the best selectivity ratio?

d. Which of the possible arrangements might be the best engineering compromise?

10-5. Find the gain-bandwidth *factor* and the selectivity ratio for two single-tuned stages tuned to give a maximally flat *delay* characteristic. (Narrow-band conditions may be assumed to prevail.)

10-6. An amplifier using 6EW6 tubes (with 6 pf of stray interstage capacitance) is to be built to give a very flat passband for telephone repeater work. Assume a band 5 Mc wide which must have a gain variation of no more than 0.5 db. The amplifier is to be made as a quadruple (i.e., four staggered stages). Compare the over-all gains obtainable from two amplifiers which would meet these specifications. The first is maximally flat with a 0.5-db bandwidth of 5 Mc. The second is an equal-ripple type with a 0.5-db ripple and a "ripple approximation band" of 5 Mc (see Fig. 9-10; the approximation band referred to is B_1).   Assume for simplicity that the capacitances on all stages including the last are equal.

10-7. An amplifier is to be a pair of stages tuned to give a maximally flat response. The lower band-edge frequency (where the relative gain is -3 db) is to be $\omega = 3$ radians/sec.   The upper band edge is at $\omega = 20$ radians/sec.

a. What is the "center" frequency ω_0 where the gain is maximum?

b. Where are the poles and zeros of the gain function of the resulting amplifier? (Find by using the lowpass-to-bandpass transformation.)

c. What are the center frequency and bandwidth of each stage?   Check your results by use of either Table 10-2 or Fig. 10-4.

10-8. An amplifier using single-tuned interstages is to be designed to give an equal-ripple response.   Four stagger-tuned stages are to be used to produce a 3-Mc bandwidth (3 db) centered at 60 Mc with a ripple of 0.1 db.   Assume that the tubes have a total capacitance $(C_{\text{out}} + C_{\text{in}} + C_w)$ of 15 pf and a g_m of 5,000 μmhos.

a. Compute the pole positions which are to be realized by the interstage networks.

b. What is the gain of the amplifier at 60 Mc?

c. What is the bandwidth over which the 0.1-db tolerance is maintained?

d. In decibels what is the increase in gain afforded by the equal-ripple case compared with a maximally flat amplifier of the same 3-db bandwidth?

10-9. A two-stage amplifier with linear-phase characteristics is desired.   Single-tuned circuits are to be used.

a. Assume that the over-all bandwidth is 10 Mc with a center frequency of 20 Mc. What are the stage resonant frequencies, and what is the Q of each stage?

b. What are the element values for each stage if $C = 15$ pf and $g_m = 5,000$ μmhos?

c. What is the over-all gain at 20 Mc?

11

The Double-tuned Interstage

In this chapter will be presented the double-tuned circuit as an alternative means of realizing amplifier gain functions, in contrast to the single-tuned interstage network employed thus far. Nothing will be added to the approximation problem: maximally flat and equal-ripple responses are still our most useful approximations to constant gain in the passband. We know what sort of complex gain functions will produce these responses, i.e., poles on a circle or on an ellipse, respectively. The task then is to find what sort of pole-zero arrangement comes out of the double-tuned circuit and how the pole coordinates are related to the circuit parameters.

The double-tuned circuit, sometimes called transformer coupling, is a logical extension from the single-tuned circuit. It represents the general process of adding more circuit complexity in exchange for improved performance. The improvement is of two kinds: a better gain-bandwidth factor and better selectivity, better in that one stage provides a two-pole response, instead of a one-pole as with the single-tuned stage.

There are several equivalent forms of the double-tuned circuit, illustrated in Fig. 11-1, the most common being the inductive coupling and the pi equivalent.

We shall confine the analytical discussion to the inductively coupled arrangement, specified by the primary inductance L_1, the secondary inductance L_2, and a coupling coefficient k, defined in the conventional way as

$$k = \frac{M}{\sqrt{L_1 L_2}} \tag{11-1}$$

The other forms of coupled circuit give identical performance if the follow-

241

ing equivalences are observed.   For the pi equivalent

$$L_{1\pi} = \frac{L_1 L_2 - M^2}{L_2 \pm M} \qquad L_{2\pi} = \frac{L_1 L_2 - M^2}{\mp M} \qquad L_{3\pi} = \frac{L_1 L_2 - M^2}{L_1 \pm M}$$

where M is the mutual inductance between the two coils L_1 and L_2.   For the autotransformer the inductances L_1 and L_2 are the open-circuit induct-

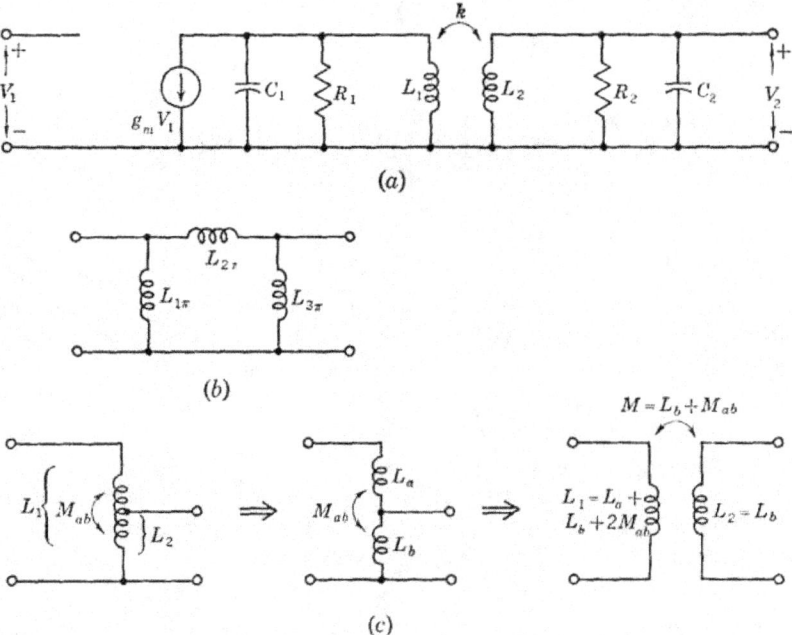

Fig. 11-1 Forms of an inductively coupled double-tuned interstage.   (a) Inductive.   (b) Pi equivalent.   (c) Autotransformer equivalent.

ances—i.e., the values measured with the opposite side of the transformer open-circuited.   The equivalences are given in the figure.   Note that a higher mutual inductance can be obtained for the autotransformer because the effective M is aided by the inductance of the lower coil, L_2.

We can further define primary and secondary resonant frequencies and Q's.   These quantities are merely definitions and are not observable resonant frequencies unless $k \to 0$.

$$\omega_1 \overset{\Delta}{=} \frac{1}{\sqrt{L_1 C_1}}$$

$$\omega_2 \overset{\Delta}{=} \frac{1}{\sqrt{L_2 C_2}}$$

(11-2)

$$Q_1 \triangleq \frac{R_1}{\omega_1 L_1}$$

$$Q_2 \triangleq \frac{R_2}{\omega_2 L_2}$$

(11-3)

We are interested in the gain of the double-tuned amplifier stage, namely, V_2/V_1 as a function of the complex variable p. This function can be obtained by direct analysis of the circuits of Fig. 11-1 with the following form of result:

$$A(p) = \frac{V_2}{V_1} = \frac{g_m k \omega_1 \omega_2}{(1 - k^2)\sqrt{C_1 C_2}} \frac{p}{p^4 + Ap^3 + Bp^2 + Cp + D}$$

(11-4)

where

$$A = \frac{\omega_1}{Q_1} + \frac{\omega_2}{Q_2} \qquad C = \frac{\omega_1{}^2 \omega_2}{Q_2(1 - k^2)}\left(1 + \frac{\omega_2 Q_2}{\omega_1 Q_1}\right)$$

$$B = \frac{\omega_1 \omega_2}{Q_1 Q_2} + \frac{\omega_1{}^2 + \omega_2{}^2}{1 - k^2} \qquad D = \frac{\omega_1{}^2 \omega_2{}^2}{1 - k^2}$$

We can see at once from Eq. (11-4) that there is one zero (at the origin) and four poles. In the cases of practical interest, the poles will appear in conjugate pairs, as depicted in Fig. 11-2. This figure suggests that the double-tuned circuit will be similar in its gain function to a pair of single-tuned circuits in cascade, with the exception (an important exception in the wideband case) that the latter would have two zeros at the origin instead of one.

There are several cases of practical importance, which we subdivide first into the narrow-band and wideband situations. Let us concentrate on the narrow-band case first. As before, we distinguish the narrow-band case by ignoring the conjugate poles (third quadrant) and any zeros at the origin. Thus Fig. 11-2 becomes Fig. 11-3.

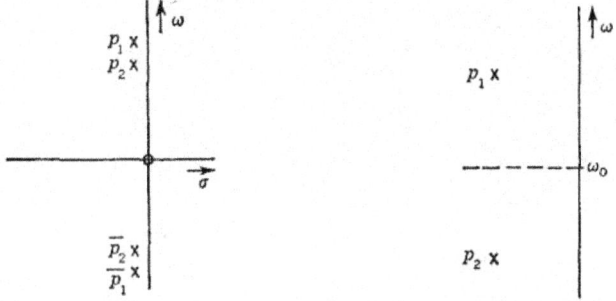

Fig. 11-2 Pole positions for the interstages of Fig. 11-1.

Fig. 11-3 Narrow-band approximation of Fig. 11-2.

Next we consider two cases: (1) *equal Q*, that is, $Q_1 = Q_2$, and (2) *one Q infinite*, for example, $Q_1 = \infty$. Because the denominator of Eq. (11-4) is a polynominal of fourth power, we cannot follow our usual procedure of factoring the denominator in general terms and then placing the resulting poles at the desired positions. Instead, to obtain a solution in closed form, we shall arbitrarily pick a set of pole positions which can be realized by a fourth-degree polynomial, form the polynomial giving the set of pole positions, and then equate the coefficients of like powers in that polynomial and Eq. (11-4). For either maximally flat or equal-ripple gain functions, the poles p_1 and p_2 in Fig. 11-3 should have equal damping (equal σ); hence let us assume the poles are located at

$$p_1 = -\sigma + j\Omega_1$$

$$p_2 = -\sigma + j\Omega_2$$

Let us further assume that the natural resonant frequencies of the transformer are equal; that is, $\omega_1 = \omega_2 = 1/\sqrt{L_1 C_1} = 1/\sqrt{L_2 C_2}$. The latter assumption may seem unduly restrictive, but even with the restrictions on ω and Q there are sufficient degrees of freedom left to realize the narrow-band case. The gain function for the assumed pole positions is

$$A(p) = \frac{p}{(p + \sigma + j\Omega_1)(p + \sigma - j\Omega_1)(p + \sigma + j\Omega_2)(p + \sigma - j\Omega_2)}$$

$$= \frac{p}{p^4 + p^3(4\sigma) + p^2(6\sigma^2 + \Omega_1{}^2 + \Omega_2{}^2) + p(4\sigma^3 + 2\sigma\Omega_1{}^2 + 2\sigma\Omega_2{}^2) + (\sigma^2 + \Omega_1{}^2)(\sigma^2 + \Omega_2{}^2)} \tag{11-5}$$

Now the coefficients of Eqs. (11-4) and (11-5) may be equated and the above assumptions included. (Let $\omega_1 = \omega_2 \stackrel{\Delta}{=} \omega_0$.)

$$A = \frac{2\omega_0}{Q} = 4\sigma \tag{11-6}$$

$$B = \frac{\omega_0{}^2}{Q^2} + \frac{2\omega_0{}^2}{1 - k^2} = 6\sigma^2 + \Omega_1{}^2 + \Omega_2{}^2 \tag{11-7}$$

$$C = \frac{2\omega_0{}^3}{Q(1 - k^2)} = 4\sigma^3 + 2\sigma\Omega_1{}^2 + 2\sigma\Omega_2{}^2 \tag{11-8}$$

$$D = \frac{\omega_0{}^4}{1 - k^2} = \sigma^4 + \sigma^2(\Omega_1{}^2 + \Omega_2{}^2) + \Omega_1{}^2\Omega_2{}^2 \tag{11-9}$$

These equations may now be solved for the pole positions in terms of the actual element values.

$$\sigma = \frac{\omega_0}{2Q} \tag{11-10}$$

$$\Omega_1 = \pm\omega_0 \sqrt{\frac{1}{1-k} - \frac{1}{4Q^2}} \tag{11-11}$$

$$\Omega_2 = \pm\omega_0 \sqrt{\frac{1}{1+k} - \frac{1}{4Q^2}} \tag{11-12}$$

Up to this point no assumptions have been made restricting the solution to narrow-band, although the pole positions have been chosen with the narrow-band situation in mind. The solution may be considerably simplified at this point by noting that in a narrow-band case Q will be high and k will be small. Hence, for $Q \geq 5$ and $k^2 \ll 1$, Eqs. (11-11) and (11-12) may be approximated by

$$\Omega_1,\ \Omega_2 \cong \pm\omega_0 \sqrt{\frac{1}{1\pm k}} \cong \pm\omega_0 \left(1 \pm \frac{k}{2}\right) \tag{11-13}$$

Therefore the gain function in the narrow-band case may be written in the following simplified form:

$$A(p) = \frac{g_m k \omega_0}{4\sqrt{C_1 C_2}} \frac{1}{(p - p_1)(p - p_2)} \tag{11-14}$$

$$p_1,\ p_2 = \frac{-\omega_0}{2Q} + j\omega_0 \left(1 \pm \frac{k}{2}\right) \qquad Q = Q_1 = Q_2 \qquad k^2 \ll 1 \tag{11-15}$$

The constant in Eq. (11-14) is modified by the effects of the neglected poles in the lower half plane and the zero at the origin [cf. the derivation of Eq. (8-6)].

The locations of the poles are shown in Fig. 11-4. Notice that, if $k = 0$, the two poles are superimposed at $-(\omega_0/2Q) + j\omega_0$; then, as k is increased, the poles separate in the vertical direction but maintain the same distance from the $j\omega$ axis. This increasing separation increases also the bandwidth (although the passband shape will vary); hence larger k is associated with larger bandwidths. Or, conversely, narrow bands require only small k, which is the justification for the assumption in Eq. (11-13) that $k^2 \ll 1$.

We now have control over the pole locations in terms of the circuit parameters ω_0, k, and Q. It is hence a straightforward task to synthesize a maximally flat or equal-ripple response. For instance, Fig. 11-5 illustrates the condition for maximal flatness with one double-tuned amplifier stage, i.e., two poles.

From Fig. 11-5 the following relationships can be seen:

$$\frac{\omega_0 k}{2} = \frac{\omega_0}{2Q} \tag{11-16}$$

$$\frac{\omega_0}{2Q} = \frac{\text{circle radius}}{\sqrt{2}} = \frac{B_r/2}{\sqrt{2}} \tag{11-17}$$

or

$$k = \frac{1}{Q} \tag{11-18}$$

$$k = \frac{1}{\sqrt{2}} \frac{B_r}{\omega_0} = \frac{1}{\sqrt{2}} \frac{B}{f_0} \qquad B_r \text{ in radians/sec, } B \text{ in cps} \tag{11-19}$$

The relationships in Eq. (11-19) are definitive for maximal flatness and are the ones usually found in the handbooks for the proper adjustment of a double-tuned circuit. The coupling coefficient k, so defined, is sometimes known as "critical," or "transitional," coupling, since it is the crossover value between a single-peaked and double-peaked amplitude response.

$$\text{Critical coupling } k_c \stackrel{\Delta}{=} \frac{1}{\sqrt{Q_1 Q_2}} = \frac{1}{Q} \tag{11-18a}$$

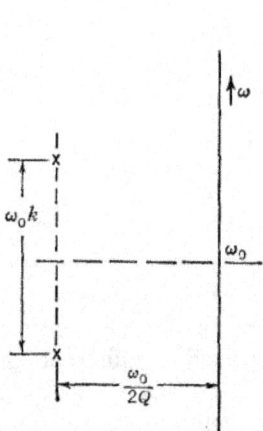

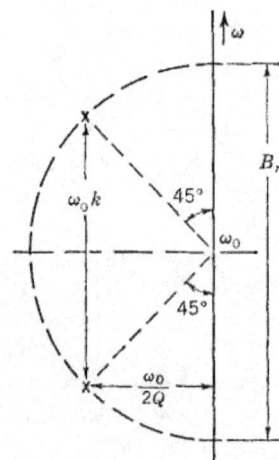

Fig. 11-4 Pole positions for a stage with $Q_1 = Q_2$.

Fig. 11-5 Pole coordinates for a maximally flat stage with $Q_1 = Q_2$.

The equal-ripple response could be similarly realized by increasing the Q in Fig. 11-5 to move the poles horizontally onto an ellipse of the proper minor diameter to yield the desired ripple magnitude.

Now that it has been shown that a desired amplitude response can be realized, what about the gain-bandwidth factor? If one takes the value of k defined by Eq. (11-19) and uses it in Eq. (11-14) to solve for the gain at band center ($p = j\omega_0$) and multiplies this by the bandwidth B (in cycles per second), which is the circle diameter B_r divided by 2π, the following obtains:

$$|A(j\omega_0)|B = \frac{g_m}{2\pi(2\sqrt{C_1 C_2})} \sqrt{2} \tag{11-20}$$

This is the gain-bandwidth product, and it contains a term dependent mainly upon the tube, namely, $g_m/2\pi(2\sqrt{C_1 C_2})$. This quantity corresponds to $g_m/2\pi(C_1 + C_2)$, which was the gain-bandwidth product of the single-tuned stage, and is very closely the same unless C_1 is very different from C_2 (the difference is only 6 per cent for a 2:1 ratio). The new form of this factor is indeed the gain-bandwidth product for the single-tuned circuit if an ideal transformer is included. The important fact is that the double-tuned circuit, with equal primary and secondary Q, is better by (at least) $\sqrt{2}$, and hence the gain-bandwidth *factor* of the circuit is $\sqrt{2}$. (Note that the GBF will be even larger for the equal-ripple case.)

The next case of interest is that in which one of the Q's is infinite. This condition can only be approximated in practice, since the primary is always loaded by the plate resistance of one tube and the secondary by the input conductance of the other. Both primary and secondary have inevitable circuit losses. Nevertheless, in wideband applications the required Q is so low that, in contrast, the unloaded Q of either primary or secondary (primary in particular, at high frequencies) will be so large in proportion that the results predicted by assuming this Q to be infinite are approximated quite closely.

Again we assume the narrow-band conditions to prevail. The gain function [Eq. (11-14)] is the correct one, except that the factors of the denominator polynomial, i.e., the poles, will be different. They are

$$p_1, p_2 = -\frac{\omega_0}{4Q_2} + j\omega_0 \left[1 \pm \frac{1}{2} \sqrt{k^2 - \left(\frac{1}{2Q_2}\right)^2} \right] \tag{11-21}$$

$$Q_1 = \infty \qquad k^2 \ll 1$$

A graphical plot of the pole location for three particular values of k will make evident the differences between this and the equal-Q case. Thus it is seen in Fig. 11-6 that, as k is increased from zero, the poles move together horizontally, then separate vertically. A particular case is that for maximal flatness, illustrated in Fig. 11-7.

$$\frac{\omega_0 k}{2} = \sqrt{2}\,\frac{\omega_0}{4Q_2} \tag{11-22}$$

$$\frac{\omega_0 k}{2} = \text{circle radius} = \frac{B_r}{2} \tag{11-23}$$

$$k = \frac{1}{\sqrt{2}\,Q_2}$$

$$= \frac{B_r}{\omega_0} \quad \text{radians/sec}$$

$$= \frac{B}{f_0} \quad \text{cps} \tag{11-24}$$

The gain-bandwidth factor for this case can be found in the same way as for the equal-Q case, namely, by substitution of the pole coordinates from Fig. 11-7 into the gain function [Eq. (11-14)] to obtain the midband gain and then by multiplying this by the bandwidth. The result is that the gain-bandwidth factor when one Q is infinite is 2.0, instead of $\sqrt{2}$ for equal Q's.

The fact that the case of $Q_1 = \infty$ is $\sqrt{2}$ times better in gain-bandwidth can be detected immediately by a comparison of Figs. 11-5 and 11-7. Suppose that we assume the circles to be the same size in both figures, thus

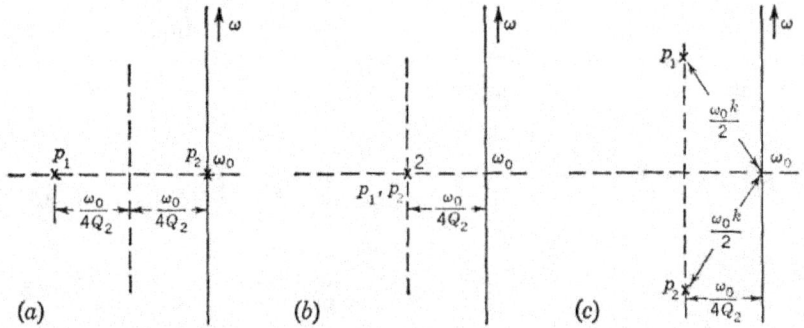

Fig. 11-6 Pole positions as k is changed ($Q_1 = \infty$). (a) $k = 0$. (b) $k = 1/2Q_2$. (c) $k > 1/2Q_2$.

giving the same bandwidth for both cases. The gain function [Eq. (11-14)] will have the same value for the fraction involving p and will differ between the two cases only in the value of k in the scale factor. Comparing the two figures reveals that k is $\sqrt{2}$ times larger in the case of $Q_1 = \infty$, and hence its gain-bandwidth factor must be larger in the same proportion.

Note that, in the $Q_1 = \infty$ case, moving the poles toward the $j\omega$ axis to achieve an equal-ripple response (Sec. 9-3) decreases the value of k as well as bringing the poles closer to the $j\omega$ axis. Hence the GBF for equal ripple is increased less than for the $Q_1 = Q_2$ case, for the same ripple.

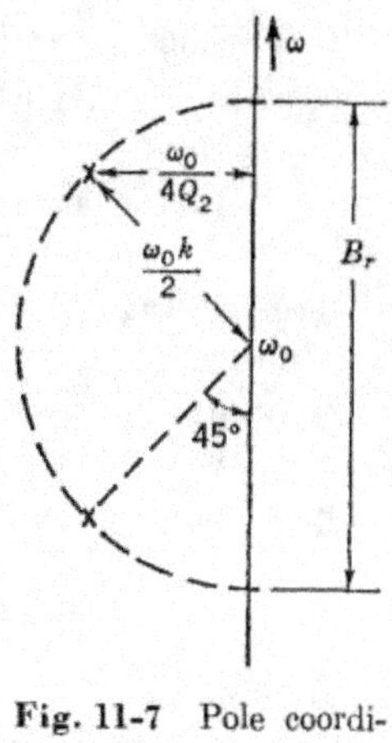

11-1 Cascading of Stages. When maximally flat, double-tuned stages are used in cascade, the bandwidth narrows. The narrowing factor is the same for any two-pole stages and hence must be the same as for staggered pairs, namely, the factor obtained from Eq. (10-3) for $n = 2$,

$$\frac{\text{Bandwidth of } m \text{ stages}}{\text{Bandwidth of one stage}} = (2^{1/m} - 1)^{\frac{1}{4}} \qquad (11\text{-}25)$$

The gain-bandwidth factor for m stages is the value of Eq. (11-25) multiplied by either $\sqrt{2}$ or 2, depending on whether $Q_1 = Q_2$ or $Q_1 = \infty$.

Fig. 11-7 Pole coordinates for a maximally flat stage with $Q_1 = \infty$.

11-2 Stagger Damping. The term "stagger damping" was coined, apparently by Wallman, to describe a form of stagger tuning using double-tuned circuits ($Q_1 = \infty$) in the narrow-band situation. The technique permits the synthesis of $2m$-pole, maximally flat responses with m double-tuned stages in cascade. The stages are tuned identically (that is, $\omega_1 = \omega_2 = \omega_0$) and have equal k's. The staggering is accomplished by varying the stage damping.

The basic principle of stagger damping can be demonstrated readily by means of Fig. 11-8, which shows a two-stage example. The four poles are shown in the proper positions to yield a maximally flat amplitude response of bandwidth B_r. Two of the poles, p_1 and p_2, are assigned to one double-tuned stage. The required k and Q for this stage are computed from Eq. (11-14) or (11-21). The other two poles, p_3 and p_4, are assigned to the second

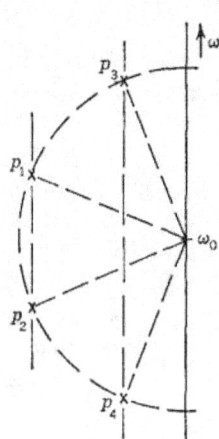

Fig. 11-8 Pole positions for a maximally flat stagger-damped amplifier (two stages) ($Q_1 = \infty$).

stage. Since the radial distance is the same, the k for each stage will be the same and the GBF for the set of m stages will be 2.0. If stages with $Q_1 = Q_2$ are used, the value of both k and Q will vary from stage to stage and computation of the GBF must take into account the different values of k since it appears in the constant multiplying Eq. (11-14). The resulting curves of stage gain vs. frequency are shown in Fig. 11-9.

The technique is obviously not limited to the four-pole case. Any even number ($2m$) of poles can be used, each pair corresponding to one double-tuned stage. The practical limits are excessive Q's and extreme precision of adjustment for the higher-order cases.

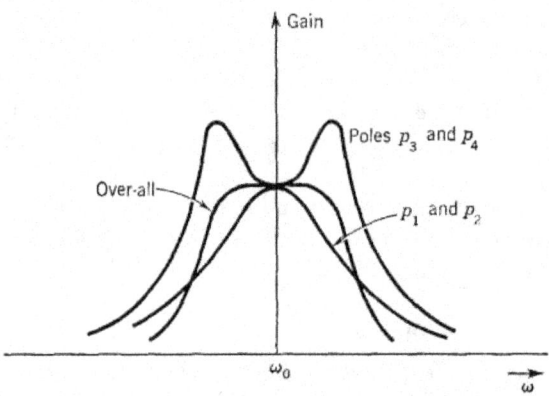

Fig. 11-9 Response of each stage corresponding to the pole positions of Fig. 11-8.

11-3   The Wideband Case. In the wideband situation, account must be taken of the zeros at the origin and the conjugate poles in the third quadrant. When these are brought into consideration, it is evident that the circular locus for the poles to give maximal flatness must be "warped" to counteract—in terms of the potential analogy—the additional charges.

The same technique of conformal mapping, or transformation of variable, can be used as in the case of single-tuned circuits. The details of the transformation must be slightly different, however, because with double-tuned circuits there is only one zero at the origin for each pair of poles in the second quadrant. The proper transformation for the inductively coupled double-tuned circuit to give a maximally flat amplitude response is given by Eq. (11-26) and illustrated by Fig. 11-10.

$$p = s^{1/2}\left(s + \frac{1}{s}\right) \tag{11-26}$$

Several features of the wideband situation can be observed from Fig. 11-10. The pole locations have been drawn to scale and the center frequency properly placed, corresponding to the frequency of maximum response. The 3-db band "edges" are the two points where the pole locus intersects the $j\omega$ axis, as before. The upper 3-db frequency is six times the lower one, thus indicating genuine wideband conditions.

The center frequency ω_0 is very closely the arithmetic mean of the upper and lower 3-db frequencies. In other words, the response curve will be arithmetically symmetrical, even for bandwidths as large as that illustrated (which, by the way, could be described by $\delta = B/f_0 = 1.43$). Moreover,

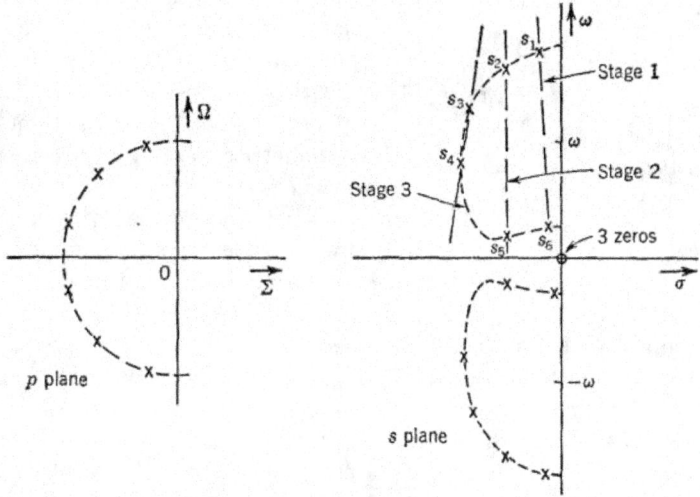

Fig. 11-10 Pole positions for the wideband case.

for bandwidths less than the one illustrated, the pole locus is very closely a circle. The proper interpretation of this situation, in terms of the potential analogy, is that the single negative charge at the origin (for each circuit) is about right to compensate for the two positive charges at a greater distance in the third quadrant. In the single-tuned case there would be two negative charges at the origin for each pair of positive charges, and this would excessively depress the potential near the origin, requiring a greater distortion of the circular locus to compensate.

The transformation given in Eq. (11-26) will not work for the equal-ripple response. The poles must be symmetrically situated about a circular locus in the p plane in order to yield a physically realizable pole set in the s plane. Hence, the elliptical locus required for equal ripple is ruled out. A suitable transformation has been devised by Trautman and Aseltine;[1] it involves elliptic functions and is not easy to use. Their study shows, however, that

[1] See D. L. Trautman and J. A. Aseltine, Equal-ripple Bandpass Amplifiers, *Univ. Calif., Los Angeles, Dept. Eng. Rept.* 51-9, August, 1951.

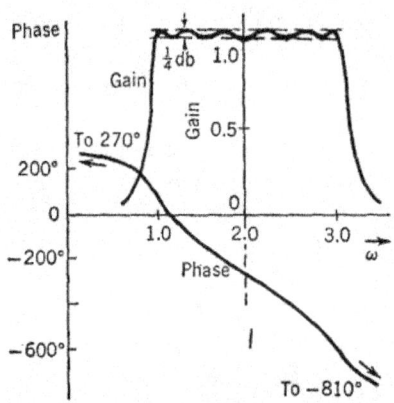

Fig. 11-11 Gain and phase response of a wideband equal-ripple double-tuned amplifier (three stages).

the pole locations are very closely on an ellipse in the s plane, even up to large bandwidths as in the maximally flat case described above. The response has arithmetic symmetry (of amplitude), as illustrated for a typical case in Fig. 11-11. This arithmetic symmetry of the amplitude response may be of advantage with a modulated signal where upper and lower sidebands must receive equal amplification. Notice, however, that the phase response is *not* arithmetically symmetrical about the center frequency.

Thus we conclude that for most practical bandwidths, say $\delta \leq 1.25$, the pole locations for the inductively coupled double-tuned circuit can be determined with adequate accuracy by simply laying out a circle or an ellipse in the s plane, essentially treating the problem as a narrow-band one. *But,* for bandwidths greater than perhaps 0.2, care must be taken in going from pole locations to circuit parameters. Some design charts have been published.[1] The results of the previous derivation of pole positions in terms of element values may be used if the final approximations are not used; i.e., the results of Eqs. (11-10) to (11-12) may be used. A similar process gives the tuning frequencies for the $Q_1 = \infty$ case. In an extremely wideband situation σ_1 will not always equal σ_2 for the poles of one interstage; to obtain this, the restriction $\omega_1 = \omega_2$ must be removed, and the result will be both staggered *tuning* and staggered *damping*.

11-4 The Capacitance-coupled Circuit. The discussion up to now has been confined to the situation where the primary and secondary circuits were inductively coupled together, either with mutual inductance or with the pi or T equivalent (Fig. 11-1). There is an alternative case which is of both theoretical and practical interest. This case is called capacitive (or capacitance) coupling and is illustrated in Fig. 11-12. The only coupling from primary to secondary is through the capacitance C_m. A coupling coefficient can be defined for the network, analogous to the inductively

[1] Maximally flat only. For one circuit, see G. E. Valley, Jr., and H. Wallman (eds.), "Vacuum Tube Amplifiers" (vol. 18, M.I.T. Radiation Laboratory Series), pp. 219–220, McGraw-Hill Book Company, Inc., New York, 1948; for two and three stages, see M. M. McWhorter, The Design, Physical Realization and Transient Response of Double-tuned Amplifiers of Arbitrary Bandwidth, *Electronics Research Lab., Stanford, Tech. Rept.* 58, February, 1953.

coupled circuit,

$$k \triangleq \frac{C_m}{\sqrt{C_1 C_2}} \tag{11-27}$$

In terms of this coupling coefficient, the gain function for the capacitively coupled double-tuned circuit is comparable in form to that of the inductively coupled circuit, as in Eq. (11-4):

$$A(p) = H \frac{p^3}{p^4 + ap^3 + bp^2 + cp + d} \tag{11-28}$$

The principal point of contrast with the inductively coupled case is that here there are three zeros at the origin, as indicated by p^3 in the numerator,

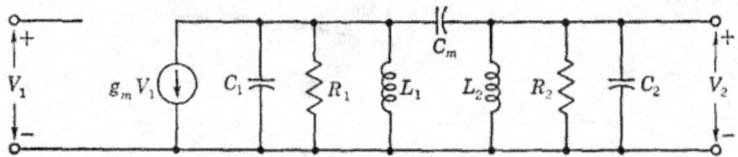

Fig. 11-12 Double-tuned capacitively coupled interstage.

instead of only one in Eq. (11-4). This has a pronounced effect on the shape of the amplitude-response curve in the wideband case.

First, however, in the narrow-band case, which we distinguish by ignoring the zeros at the origin and also the conjugate poles in the third quadrant (Fig. 11-13), there is essentially no difference between capacitance and inductance coupling. The variation of pole locations with k and Q, the gain-bandwidth factor, etc., are all the same in the narrow-band case.

But, in the wideband case, matters are far from equivalent. Speaking in terms of the potential analogy, the three negative charges at the origin

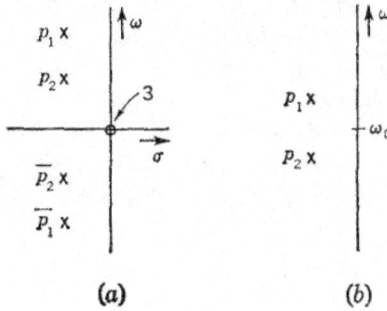

(a)   (b)

Fig. 11-13 Pole positions for a double-tuned capacitively coupled interstage. (a) Wideband. (b) Narrow band.

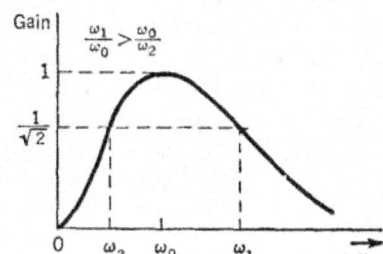

Fig. 11-14 Response of a double-tuned capacitively coupled interstage.

cause the potential on the low-frequency side to fall much more rapidly with frequency than it does on the high-frequency side of band center. The amplitude response thus appears as in Fig. 11-14.

Moreover, the gain-bandwidth factor of the capacitance-coupled circuit is highly unfavorable in wideband situations. For comparison with the inductively coupled case, there are plotted in Fig. 11-15 the curves of gain-bandwidth factor for $Q_1 = \infty$.

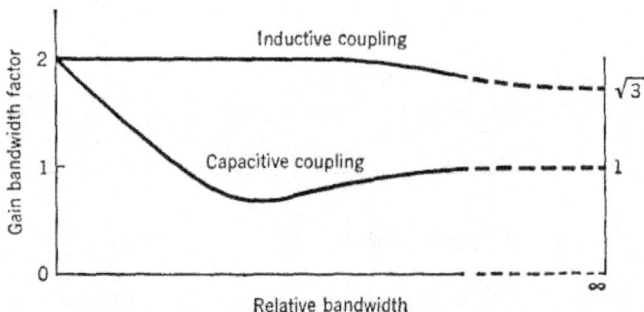

Fig. 11-15   Comparison of GBF for inductive and capacitive coupling.

11-5   The Autotransformer.   Another form of double-tuned interstage, which is really a special case of inductive coupling, is the autotransformer shown in Fig. 11-1c. While this scheme is well known at 60 cps, it has not received wide employment at radio frequencies. It nevertheless is quite applicable, and the design can be straightforward.[1] It has particular advantage in wideband double-tuned amplifier stages with large bandwidth and large ratio C_1/C_2; in fact, it turns out conveniently that the autotransformer is physically realizable in those regions of operating conditions where the pi equivalent is not (because of one or more negative elements).[2]

11-6   Selectivity Ratio.   The selectivity ratio of one maximally flat double-tuned stage is the same as that of the maximally flat staggered pair. Similarly, the selectivity ratio of cascades of identical stages is the same as for cascades of pairs. Hence, both Eq. (10-3) and Fig. 10-10 apply, provided that one takes values only for $n = 2$.

Stagger damping or stagger tuning in the wideband case can also be studied from Fig. 10-10. For a maximally flat pair of double-tuned stages take $n = 4$; for a maximally flat triple take $n = 6$; etc.

[1] W. A. Edson, The Single-layer Solenoid as an RF Transformer, *Proc. IRE*, vol. 43, pp. 932–936, August, 1955.

[2] M. M. McWhorter and J. M. Pettit, The Design of Stagger-tuned Double-tuned Amplifiers for Arbitrarily Large Bandwidth, *Proc. IRE*, vol. 43, pp. 923–931, August, 1955.

PROBLEMS

11-1. An amplifier using double-tuned stages with $Q_1 = \infty$ is to be designed. As an aid to the design, graphs similar to Fig. 10-9 are desired, but for double-tuned stages.

a. On a single sheet of three-cycle semilogarithmic graph paper construct curves for one, two, and three identical stages; one, two, and three staggered pairs; and one staggered triple. [The normalized bandwidth will be $B/(g_m/4\pi\sqrt{C_1 C_2})$.]

b. Assume that the amplifier is to use a minimum number of tubes, $g_m = 5,000$ μmhos, $C_{in} = C_{out} = 5$ pf, gain = 75 db (over-all), the over-all bandwidth is 5×10^7 radians/sec, and the band center is $\omega_0 = 10^9$ radians/sec. Determine what amplifier configuration uses the minimum number of tubes, and find the stage tuning frequencies, Q's, and k's for each stage of the amplifier.

11-2. Derive Eq. (11-21) for the $Q_1 = \infty$ case, using the same method as that outlined for the derivation of Eq. (11-13), the $Q_1 = Q_2$ case.

11-3. A two-stage amplifier using identical double-tuned circuits ($Q_1 = \infty$) is required. The amplifier is to be adjusted so that the *over-all* response has 0.5 db ripple (i.e., 20 times the logarithm of the ratio of peak to valley gain is 0.5).

a. What is the gain-bandwidth factor of the amplifier? (Assume that $C_1 = C_2$.)

b. Compare this amplifier with one using two identical single-tuned circuits adjusted to give the equal-ripple response specified above. Calculate the ratio of gains obtained for the same bandwidth, the ratio of bandwidths obtained for the same gain, and the selectivity ratios of the two amplifiers.

c. A line representing this double-tuned amplifier can be drawn on Fig. 10-9. State clearly where this line should be drawn; i.e., give the slope and the intercept with the 0-db axis.

d. In terms of the fractional bandwidth (assume that this is small), find the values of k and Q_2 which should be used for each stage.

11-4. Compare the gain-bandwidth factors for stages giving an equal-ripple response with 0.1 db ripple, but in one case having $Q_1 = Q_2$ and in the other case having $Q_1 = \infty$. Why is the percentage change in the gain-bandwidth factor from the corresponding GBF's for maximally flat stages different in the two cases?

11-5. Design the interstages for a two-stage stagger-damped amplifier using double-tuned circuits. The over-all bandwidth is to be 5 Mc with a center frequency of 60 Mc. The response is to be equal ripple with 0.1 db ripple. 6AK5 tubes are to be used with capacitances of $C_1 = 6.5$ pf and $C_2 = 6$ pf, and $g_m = 5,000$ μmhos.

a. Make a table giving the values of ω_1, ω_2, L_1, L_2, k, and Q for each stage.

b. What is the gain of the two stages at $\omega = \omega_0$?

11-6. Determine the circuit parameters k and Q in terms of the over-all amplifier bandwidth and center frequency for the three stages of a stagger-damped double-tuned maximally flat amplifier. Assume that the bandwidth is less than 10 per cent of the center frequency and that the primary and secondary Q's of each interstage are equal. Sketch the gain-frequency curve for each stage. (Do not calculate the curves, but try to visualize them from the pole-zero diagram.)

11-7. The gain-bandwidth factor for maximally flat stagger-damped double-tuned stages is the same as for a single maximally flat stage (i.e., 2) if $Q_1 = \infty$. However, if $Q_1 = Q_2$, then the GBF for the stagger-damped stages will not be equal to the GBF for a single maximally flat stage. Explain why this is true. Which will have the greater GBF, the single stage or the stagger-damped stages?

11-8. A given amplifier is to have three stages and an over-all bandwidth of 10 Mc. The tubes to be used have a gain-bandwidth product of 100 Mc (with an allowance for stray capacitances).   Make a table giving the gain attainable in this amplifier if the following types of amplifier are used:

   a. Single-tuned, synchronously tuned stages
   b. A single-tuned, maximally flat triple
   c. Double-tuned identical stages ($Q_1 = Q_2$)
   d. A double-tuned maximally flat stagger-damped triple ($Q_1 = \infty$)

12

The Feedback Pair

Among the available amplifier circuits for realizing a maximally flat or equal-ripple response, while at the same time providing a high gain-bandwidth factor, are a class of circuits employing feedback.[1] The use of feedback is for the purpose of increasing the gain-bandwdith product with simple circuits, and not for the usual objective of high stability of gain. The amount of feedback introduced is usually too small to make any significant difference in stability.

12-1 General Design Relations for a Feedback Pair. A very useful and indeed the simplest example of this kind of feedback amplifier is the "feedback pair," illustrated in Fig. 12-1. Notice that either a bandpass

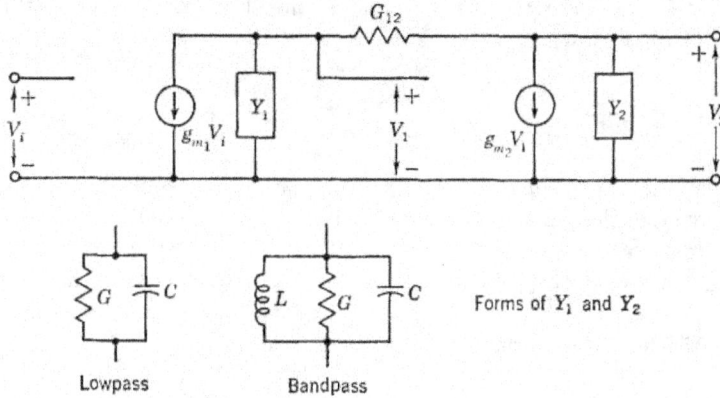

Fig. 12-1 Basic circuit of the feedback pair.

[1] See G. E. Valley, Jr., and H. Wallman (eds.), "Vacuum Tube Amplifiers" (vol. 18, M.I.T. Radiation Laboratory Series), chap. 6, McGraw-Hill Book Company, Inc., New York, 1948; H. N. Beveridge, Broadband Feedback Amplifiers, *IRE Conv. Record*, pt. 5, pp. 52–56, March, 1953.

or a lowpass amplifier can be produced by proper choice of Y_1 and Y_2. The feedback path can, in general, contain a variety of networks, but for the case at hand it consists of only a resistor having a conductance G_{12}, connected from the plate of the second tube back to the plate of the first tube (grid of second tube).

$$\text{Gain } A(s) = \frac{V_2}{V_i} = \frac{g_{m1}(g_{m2} - G_{12})}{(G_{12} + Y_1)(G_{12} + Y_2) + G_{12}(g_{m2} - G_{12})}$$

$$= \frac{g_{m1}(g_{m2} - G_{12})}{Y_1 Y_2 + G_{12}(Y_1 + Y_2) + G_{12} g_{m2}} \tag{12-1}$$

A straightforward nodal analysis yields the gain function given in Eq. (12-1). Consider the bandpass case where

$$Y_1 = Y_2 = G + sC + \frac{1}{sL}$$

$$= G\left[1 + Q\left(\frac{s}{\omega_0} + \frac{\omega_0}{s}\right)\right]$$

$$= G(1 + Qp) \tag{12-2}$$

where $\quad p \triangleq \dfrac{s}{\omega_0} + \dfrac{\omega_0}{s} \qquad Q \triangleq \dfrac{\omega_0 C}{G} \qquad$ and $\qquad \omega_0 \triangleq \dfrac{1}{\sqrt{LC}}$

Substituting these values for Y_1 and Y_2 into Eq. (12-1) gives for the gain in terms of the lowpass variable p

$$A(p) = \frac{g_{m1}(g_{m2} - G_{12})}{p^2 G^2 Q^2 + p(2G^2 Q + 2G_{12} GQ) + G^2 + 2G_{12} G + G_{12} g_{m2}} \tag{12-3}$$

This function has only poles in the finite part of the p plane; therefore we may arrange the poles as before to give a desired response. However, instead of then transforming the pole locations back into the s plane, we may, instead, discover the *p-plane* coordinates in terms of the element values G, L, C, etc. Thus we do not need to perform the reverse transformation since all the data we need will be present in the p plane. The poles of Eq. (12-3) lie at

$$p_1, p_2 = \frac{1}{Q}\left(-\frac{G + G_{12}}{G} \pm \frac{G_{12}}{G}\sqrt{1 - \frac{g_{m2}}{G_{12}}}\right) \tag{12-4}$$

$$= \frac{-(G + G_{12})}{\omega_0 C} \pm j\frac{G_{12}}{\omega_0 C}\sqrt{\frac{g_{m2}}{G_{12}} - 1} \tag{12-5}$$

The pole locations are as shown in Fig. 12-2 if $g_{m2} > G_{12}$. These two poles produce four complex poles and a zero at the origin in the s plane, as shown in Fig. 12-3. If we wish a maximally flat response in the p and s planes, we place the poles so that the real and imaginary components are equal in the

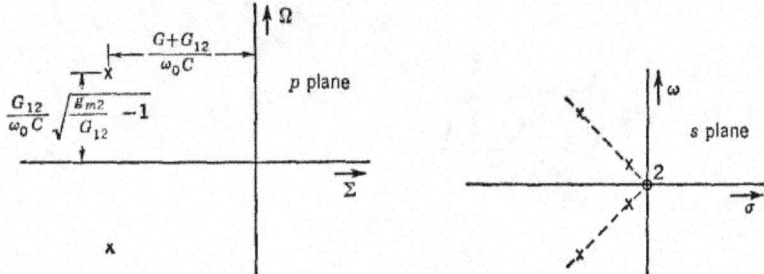

Fig. 12-2 Pole locations in the p plane.   Fig. 12-3 Pole locations in the s plane corresponding to Fig. 12-2.

p plane. The radial distance from the origin to the pole is then the normalized bandwidth, B_r/ω_0. In terms of the pole coordinates, the bandwidth is

$$B_r = \sqrt{2}\,\omega_0\,\frac{G_{12} + G}{\omega_0 C} = \sqrt{2}\,\omega_0\,\frac{G_{12}}{\omega_0 C}\sqrt{\frac{g_{m2}}{G_{12}} - 1} \tag{12-6}$$

Solving Eq. (12-6) for the value of G_{12} necessary to give the required bandwidth, we obtain

$$G_{12} = \frac{g_{m2}}{2}\left[1 \pm \sqrt{1 - \frac{2(B_r C)^2}{g_{m2}{}^2}}\right] \tag{12-7}$$

The negative sign in Eq. (12-7) is usually taken, since larger gain results. For values of $B_r C/g_{m2} \ll 1$ an approximate value for G_{12} may be obtained by using the binomial expansion for the square root,

$$G_{12} \cong \frac{(B_r C)^2}{2g_{m2}} \qquad \frac{B_r C}{g_{m2}} \ll 1 \tag{12-8}$$

The value of loading conductance G may be found from Eq. (12-6) also,

$$G = \frac{B_r C}{\sqrt{2}} - G_{12} \tag{12-9}$$

The values of L and C are such as to resonate at ω_0, of course. The resulting amplifier will be a maximally flat bandpass amplifier with center frequency ω_0 and bandwidth B_r. This has been obtained without ever discovering the s-plane pole locations. The response shape will be identical to a maximally flat stagger-tuned pair, but the gain-bandwidth factor will be slightly less (see Sec. 12-3).

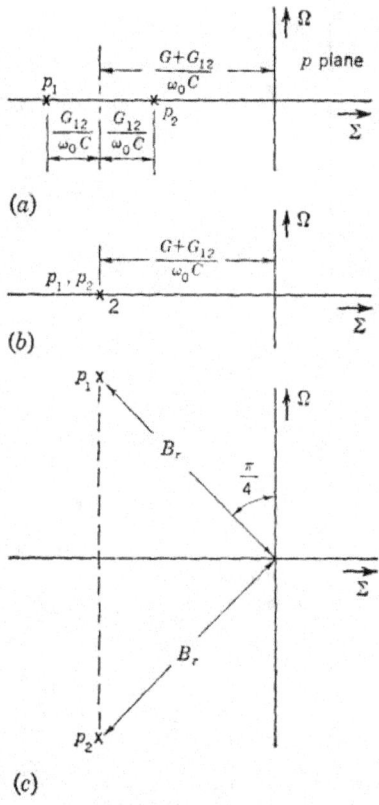

(a)

(b)

(c)

Fig. 12-4 Pole positions of the feedback pair as g_{m2} is varied. (a) $g_{m2} = 0$. (b) $g_{m2} = G_{12}$. (c) $g_{m2} = G_{12} + \dfrac{(G + G_{12})^2}{G_{12}}$ (maximally flat).

The g_m of the second stage has an important effect upon the response shape; therefore g_{m2} may not be varied to control gain alone but may instead be varied to change the passband shape. The pole locations for different values of g_{m2} are shown in Fig. 12-4. Note from Eq. (12-3) that gain is produced even when $g_{m2} = 0$, as in Fig. 12-4a, but zero gain results for $g_{m2} = G_{12}$, as in Fig. 12-4b.

12-2 Bandwidth Switching. A unique feature of the feedback pair is the possibility of a two-position bandwidth switch, involving no modification of the high-frequency circuits but only a d-c change. The bias on the second tube is switched between the conditions of normal plate current for the WIDE and zero plate current for the NARROW positions.

There are enough degrees of freedom in the circuit parameters [1] to permit specification of:

1. Maximally flat response in the WIDE position
2. Bandwidth B_w in the WIDE position
3. Ratio of gains between the two positions
4. Ratio of bandwidths between the two positions

General design equations are not available, but a large number of cases are presented by Valley and Wallman. As an example, a specification of a bandwidth ratio of 4:1, with equal gains in WIDE and NARROW positions, leads to the following requirements for a given radian bandwidth B_w:

$$9.20 \, \frac{G_1}{C} = 1.97 \, \frac{G_2}{C} = 2.49 \, \frac{G_{12}}{C} = 0.746 \, \frac{g_{m2} - G_{12}}{C} = B_w \qquad \text{radians/sec}$$

$$(12\text{-}10)$$

For this example, the pole locations (in the p plane) prove to be as follows:

WIDE

$$\frac{(-0.707 + j0.707)B_w}{\omega_0}$$

$$\frac{(-0.707 - j0.707)B_w}{\omega_0}$$

NARROW

$$\frac{-1.158B_w}{\omega_0}$$

$$\frac{-0.260B_w}{\omega_0}$$

It is of interest to note that the design requirements specify the g_m of the tube for a given capacitance C. This may mean operating at reduced g_m in order to achieve the switching benefits.

12-3 Gain-Bandwidth Factor. The feedback pair is similar to the staggered pair in many respects, including gain-bandwidth factor. This factor must always be slightly less than for the staggered pair, because of the equivalent reduction of the g_m of the second tube, as can be seen from the $g_{m2} - G_{12}$ term in the scale factor of Eq. (12-3). For the bandwidth-switching example just described, it turns out that the gain-bandwidth factor in the WIDE position is 0.88.

---

[1] The two load conductances G_1 and G_2 cannot in general be identical as they are in Eq. (12-2).

12-4   Selectivity Ratio. This parameter depends upon the nature of the gain function, e.g., maximally flat, and not upon the circuit which produces it. Hence, a maximally flat feedback pair has the same selectivity ratio as a staggered pair or double-tuned stage.

PROBLEMS

12-1. Find an expression similar to Eq. (12-7) which gives the value of G_{12} necessary to obtain a prescribed equal-ripple response. The expression should be a function of B_r, C, g_{m2}, and $\tanh a$. (For the latter see Chap. 9.)

12-2. A feedback pair is desired which has a bandwidth B_w with tube 2 operating that is approximately five times the bandwidth B_n obtained with tube 2 off ($g_{m2} = 0$). The passband in the WIDE position is to be maximally flat. Assume that identical plate networks are used (that is, $Y_1 = Y_2$).

a. Find the parameters $G = G_1 = G_2$, G_{12}, and g_{m2} in terms of B_w and C.

b. Find numerical values for the preceding parameters and the tuning inductors if $B_w = 5$ Mc, $f_0 = 15$ Mc, and $C = 20$ pf.

c. Sketch the pole-zero diagram in the s plane which results when $g_{m2} = 0$.

12-3. A feedback pair is to have a bandwidth of 25 Mc and a center frequency of 60 Mc. The tubes to be employed are Mullard E180F pentodes with the specifications given below. One power supply may be used with any desired value of voltage. Assume that the stray capacitance is 5 pf per stage.

$E_f = 6.3$ volts	$E_{c3} = 0$ volt
$I_f = 0.3$ amp	$E_{c1} = -1$ volt
$C_{\text{in}} = 7.9$ pf	$I_{g2} = 3.0$ ma
$C_{\text{out}} = 2.9$ pf	$I_b = 13$ ma
$C_{\text{in}} = 11.2$ pf	$g_m = 16.5$ ma/volt
(at $I_b = 13$ ma)	Input damping = 6 kilohms
$E_b = 190$ volts	(measured at 50 Mc)
$E_{c2} = 160$ volts	

a. Give a complete schematic diagram of the resulting feedback pair, assuming that these stages are the third and fourth stages of an eight-stage amplifier. Include all components with typical or calculated values. (The bypass capacitors have no uniquely determinable value, but reference to the first part of Chap. 15 may prove useful in determining a reasonable value.)

b. Find the center-frequency gain and over-all bandwidth of four such feedback pairs.

c. What is the approximate bandwidth of the pair if the plate current to the second tube is cut off?

13

Noise in Amplifier Circuits

13-1 Circuit Noise. In a broad sense "noise" could be defined as any current or voltage not the signal. Thus, if the "signal" is a single-frequency sine wave, any currents having frequency components at other than this frequency would be classed as noise. Hence an imperfectly filtered power supply might introduce harmonics of 60 cycles (hum) into the output; however, such spurious signals could always be removed by better filtering of the supply. Another form of such a signal, which cannot be removed by such filtering, is hum originating from the a-c operation of tube heaters. Even such noise could, in principle, be completely removed at the expense of a d-c heater supply. Other forms of noise which originate *inside* an amplifier and which can be reduced or eliminated by better environment include microphonic noises caused by tube-element vibration and in some cases the vibration of passive elements, particularly capacitors sustaining a d-c voltage. (Note that the transistor is superior in eliminating most of these mechanical noises.)

Another type of noise, which is perhaps more troublesome, originates *outside* the amplifier. For example, static and interfering signals are, in the broad sense, both noise. These types of noise, however, are really a *system* problem in that decreasing the noise cannot be achieved by changing the amplifier itself (except by providing the proper band shape to pass the desired signal and reject as much interference as possible).

The amplifier itself does introduce, however, a noise signal which determines the ultimate limit to the sensitivity of a system when noise sources like the preceding have been eliminated. This noise, which is particularly troublesome in wideband high-gain amplifiers, has its origin in the input stages of the amplifier. The presence of this natural input noise sets a limit on the gain which can usefully be attained. If the gain of the amplifier is sufficient to produce a large noise signal in the output, then adding more gain may only cause the output to saturate on noise and will not increase the

263

sensitivity of the amplifier to weak signals.   Indeed the signal-to-noise ratio in a high-gain amplifier is essentially independent of the gain of the amplifier and is determined primarily by the design of the input amplifier stage. (This assumes that those sources of noise which can be reduced in the amplifier have been made negligible.)

The noise which will be discussed here is properly known as *random noise*, hereafter referred to simply as *noise*, and has its origin in the motion of electrons.   In an electrical conductor at a temperature above absolute zero there is always thermal agitation that produces motion of the electrons; this motion manifests itself as a minute electric current which can be detected with sufficient amplification.   The amount of noise can be forecast from theoretical consideration of the thermodynamics involved.[1] The name usually given to this noise is *thermal noise*.   Because of the large number of charges in any conductor, moving at random with high velocities made possible by their small mass, the resulting currents are random in their amplitude and phase and contain all frequencies of any practical interest up to thousands of megacycles.   Accordingly, it is not possible to describe the amount of noise by the usual quantities of amplitude of waveform or of frequency component.   Even a description of the frequency distribution, or power spectrum, tells little, for this spectrum is uniform for all practical frequencies.   There is one useful quantity, however, which provides a measure that can be compared with ordinary signal quantities, namely, the average power, or the mean square, of the current or voltage (or the rms) measured in a band of frequencies.

The basic law for the amount of thermal noise that would appear as an open-circuit voltage across a conductor of resistance R, and summed as a mean square across a bandwidth B, is as follows:

$$\overline{v_n^2} = 4kTBR \qquad \text{volts}^2 \tag{13-1}$$

where k = Boltzmann's constant $(1.37 \times 10^{-23}$ watt-sec/deg$)$

   T = absolute temperature, °K (°C + 273.1)

   B = noise bandwidth, cps

   R = resistance, ohms

   $4kT = 1.6 \times 10^{-20}$ for $T = 293°K$

The noise equivalent circuit shown in Fig. 13-1 indicates the mean-square noise voltage $\overline{v_n^2}$ in series with a noiseless resistor R.   The Norton equivalent is also shown in Fig. 13-1 and has a current generator of mean-square value equal to

$$\overline{i_n^2} = 4kTBG \tag{13-1a}$$

[1] The experimental discovery is usually credited to J. B. Johnson, Thermal Agitation of Electricity in Conductors, *Phys. Rev.*, vol. 32, p. 97, July, 1928; the first analysis is due to H. Nyquist, Thermal Agitation of Electronic Charge in Conductors, *Phys. Rev.*, vol. 32, p. 110, July, 1928.   An excellent tutorial article on noise is given by J. R. Pierce, Physical Sources of Noise, *Proc. IRE*, vol. 44, p. 601, May, 1956.

Note that the available power from either representation is

$$P_{\mathrm{av}} = kTB \tag{13-2}$$

A sufficiently sensitive instrument responding to mean-square currents or voltages, such as a thermocouple or a bolometer, if preceded by a noise-free amplifier with bandwidth B and with an infinite input impedance, would give a reading proportional to Eq. (13-1). The resistance is assumed constant over the passband, which is assumed to have a rectangular shape of width B.

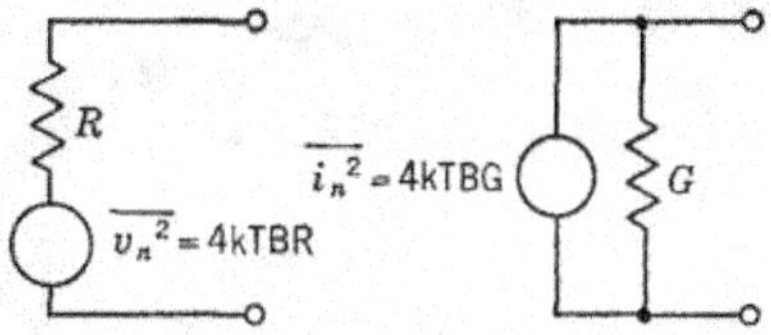

Thermal noise sets an ultimate limit on how weak a signal one can amplify and perceive at the output of the amplifier. Such a signal will always originate in a source having an internal resistance, and hence the

$\overline{i_n{}^2} = 4kTBG$ $\overline{v_n{}^2} = 4kTBR$

Fig. 13-1 Equivalent circuits for thermal noise.

signal will be accompanied by noise. Even a perfect amplifier, introducing no additional noise, cannot reduce the thermal noise generated in the signal source.

The practical fact is that the amplifier will introduce a certain amount of additional noise. It will be our purpose here to evaluate the amount of additional noise and later to see what can be done to minimize it.

13-2 Tube Noise. The added noise in a vacuum-tube amplifier comes from the tubes, particularly the first stage. The electron stream in the tube is not a perfect fluid but consists of discrete particles. These particles, the electrons, arrive at the anode of the tube with random phase and combine to produce a spectrum of frequency components of current extending from d-c to extremely high frequencies—as high as are encountered in any electronic system. The spectrum is uniform up to frequencies at which transit time becomes important. The noise current can again be defined quantitatively by its mean-square value.

There are three principal classes of tube noise occurring in conventional amplifier tubes. These are shot noise, partition noise, and induced grid noise. In any given amplifier the three contributions can be evaluated.

Shot noise is the result of the random arrival of electrons at the anode of a tube. The simplest situation is that of a temperature-limited diode in which the mean-square noise current in a bandwidth B is [1]

$$\overline{i_n{}^2} = 2qI_{\mathrm{dc}}B \tag{13-3}$$

where q = electron charge (1.60×10^{-19} coulomb)

    I_{dc} = d-c component of diode current, amp

---

[1] For transit-time correction, see D. B. Fraser, Noise Spectrum of Temperature-limited Diodes, *Wireless Engr.*, vol. 26, pp. 129–131, April, 1949.

Temperature-limited operation is of little use in amplifiers, although a diode used in this manner provides a convenient noise source for measurement purposes. When tubes are space-charge-limited, the noise current is less and is usually expressed by the following relationship:

$$\overline{i_n{}^2} = 4kTBR_{eq}g_m{}^2 \tag{13-4}$$

where
$$R_{eq} \cong \frac{2.5}{g_m} \quad \text{triodes} \tag{13-5a}$$

$$R_{eq} \cong \frac{I_b}{I_b + I_{c2}} \left(\frac{2.5}{g_m} + \frac{20I_{c2}}{g_m{}^2} \right) \quad \text{pentodes} \tag{13-5b}$$

where I_b = d-c plate current, amp

I_{c2} = d-c screen current, amp

g_m = grid-plate transconductance, mhos (use value appropriate to pentode or triode operation)

The pentode case requires special explanation. The second of the two terms in Eq. (13-5b) for R_{eq} is not truly shot noise but is due to *partition noise*. In a pentode, as opposed to a diode or triode, the current leaving the cathode does not all go to the anode or plate. A portion goes to the screen grid, but not an absolutely constant portion. There is a small random variation in the division, or "partition," of the total current which introduces an additional set of noise components (partition noise) in the plate current.

The custom of expressing shot and partition noise in terms of the equivalent noise resistor has arisen from the fact that, if the proper resistor generating thermal noise as given by Eq. (13-1) were connected between grid and cathode of a tube having a given g_m but idealized by having no noise, there would be amplified thermal noise in the plate current exactly equal to the shot noise appearing in the actual tube.

The final source of noise that is important in wideband amplifiers is called *induced grid noise*. The random flow of electrons in the tube current can induce currents in the grid circuit as the electrons pass the grid of the tube. The amount of this noise varies with frequency squared, as does the input conductance of the tube; hence the usual expression for induced grid noise is in terms of the input conductance as follows,

$$\overline{i_n{}^2} = 5(4kTG_rB)$$

$$= \text{mean-square noise in the grid} \tag{13-6}$$

where G_r = input conductance due to electron stream (not due to feedback such as grid-to-plate C or cathode circuit L)

5 = empirical factor typical of oxide-cathode amplifier tubes

The several noise sources associated with the vacuum tube and its input circuit are shown in Fig. 13-2. The two generators $\overline{i_{n2}^2}$ and $\overline{i_{n3}^2}$ are caused by the tube; the generator $\overline{i_{n1}^2}$ is due to the input-circuit conductance. The

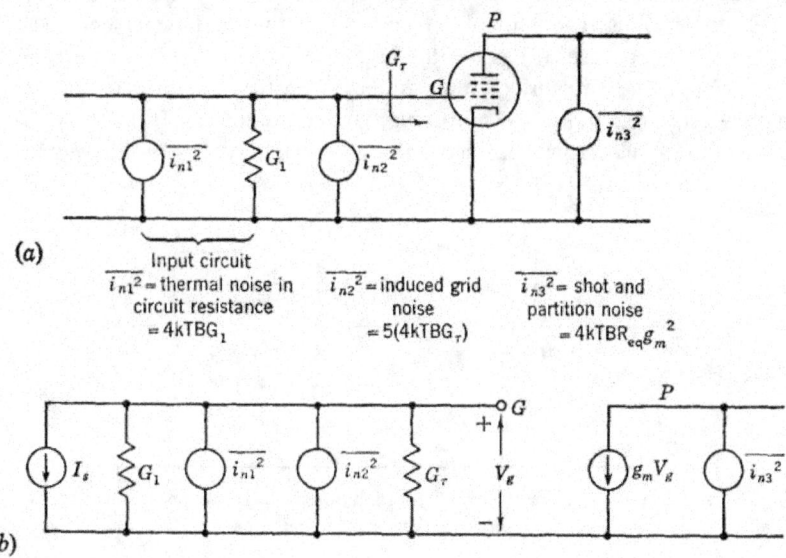

(a)

$\overline{i_{n1}^2}$ = thermal noise in circuit resistance = 4kTBG₁

$\overline{i_{n2}^2}$ = induced grid noise = 5(4kTBG_r)

$\overline{i_{n3}^2}$ = shot and partition noise = 4kTBR_{eq}g_m^2

(b)

Fig. 13-2 (a) Schematic representation of the noise sources in a vacuum tube (pentode or triode). (b) Equivalent circuit.

bandwidth B is the noise bandwidth over which the noise is being observed and is often largely determined by subsequent circuits.

For many situations it is more convenient to transform the noise generator in the plate circuit over to the grid circuit so that all the noise-producing elements are in the same part of the circuit. This may be accomplished by dividing the current $\overline{i_{n3}^2}$ by g_m^2, giving an equivalent noise voltage at the grid which is the same as that produced by a resistor of value R_{eq}. The result is shown in Fig. 13-3. (Note that the actual grid connection is shown

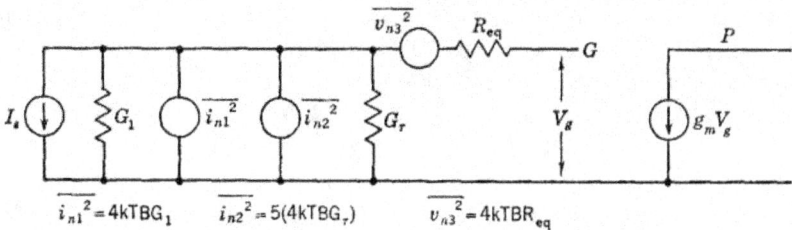

$\overline{i_{n1}^2} = 4kTBG_1$ $\overline{i_{n2}^2} = 5(4kTBG_r)$ $\overline{v_{n3}^2} = 4kTBR_{eq}$

Fig. 13-3 Equivalent circuit for tube noise with the shot-noise generator transformed into an equivalent noisy resistor R_{eq}.

to the left of R_{eq} and that the actual grid voltage is always v_1 plus the noise voltage appearing across R_{eq}.)

13-3 Transistor Noise. The noise in a transistor is produced by several processes. In the case of an *NPN* transistor,[1] these are (1) noise due to electrons going from emitter to collector, (2) noise due to electrons going from emitter to base, (3) noise due to electrons injected into the base and returning to the emitter, (4) noise due to electrons trapped in the emitter space-charge region and recombining with holes coming from the base, (5) noise due to electrons trapped in the emitter space-charge region and

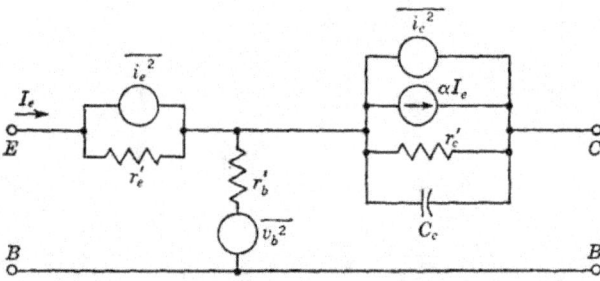

Fig. 13-4 An equivalent for the transistor including noise sources. (*From E. G. Nielsen, Behavior of Noise Figure in Junction Transistors, Proc. IRE, vol. 45, pp. 957–963, July, 1957.*)

returning to the emitter after being detrapped thermally, and (6) thermal noise due to the ohmic base resistance r_b'.

At low frequencies, typically below a few kilocycles, an additional noise source due to surface-leakage effects becomes prominent. The noise typically varies approximately as $1/f$ and is usually called "excess noise" or simply "$1/f$ noise."

The effect of all these noise sources may be represented by the addition of three noise generators to the T equivalent circuit. One current generator appears across the emitter junction, and one appears across the collector junction. These two generators are partially correlated. The third generator is a voltage generator in series with the ohmic base resistance and is uncorrelated with either of the current generators.[2] Such a complicated equivalent circuit can be considerably simplified and still retain reasonable accuracy up to the alpha-cutoff frequency of the transistor.

The simplified noise equivalent circuit given by Nielsen is shown in Fig. 13-4, in which the noise at the emitter is represented by a current

---

[1] A. van der Ziel, Shot Noise in Transistors (letter), *Proc. IRE*, vol. 48, pp. 114–115, January, 1960; and A. van der Ziel and A. G. T. Becking, Theory of Junction Diode and Junction Transistor Noise, *Proc. IRE*, vol. 46, pp. 589–594, March, 1958.

[2] A. van der Ziel, Shot Noise in Junction Diodes and Transistors, *Proc. IRE*, vol. 43, pp. 1639–1646, November, 1955.

generator of value

$$\overline{i_e^2} = 2qI_EB \tag{13-7}$$

where q = electron charge

$\quad I_E$ = d-c emitter current, amp

This is a current generator giving full shot noise [cf. Eq. (13-3)] in parallel with the emitter resistance $r_e' = kT/qI_E$.

Across the collector junction another noise current generator appears which is assumed *uncorrelated* [1] with the emitter noise generator. The value of the noise current produced by the collector generator is

$$\overline{i_c^2} = 2qI_C \left(1 - \frac{|\alpha|^2}{\alpha_0} \right) B \tag{13-8}$$

where I_C = d-c collector current, amp

$\quad \alpha$ = common-base current gain (a function of frequency)

$\quad \alpha_0$ = α at zero frequency

The collector current noise generator is not independent of frequency as the emitter current generator was assumed to be. At low frequencies $\overline{i_c^2}$ is considerably less than full shot noise but increases with frequency since $|\alpha|$ is a decreasing function of frequency.

The remaining noise generator, $\overline{v_b^2}$, simply gives the thermal noise generated in the ohmic base resistance,

$$\overline{v_b^2} = 4kTr_b'B \tag{13-9}$$

The resulting equivalent circuit is the same as that shown in Fig. 2-8 and in Fig. 2-20 with the addition of the equivalent noise generators. Note that the emitter capacitance $1/\omega_\alpha r_e'$ and the reverse transfer parameter μ_{ec} are being neglected. If α is approximated by the expression

$$\alpha = \frac{\alpha_0}{1 + jf/f_\alpha} \tag{13-10}$$

then Eq. (13-8) may be written in terms of frequency,

$$\overline{i_c^2} = 2qI_C(1 - \alpha_0) \frac{\left[1 + \left(\frac{f}{\sqrt{1 - \alpha_0} f_\alpha} \right)^2 \right] B}{1 + (f/f_\alpha)^2} \tag{13-11}$$

From Eq. (13-11) $\overline{i_c^2}$ is seen to have a minimum value at low frequencies of

$$\overline{i_c^2} = 2qI_CB \, (1 - \alpha_0) \qquad f \to 0 \tag{13-12}$$

---

[1] The fact that the two generators are uncorrelated has the practical effect that the noise power, say at the output of the device, may be computed by calculating the power due to each noise generator *separately* and summing the two powers to obtain the true total power.

The magnitude of $\overline{i_c^2}$ as a function of frequency is shown in Fig. 13-5, in which $\overline{i_c^2}$ is seen to start increasing at a frequency somewhat greater than the beta-cutoff frequency, $(1 - \alpha_0)f_\alpha$.

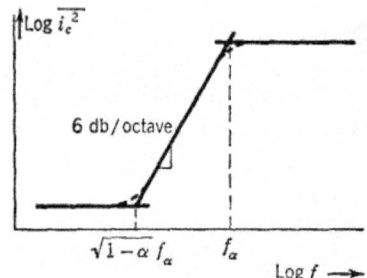

Fig. 13-5   Frequency dependence of the collector noise generator.

For some applications the Thévenin equivalent of the noise generators is more convenient.   Such a circuit as is shown in Fig. 13-6 has equivalent noise generators of the following values,

$$\overline{v_b^2} = 4kTr_b'B \tag{13-9}$$

$$\overline{v_e^2} = \overline{i_e^2}r_e'^2 = (2qI_EB)\left(\frac{kT}{qI_E}\right)r_e' = 2kTr_e'B \tag{13-13}$$

$$\overline{v_c^2} = 2qI_C(1 - \alpha_0)F(f)|Z_c|^2B$$

where

$$F(f) \triangleq \frac{1 + \left(\dfrac{f}{f_\alpha\sqrt{1 - \alpha_0}}\right)^2}{1 + (f/f_\alpha)^2}$$

$$Z_c = \frac{1}{1/r_c' + j\omega C_c}$$

$$\overline{v_c^2} = \frac{2kT\alpha_0}{r_e'}(1 - \alpha_0)|Z_c|^2BF(f) \tag{13-14}$$

(assuming that $I_C = |\alpha_0 I_E|$).

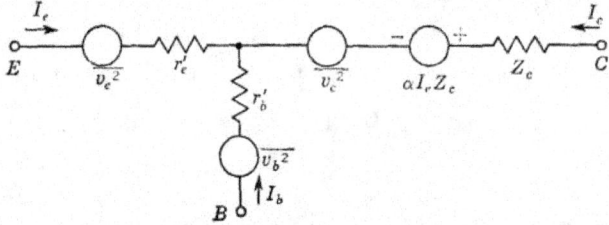

Fig. 13-6   Thévenin equivalent of Fig. 13-4.

Using the representation of Fig. 13-4 or that of Fig. 13-6, we have a quite accurate equivalent circuit for analyzing the noise behavior from frequencies above a few hundred cycles up to the alpha-cutoff frequency.

13-4 Calculation of Amplifier Noise Factor (Vacuum-tube Amplifier). The *noise factor* (or "noise figure") is a measure of the relative magnitudes of signal and noise powers to be expected in the output of an amplifier compared with these same powers in the output of an "ideal noise-free" amplifier. From a knowledge of the noise sources within an

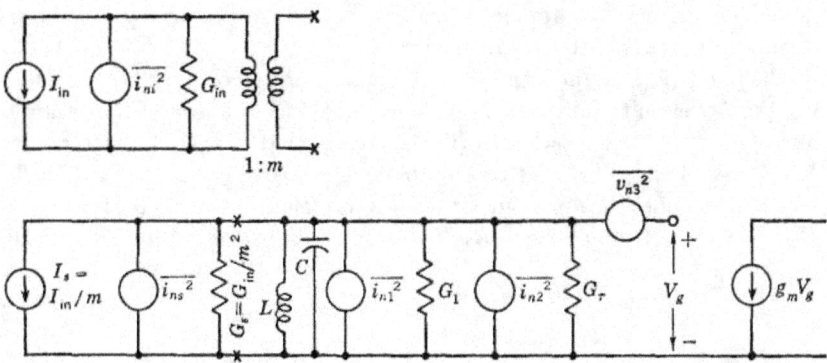

Fig. 13-7 Equivalent circuit of the input stage of a vacuum-tube amplifier. [*Adapted from G. E. Valley, Jr., and H. Wallman (eds.), "Vacuum Tube Amplifiers" (vol. 18, M.I.T. Radiation Laboratory Series), pp. 682–694, McGraw-Hill Book Company, Inc., New York, 1948; and M. T. Lebenbaum, Design Factors in Low-noise-figure Input Circuits, Proc. IRE, vol. 38, pp. 75–80, January, 1950.*]

amplifier and the manner in which the signal and noise are amplified, we can make any desired calculations.

Let us set out to evaluate how much signal and how much noise will be present in the output of an amplifier. The signal, of course, originates in the signal source connected to the input terminals, as does also a certain amount of thermal noise in the internal resistance of the source. In the first tube of the amplifier there is added some more noise: shot, partition, and induced grid noise. This first tube amplifies the signal and also any noise appearing in its grid circuit. In most cases the amplification is sufficient to make the amplified noise of the first tube much greater than the noise contributed by the second tube. We shall assume this to be the case in our initial analysis.

The circuit to be considered is shown in Fig. 13-7, in which the tube is represented as shown in Fig. 13-3. For the example, a single-tuned circuit is shown in the grid of the first tube, although in some circumstances a double-tuned circuit would prove advantageous. We shall want to evaluate the

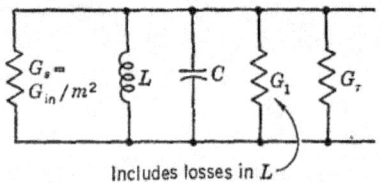

Fig. 13-8 Circuit for computing the input bandwidth.

bandwidth of the circuit, because it may have to fit in the circuits of the other stages to provide stagger tuning. Then, in order to demonstrate the effects of matching (or mismatching) the generator impedance to the input impedance of the amplifier, there has been shown an ideal transformer in the circuit; this can be approximated in practice by tapping down on the inductor L or by use of the double-tuned circuit already mentioned.

The bandwidth of the input circuit can be readily computed by considering the LC circuit and the loading represented by the sum of the conductances. Notice that this bandwidth is merely that of the input circuit, *not* the over-all bandwidth B of the complete amplifier (several stages, usually) which determines the noise output. The equivalent input circuit is shown in Fig. 13-8, and the bandwidth B_1 of this circuit is expressed by Eq. (13-15).

$$B_1 = \frac{G_s + G_1 + G_r}{2\pi C} \tag{13-15}$$

= 3 db bandwidth of input circuit only

The bandwidth B_1 can, in principle, be adjusted either by varying the transformation ratio m or by adding to the conductance G_1. Before saying which is preferable, let us examine the noise situation.

In order to combine all the noise sources directly, it is convenient to transform the voltage source $\overline{v_{n3}^2}$ to a current source in parallel with the other currents, as shown in Fig. 13-9. The value of this current generator is

$$\overline{i_{n3}^2} = \overline{v_{n3}^2}(G_s + G_1 + G_r)^2$$

$$= 4kTBR_{eq}(G_s + G_1 + G_r)^2 \tag{13-16}$$

We now have all the noise sources together, and with them the signal source, so that some manner of comparing them can now be chosen. Because of the added noise sources in the first tube, the output signal-to-

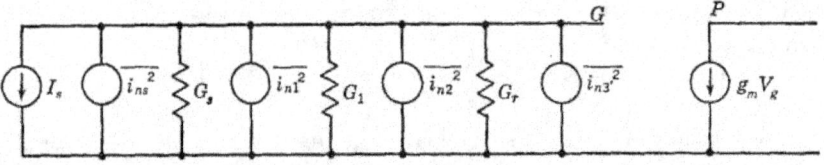

Fig. 13-9 Equivalent circuit for the calculation of F.

noise ratio will be worse than at the input; therefore it would seem reasonable to compare the signal-to-noise ratio at the input of the amplifier with that at the output. If the input source is at the standard temperature of 290°K (which approximates room temperature), the ratio of input signal-to-noise ratio to output signal-to-noise ratio is defined as the *noise factor*.

Since the noise is a measurable quantity only in a mean-square sense, let us take as our signal-to-noise ratio S/N the square of the signal current divided by the mean-square noise current. Then from the preceding definition of noise factor we obtain

$$\text{Input } S/N \text{ ratio} \triangleq \frac{\overline{I_{\text{in}}^2}}{\overline{i_{ni}^2}} = \frac{\overline{I_s^2}}{\overline{i_{ns}^2}} = \frac{\overline{I_s^2}}{4kTBG_s} \tag{13-17}$$

$$\text{Output } S/N \text{ ratio} \triangleq \frac{\overline{I_s^2}}{\overline{i_{ns}^2} + \overline{i_{n1}^2} + \overline{i_{n2}^2} + \overline{i_{n3'}^2}}$$

$$= \frac{\overline{I_s^2}}{4kTB[G_s + G_1 + 5G_r + R_{\text{eq}}(G_s + G_1 + G_r)^2]} \tag{13-18}$$

$$\text{Noise factor}^1 \; F \triangleq \frac{\text{input } S/N \text{ ratio}}{\text{output } S/N \text{ ratio}} \tag{13-19}$$

$$= 1 + \frac{G_1}{G_s} + \frac{5G_r}{G_s} + \frac{R_{\text{eq}}}{G_s}(G_s + G_1 + G_r)^2 \tag{13-20}$$

(assumes that T_s, the source temperature, is 290°K).

Notice that the minimum value of the noise factor is unity, corresponding to an amplifier in which there are no additional noise sources beyond the generator. Anything that can be done to decrease G_1, G_r, or R_{eq} will improve the noise factor. The latter two are the principal noise contributors, and they depend upon the choice of tube.

Once the tube choice has been made, what about the circuit design? For instance, what about the value of transformation ratio m? The noise factor is a function of m, since G_s is G_{in}/m^2, and the source impedance (conductance) G_{in} is usually fixed in any situation, rather than the transformed conductance G_s. One can find an optimum value of m by determining the minimum noise factor F in Eq. (13-20) in the usual way of differentiating

---

[1] On the assumption that the amplifier is linear, this is the same as the noise-factor definition given in IRE Standards on Methods of Measuring Noise in Linear Twoports, 1959, *Proc. IRE*, vol. 48, p. 61, January, 1960.

with respect to m and equating the derivative to zero.   For minimum noise factor

$$m_{opt} = \left[\frac{G_{in}^2 R_{eq}}{G_1 + 5G_r + R_{eq}(G_1 + G_r)^2} \right]^{\frac{1}{4}} \qquad (13\text{-}21a)$$

In some cases the value of the source conductance itself may be altered to provide the proper value for minimum noise factor.   The optimum source conductance is

$$G_{s,opt} = \sqrt{\frac{G_1 + 5G_r}{R_{eq}} + (G_1 + G_r)^2} \qquad (13\text{-}21b)$$

Both Eqs. (13-21a) and (13-20) may be simplified for the special case of negligible input circuit loss ($G_1 \approx 0$) and $R_{eq}G_r \ll 5$.   If we define a new parameter $G_n \triangleq 5G_r$ which varies as frequency squared, we may rewrite Eq. (13-21b) in the simpler form

$$G_{s,opt} \cong \frac{f_x}{f_0} \sqrt{\frac{G_n(f_0)}{R_{eq}}} \qquad (13\text{-}21c)$$

In this equation $G_n(f_0)$ is the value of G_n at the frequency f_0, while f_x is the operating frequency.   With the same approximations the optimum value of F may be found from Eq. (13-20) by substituting $G_{s,opt}$ for G_s.   The resulting minimum value of noise factor is found to be

$$F_{min} \cong 1 + \frac{2f_x}{f_0} \sqrt{R_{eq}G_n(f_0)} \qquad (13\text{-}20a)$$

From Eq. (13-20a) it is seen that the minimum noise factor will increase approximately linearly with operating frequency, although the equation must be used with care both at very low and at very high frequencies.

Table 13-1 summarizes the noise parameters for some receiving tubes.   A small value of $R_{eq}G_n$ is, of course, desirable to obtain a low noise factor. Note that both the 7077 and 416A are relatively special, close-spaced microwave triodes which give exceptionally low $R_{eq}G_n$.

The value of m to give minimum noise factor may not be the proper value for bandwidth, as determined from Eq. (13-15), or even the proper value to give maximum signal voltage at the grid of the tube; i.e., for maximum signal voltage at grid

$$m = \sqrt{\frac{G_{in}}{G_1 + G_r}} \qquad (13\text{-}22)$$

Accordingly, some compromise must be made, the decision usually resting upon which performance feature is the most important.   Usually the maximum grid voltage is of the least consequence, because gain added later

Table 13-1 *

Tube type	g_m, μmhos	R_{eq}, ohms	G_n at 100 Mc, μmhos	$R_{eq}G_n$
416A	50,000	~100	106	0.011
6201	5,500	600	400	0.238
6688	18,500	120	1,440	0.172
7077	9,000	350	170	0.061
6AK5 (pentode)	5,000	1,880	670	1.260
6AK5 (triode)	6,600	385	670	0.258
6AM4	9,800	260	740	0.193
6AN4	10,000	250	680	0.171
6BC4	10,000	260	670	0.173
6BC8	6,200	600	400	0.238
6BK7A	9,300	240	640	0.155
6BN4	6,800	420	480	0.203
6BQ7A	6,400	435	360	0.156
6BZ7	6,800	490	430	0.212
6CE5	9,000	650	1,480	0.964
2CY5	8,000	525	790	0.415
PC86	. . . . .	170	880	0.150

* Unless otherwise indicated, all pentodes are triode-connected.

can make up for the inefficiency in the first stage. The noise factor, on the other hand, is determined at the input, and later stages can do nothing to improve it.

The problem of greatest practical difficulty is that occurring when the chosen value of m to give minimum noise factor according to Eq. (13-21a) yields a bandwidth which is too narrow. The designer must then choose among three possible alternatives: (1) increase G_s by changing m, (2) increase G_1 by adding shunt resistance or by introducing feedback to raise the input conductance of the tube, (3) or adopt a different circuit altogether. The first two alternatives increase the noise factor above the minimum value, although the first one is usually less harmful than the second because the F as a function of G_s has a broad minimum. The third alternative is best; the most commonly used alternative circuit is the double-tuned one, which, for the same conductance G_s, will yield up to twice the bandwidth of the circuit in Fig. 13-7.

The noise associated with the second stage in the amplifier will make an appreciable contribution to the over-all noise factor if the gain of the first stage is not high. A given situation can be evaluated by transforming the noise of the second tube backward through the first tube to the input grid circuit as was done with the plate shot noise. Thus the mean-square noise current in the grid circuit of the second tube can be divided by g_m^2 to give an equivalent mean-square grid voltage, which in turn can be transformed to a current source for adding in with the other current generators. An alternative approach to this contribution of the second stage is that of computing the over-all noise factor of two networks in cascade, by using relationships presented in Chap. 14.

13-5   Noise Factor of a Transistor Stage. The noise factor of a transistor stage may be found in a manner similar to that used to find the noise factor of a vacuum-tube input stage. The equivalent circuit which

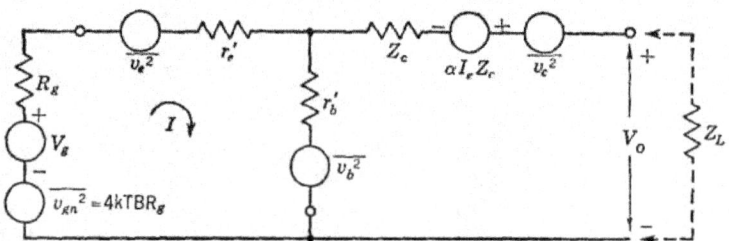

Fig. 13-10 Circuit for calculation of F for a common-base stage. (NOTE: V_o is the open-circuit output voltage.)

will be analyzed is shown in Fig. 13-10 and is the same as that of Fig. 13-6 with a signal source added. The circuit is shown in the common-base connection, but, as will later be shown, the results obtained are the same as for the common-emitter connection. The method to be used is, briefly, to calculate the mean-square open-circuit voltage (both signal and noise) appearing at the output of the transistor. The power delivered to the load Z_L is then a constant times this mean-square voltage—the constant depending only upon Z_L and the output impedance of the transistor. (The noise in Z_L is assumed small compared with the amplifier noise currents.)

A convenient way to make the calculation is to compute the mean-square output voltage caused by each generator in Fig. 13-10. To make the calculation, an arbitrary polarity must be temporarily assigned to the noise generators so that their contribution can be properly computed. As an example, assume that the generator v_b is positive on top. The current in the input loop due only to v_b is then

$$I = \frac{-v_b}{R_g + r_b' + r_e'} \qquad V_g = v_{gn} = v_e = v_c = 0 \qquad (13\text{-}23a)$$

The open-circuit output voltage due to v_b is then

$$v_o = \frac{-v_b(\alpha Z_c - R_g - r_e')}{R_g + r_b' + r_e'} \qquad (13\text{-}23b)$$

and the mean-square value of the output voltage is

$$\overline{v_o^2} = \frac{\overline{v_b^2}\,|\alpha Z_c - R_g - r_e'|^2}{(R_g + r_b' + r_e')^2} \tag{13-23c}$$

In a similar manner, the mean-square voltages due to $\overline{v_{gn}^2}$, $\overline{v_e^2}$, and $\overline{v_c^2}$ may be found and added together to give the total $\overline{v_o^2}$ (note that all the generators are uncorrelated).

$$\overline{v_o^2} = \left(\overline{v_{gn}^2}\,|\alpha Z_c + r_b'|^2 + \overline{v_e^2}\,|\alpha Z_c + r_b'|^2 \right.$$
$$\left. + \overline{v_b^2}\,|\alpha Z_c - r_e' - R_g|^2\right)\frac{1}{(R_g + r_e' + r_b')^2} + \overline{v_c^2} \tag{13-24}$$

This equation may be simplified by noting that usually $Z_c \gg r_b'$ and $Z_c \gg R_g + r_e'$. Therefore Eq. (13-24) may be approximated by

$$\overline{v_o^2} = (\overline{v_{gn}^2}\,|\alpha Z_c|^2 + \overline{v_e^2}\,|\alpha Z_c|^2 + \overline{v_b^2}\,|\alpha Z_c|^2)\frac{1}{(R_g + r_e' + r_b')^2} + \overline{v_c^2}$$
$$\tag{13-25}$$

We may now rewrite the definition of noise factor to use more conveniently here.

$$F = \frac{(S/N)_{\text{in}}}{(S/N)_{\text{out}}} \tag{13-26}$$

(power ratio at $T_{\text{source}} = 290°\text{K}$). The output noise power into Z_L from the stage is

$$P_L = \frac{\overline{v_o^2}}{|Z_o + Z_L|^2}\,\text{Re}\,Z_L = K_1\overline{v_o^2} \tag{13-27}$$

where Z_o is the output impedance looking into the collector and base terminals. Therefore, if V_{og} is the open-circuit output voltage due to the source V_g, then the output S/N is

$$S/N_{\text{out}} = \frac{V_{og}^2 K_1}{\overline{v_o^2}K_1} = \frac{GV_g^2 K_1}{\overline{v_o^2}K_1}$$
$$= \frac{GV_g^2}{\overline{v_o^2}} \tag{13-28}$$

The gain G is the square of the open-circuit voltage gain. The input S/N is a similar ratio of voltages squared.

$$S/N_{\text{in}} = \frac{V_g^2 K_2}{\overline{v_{gn}^2}K_2} = \frac{V_g^2}{\overline{v_{gn}^2}} \tag{13-29}$$

Hence the noise factor may be written as

$$F = \frac{V_g^2 / \overline{v_{gn}^2}}{G V_g^2 / \overline{v_o^2}} = \frac{\overline{v_o^2}}{G \overline{v_{gn}^2}} \qquad (13\text{-}30)$$

The quantity $G\overline{v_{gn}^2}$ is simply the first term in Eq. (13-24) since this term is the open-circuit voltage due to the generator $\overline{v_{gn}^2}$. Hence the noise factor is obtained by dividing Eq. (13-24) by its first term, resulting in

$$F = 1 + \frac{\overline{v_e^2}}{\overline{v_{gn}^2}} + \frac{\overline{v_b^2}}{\overline{v_{gn}^2}} + \frac{\overline{v_c^2}(R_g + r_b' + r_e')^2}{\overline{v_g^2} \,|\alpha Z_c|^2} \qquad (13\text{-}31)$$

If the values previously found for each of these generators are incorporated into this equation, we obtain the desired equation for common-base noise figure.

$$F_b = 1 + \frac{r_e'}{2R_g} + \frac{r_b'}{R_g} + \frac{(1 - \alpha_0)(R_g + r_e' + r_b')^2}{2\alpha_0 r_e' R_g}\left[1 + \left(\frac{f}{f_\alpha \sqrt{1 - \alpha_0}}\right)^2\right]$$

$$(13\text{-}32)$$

The noise factor as given also applies for the common-emitter transistor connection, because the noise factor depends only upon a set of open-circuit voltages [Eqs. (13-30), (13-31)]. The open-circuit output voltages do not depend upon whether the emitter or base terminal is considered common. The output impedance depends upon the common terminal; hence the actual power *delivered* to a given Z_L does depend upon the circuit configuration. The common-collector noise factor is slightly different (see Prob. 13-7).

The noise factor as given by Eq. (13-32) may be plotted as a function of R_g for typical values of r_e, r_b', and α_0, as shown in Fig. 13-11. The noise

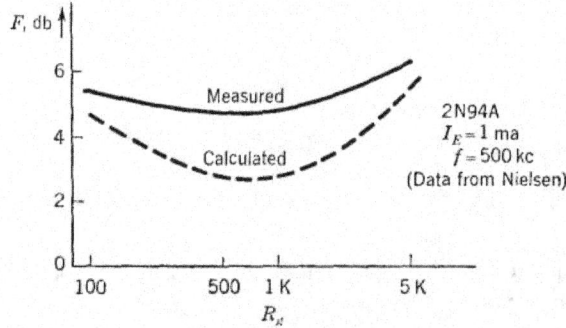

Fig. 13-11 Noise factor as a function of source resistance for a typical common-base or common-emitter stage.

factor has a definite minimum for a particular value of R_g, as is evident from the figure. The optimum value of R_g may be found more easily if the noise factor is expressed in the form

$$F = 1 + \frac{K_1}{R_g} + \frac{(K_2 + R_g)^2 K_3}{R_g} \qquad (13\text{-}33)$$

where

$$K_1 = r_b' + \frac{r_e'}{2}$$

$$K_2 = r_b' + r_e'$$

$$K_3 = \frac{(1 - \alpha_0)[1 + (f/f_\alpha \sqrt{1 - \alpha_0})^2]}{2\alpha_0 r_e'}$$

We differentiate Eq. (13-33) with respect to R_g and set the derivative equal to zero.

$$\frac{\partial F}{\partial R_g} = \frac{-K_1}{R_g{}^2} + K_3 \frac{2R_g(K_2 + R_g) - (K_2 + R_g)^2}{R_g{}^2} = 0 \qquad (13\text{-}34)$$

Solving for the R_g to obtain the minimum value of F ($= F_{\min}$), we get

$$R_{g,\text{opt}} = \sqrt{K_2{}^2 + \frac{K_1}{K_3}} \qquad (13\text{-}35)$$

Thus the optimum value of source impedance is seen to decrease with increasing frequency. The optimum value of R_g at low frequencies ($f \ll f_\alpha \sqrt{1 - \alpha}$) is given by the expression

$$R_{g,\text{opt}} = r_b' \sqrt{1 + \frac{(r_e'/r_b')^2 + 2r_e'/r_b'}{1 - \alpha}} \qquad (13\text{-}36)$$

This value of R_g is near the input impedance of the transistor in the common-emitter configuration; hence, nearly the maximum available gain is obtained when the optimum source resistance for noise is used. In the common-base configuration the optimum source resistance is the same, but the source to give maximum available gain is much lower. Hence using $R_{g,\text{opt}}$ in a common-base stage gives a lowered stage gain and makes the noise contribution of the following stage important. A similar mismatched condition also exists in using a common-collector stage with $R_{g,\text{opt}}$ except that the source giving the largest available power gain with a common-collector stage is greater than $R_{g,\text{opt}}$. Consequently, the common-emitter stage is the most useful as an input stage because the largest stage gain can be achieved with the proper source resistance for a low noise factor.

If the source impedance is given the optimum value, as in Eq. (13-35), then the value of F_{min} is

$$F_{min} = 1 + 2K_3 \left(\sqrt{K_2{}^2 + \frac{K_1}{K_3}} + K_2 \right) \qquad (13\text{-}37)$$

The shape of the curve of F_{min} as a function of frequency is shown approximately in Fig. 13-12. The noise factor is constant in the medium-frequency range but begins to increase at frequencies of the order of $f_\alpha \sqrt{1 - \alpha_0}$. The dashed curve showing an increase in F at low frequencies indicates the effect of $1/f$ noise, which is not included in the noise equivalent

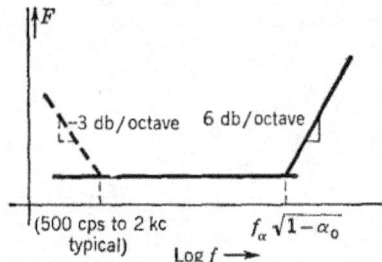

Fig. 13-12 Typical behavior of transistor noise factor vs. frequency.

circuit. The frequency at which $1/f$ noise becomes important unfortunately varies widely, often even among transistors of one type.

Some general conclusions may be drawn concerning the operation of transistors where low F is required. The use of $R_{g,\text{opt}}$ is desirable although the minimum in F is a broad one as R_g is varied. The value of $R_{g,\text{opt}}$ may by altered by adjusting r_e' by changing I_E. If high values of source impedance must be used, then very low values of I_E may be valuable;[1] however, the lowest possible value of F is usually obtained with R_g in the range of 500 to 2,000 ohms. In general, a transistor with small $(1 - \alpha_0)$ is desirable, and if operation at high frequencies is necessary, a high f_α. For operation at very low frequencies selection of an optimum transistor is useful to minimize $1/f$ noise. So-called "low-noise" transistors may not have low $1/f$ noise unless they are specifically selected for this frequency range. Feedback arrangements including the common-collector connection have very little effect on changing the optimum source impedance for low F, although the feedback may have great effect on the input impedance of the amplifier. The noise figure as predicted by the preceding equations is usually slightly lower than that actually realized, as is shown in Nielsen's article. Typically the error in F_{min} is of the order of 1 to 2 db.

[1] A. E. Bachman, Rausharmer Transistorverstarker mit hoher Eingangsimpedanz, *Arch. Elek. Uebertragung*, April, 1958, pp. 331–333.

PROBLEMS

13-1. A multistage amplifier has a 6AK5 pentode in the first stage, as shown in Fig. P13-1. The center frequency of the amplifier is 100 Mc, resulting in the following parameters:

$$G_r = 130 \ \mu\text{mhos}$$

$$R_{eq} = 1,900 \ \text{ohms}$$

$$g_m = 5,000 \ \mu\text{mhos}$$

$$C_{in} = 8 \ \text{pf (including wiring capacity)}$$

$$G_1 = 0$$

a. Select a value of m so that the input-circuit bandwidth is 20 Mc. Assume that the source resistance is 50 ohms.

b. Compute the percentage of total noise power at the amplifier output due to each of the following: (1) thermal noise from the source; (2) shot and partition noise; (3) induced grid noise. (Assume that all noise originates in the source and first stage.)

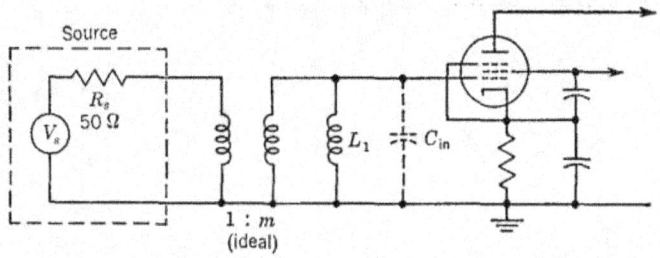

Fig. P13-1

c. Find the noise factor of the amplifier (in decibels).

d. What is the minimum F which could be achieved, disregarding bandwidth?

e. If the maximum Q which can be attained in an inductor of a given size is 100, what increase (in decibels) in the noise factor found in (c) and (d) results?

13-2. What is the minimum noise factor which may be obtained with the pentode of Prob. 13-1 if the frequency of operation is 10 Mc and the maximum attainable inductor Q is 250? By how many decibels may the noise factor be decreased if the pentode is triode-connected and the source again adjusted for optimum noise factor? Assume that the 6AK5 triode noise parameters are the same as for the pentode except that $R_{eq} = 385$ ohms. (Do not forget that G_r in Prob. 13-1 is specified at 100 Mc.)

13-3. To show that the noise factor has a relatively broad minimum at $G_{s,\text{opt}}$, compute $F_{\min}$ for a 6AK5 pentode at 30 Mc. Also compute F for source conductances of $2G_{s,\text{opt}}$ and $\frac{1}{2}G_{s,\text{opt}}$. Roughly plot the resulting F's as a function of G_s. (Make use of the 6AK5 data from Prob. 13-1.)

13-4. Show that the noise figure of a triode in the grounded-grid connection is

$$F = 1 + \frac{G_1}{G_s} + \frac{5G_r}{G_s} + \left(\frac{\mu}{\mu+1}\right)^2 \frac{R_{eq}}{G_s}(G_s + G_1 + G_r)^2$$

13-5. Assume that a common-emitter connected transistor has the following parameters while operating with $I_E = 1$ ma and $V_{CE} = 3$ volts:

$$r_e' = 25 \text{ ohms} \qquad\qquad r_c' = 8 \text{ megohms}$$

$$r_b' = 100 \text{ ohms} \qquad\qquad f_\alpha = 5 \text{ Mc}$$

$$\alpha_0 = 0.98$$

Find the optimum source impedance and minimum noise factor for this transistor operating at 300 kc.

13-6. Assuming that the theoretical relationship between I_E, the d-c emitter current, and r_e' holds (see Chap. 2), plot a curve of minimum noise factor versus I_E and a curve giving the source resistance for this minimum noise factor versus I_E. Assume that r_b' is independent of I_E and equal to 100 ohms; $\alpha_0/(1 - \alpha_0)$ is given for a particular unit by Fig. 2-10. Plot the results on log-log graph paper for the range 100 μa $\leqq I_E \leqq$ 10 ma.

13-7. Show that the noise factor in the common-collector configuration (F_c) is given by

$$F_c = 1 + \frac{r_b'}{R_g} + \frac{r_e'}{2R_g} + \frac{\alpha_0(1 - \alpha_0)[1 + (f/f_\alpha\sqrt{1 - \alpha_0})^2](R_g + r_b')^2}{2r_e'R_g[1 + (f/f_\alpha)^2]}$$

Discuss the behavior of F_c as a function of frequency.

14

Some General Noise-factor Relationships

14-1 Available Power and Available Power Gain. In the last chapter there was derived a figure of merit which we called the noise factor. It was defined in a reasonably obvious fashion from a comparison of mean-square noise and signal currents. There is a more general definition, based on the concept of "available power," which is very useful when it comes to the analysis or measurement of systems which cannot be reduced to such simple circuits as those of Chap. 13.

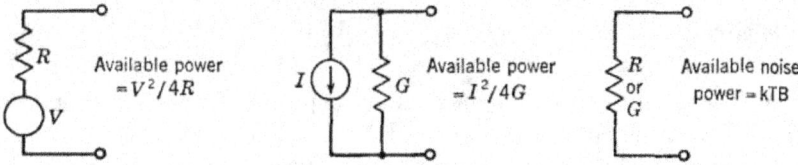

Fig. 14-1 Available power from various sources.

Available power from a generator or active network is a measure of the active portion of such a network, or its energy source. It is an alternative parameter to the more frequently used open-circuit voltage or short-circuit current. It is the power that would be delivered to a matched load and hence is the maximum power available from the generator, or simply the "available power" (see Fig. 1-1). Examples are given in Fig. 14-1.

One feature of the available-power concept is the simple expression for thermal noise, always kTB regardless of the value of R or G. It might be well to point out, however, that this power could not be measured, since the matched load (also R or G) would be generating an equal amount of

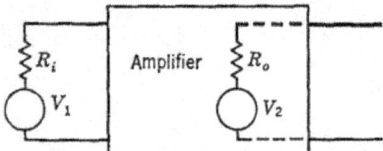

Fig. 14-2   Circuit for definition of available power gain.

thermal noise—if it were at the same temperature—and hence the net power flow would be zero. If the load were an idealized noise-free resistor, then kTB could flow to it.

Available power is a perfectly practical quantity, however, for signal sources, and indeed noise sources if the noise level is high. In fact, at microwave frequencies the available power can readily be measured by means of a matching network and a power-measuring device such as a bolometer, whereas open-circuit voltage or short-circuit current are not only impossible to measure but even ambiguous to define (in waveguides, for instance).

Once we have defined available power, we can also introduce the term *available power gain*. If a generator has an available power P_1 and to it is connected an amplifier, there will be a new available power at the output terminals of the amplifier (see Fig. 14-2). The ratio of the output available power P_2 to the input available power from the generator P_1 is defined as the available power gain $\mathcal{G}$.

Available power P_1 from generator
$$P_1 = \frac{V_1{}^2}{4R_i}$$

Available power P_2 from amplifier
$$P_2 = \frac{V_2{}^2}{4R_o}$$

Available power gain
$$\mathcal{G} \overset{\Delta}{=} \frac{P_2}{P_1} = \left(\frac{V_2}{V_1}\right)^2 \frac{R_i}{R_o} \tag{14-1}$$

Notice that this kind of gain does not depend upon the load connected to the amplifier, because it is a measure of the available rather than the actual power. Moreover, the input resistance of the amplifier does not appear explicitly, although it does influence the magnitude of V_2 for a given V_1.

The definition of available power gain can be contrasted with other gain definitions which are commonly used.[1] These are illustrated in Fig. 1-1.

14-2 Noise Factor in Terms of Available Power. The quantities available power and available power gain are remarkably useful in the manipulation of noise-factor relationships and are the key to successful circuit design for low noise. Let us redefine the noise factor in the way that it is generally given by workers in this technical field.

[1] IRE Standards on Electron Tubes: Definitions of Terms, 1950; F. E. Terman and J. M. Pettit, "Electronic Measurements," 2d ed., pp. 311–322, McGraw-Hill Book Company, Inc., New York, 1952.

$$F \triangleq \frac{\text{(available input signal)}/\text{(available input noise)}}{\text{(available output signal)}/\text{(available output noise)}} \qquad (14\text{-}2)$$

$$F = \frac{\text{output } S/N \text{ of ideal amplifier}}{\text{output } S/N \text{ of actual amplifier}} \qquad (14\text{-}2a)$$

The evaluation is customarily restricted to the case where the only noise at the input is thermal noise in the signal source resistance; hence

$$F = \frac{\text{(available input signal)}/kTB}{\text{(available output signal)}/\text{(available output noise)}}$$

$$= \frac{\text{(available output noise)}/\mathcal{G}}{kTB} \qquad (14\text{-}3)$$

where $\mathcal{G}$ = available power gain

T = standard temperature, usually 290°K

The utility of the new definitions is apparent in Eq. (14-3), where the numerator can be interpreted as an equivalent available noise at the input, which is to be compared with kTB to define the noise figure. Moreover, the equation suggests a possible measurement technique. We could measure the available output noise with a matched mean-square instrument; we could measure $\mathcal{G}$ at some convenient level with a conventional signal generator, and we could compute kTB.

If the new definition of F in Eq. (14-2) is applied to the circuit of Chap. 13, the same relationships used there in Eqs. (13-17) to (13-20) would result.

14-3 Noise Factor of Networks in Cascade. Finally, a unique and valuable feature of the new definitions is a systematic technique for handling the noise evaluation of two or more networks in cascade. Suppose that we have two amplifiers (or other networks, for that matter) connected in cascade, as in Fig. 14-3, with noise figures and available power gains (or losses) which have been separately evaluated in accordance with

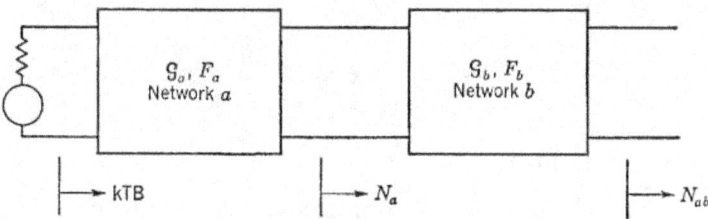

Fig. 14-3 Circuit for determining the noise factor of cascaded networks. (The rectangular bandwidth B either is the same for both networks or else is determined by networks following a and b.)

Eqs. (14-1) and (14-3). The question is: What is the noise figure F_{ab} of the combination as shown in Fig. 14-3?

$$N_a = F_a \mathcal{G}_a kTB \tag{14-4}$$

$$N_{ab} = F_{ab} \mathcal{G}_a \mathcal{G}_b kTB \tag{14-5}$$

Consider network b tested separately (Fig. 14-4).

$$F_b = \frac{N/\mathcal{G}_b}{kTB} \tag{14-6}$$

or

$$N = F_b \mathcal{G}_b kTB \tag{14-7}$$

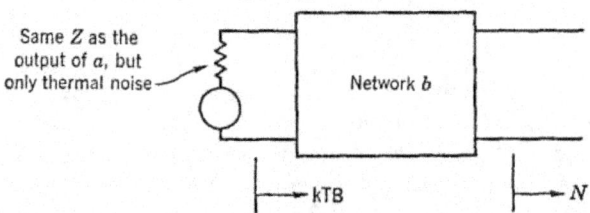

Same Z as the output of a, but only thermal noise

Network b

$\longmapsto kTB$ $\longmapsto N$

Fig. 14-4 Determination of noise output of second stage alone.

On the premise that noise *powers* add,

$$\text{Portion of } N \text{ due to source} = \mathcal{G}_b kTB \tag{14-8}$$

$$\text{Portion of } N \text{ due to noise originating within network } b = N - \mathcal{G}_b kTB$$

$$= F_b \mathcal{G}_b kTB - \mathcal{G}_b kTB$$

$$= (F_b - 1)\mathcal{G}_b kTB \tag{14-9}$$

Thus in Fig. 14-3

$$N_{ab} = \underset{\underset{\substack{\text{Source} \\ \text{noise} \\ \text{for } b}}{\uparrow}}{N_a \mathcal{G}_b} + \underbrace{(F_b - 1)\mathcal{G}_b kTB}_{\substack{\text{Noise} \\ \text{from} \\ \text{within } b}} \tag{14-10}$$

$$N_{ab} = F_a \mathcal{G}_a \mathcal{G}_b kTB + (F_b - 1)\mathcal{G}_b kTB \tag{14-10a}$$

Combine Eqs. (14-10a) and (14-5),

$$F_{ab} \mathcal{G}_a \mathcal{G}_b kTB = F_a \mathcal{G}_a \mathcal{G}_b kTB + (F_b - 1)\mathcal{G}_b kTB$$

$$F_{ab} = F_a + \frac{F_b - 1}{\mathcal{G}_a} \tag{14-11}$$

This relationship is an important one. It shows that the over-all noise figure is essentially that of the first network if the available power gain of that network is high. If it is not high, the relationship gives the contribution of the noise figure of the second network, namely, the second term on the right-hand side of the equation.[1]

For three networks in cascade, the over-all noise figure has a similar expression,

$$F_{abc} = F_a + \frac{F_b - 1}{\mathcal{G}_a} + \frac{F_c - 1}{\mathcal{G}_a \mathcal{G}_b} \tag{14-12}$$

14-4 Noise in Terms of Equivalent Temperature. The emphasis in this treatment of amplifier circuit noise is appropriate to those applications where the signal source is at room temperature, which is conveniently taken to have a standard value of 290°K. In some applications, such as receivers for radio astronomy, the source noise is at a much lower temperature. It is then more common in making quantitative comparisons of amplifiers to use *equivalent input noise temperature*[2] T_e, which is related to noise factor F by

$$T_e = 290(F - 1) \qquad °\text{K} \tag{14-13}$$

An ideal noiseless amplifier ($F = 1$, or 0 db) would have $T_e = 0°$, while an amplifier with $F = 2$, or 3 db, would give $T_e = 290°\text{K}$.

14-5 Equivalent Noise Bandwidth. The assumption of a passband of rectangular shape and width B is neither practical nor necessary for any amplifier in whose output noise is observed. Thermal noise from a resistor connected to the input, for example, is observed at the output as the integration of

$$kT\mathcal{G}(f)\, df \tag{14-14}$$

where df = incremental bandwidth

$\mathcal{G}(f)$ = available power gain at frequency f

$$\text{Available output noise (power)} = kT \int_0^\infty \mathcal{G}(f)\, df \tag{14-15}$$

$$= kT \underbrace{\left[\frac{1}{\mathcal{G}(f_0)} \int_0^\infty \mathcal{G}(f)\, df \right]}_{B} \mathcal{G}(f_0)$$

$$= kTB\mathcal{G}(f_0) \tag{14-15a}$$

where B = equivalent noise bandwidth $= \dfrac{1}{\mathcal{G}(f_0)} \displaystyle\int_0^\infty \mathcal{G}(f)\, df$ (14-16)

---

[1] The derivation above follows that of H. T. Friis, Noise Figures of Radio Receivers, *Proc. IRE*, vol. 32, pp. 419–429, July, 1944.

[2] IRE Standards on Methods of Measuring Noise in Linear Twoports, 1959, sec. 5.

The equivalent noise bandwidth can be thought of as the width of an equivalent rectangular passband having a height $\mathcal{G}(f_0)$ and the same total output noise as the actual passband (see Fig. 14-5).

The equivalent noise bandwidth is approximated fairly well by the 3-db bandwidth in most practical cases. As an example, consider a maximally flat triple, where the voltage gain has been shown to vary with frequency (normalized) as $1/\sqrt{1 + x^6}$. The 3-db bandwidth is 2.0. The available

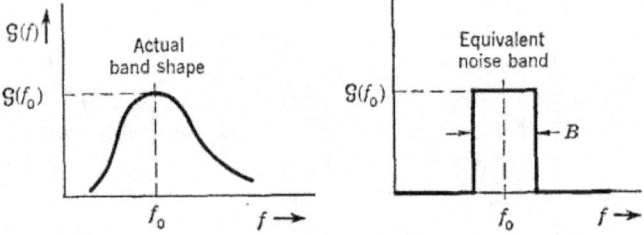

Fig. 14-5 The meaning of *equivalent noise bandwidth*.

power gain can be expected to vary as the square of this voltage gain, and thus

$$\frac{\text{Noise bandwidth}}{\text{3-db bandwidth}} = \frac{\displaystyle\int_{-\infty}^{\infty} [\mathcal{G}(x)\,dx]/\mathcal{G}(0)}{2} \qquad (14\text{-}17)$$

Since $\mathcal{G}(0) = 1$ and $\mathcal{G}(x)$ is even,

$$\frac{B_{\text{noise}}}{B_{\text{3-db}}} = \frac{2\displaystyle\int_{0}^{\infty} \mathcal{G}(x)\,dx}{2}$$

$$= \int_{0}^{\infty} \frac{dx}{1 + x^6}$$

$$= \frac{\pi/6}{\sin(\pi/6)} = 1.045$$

$$\cong 1 \qquad (14\text{-}18)$$

On the other hand, the noise bandwidth of a single, single-tuned stage is found to be quite different from the 3-db bandwidth; that is, $B_{\text{noise}}/B_{\text{3-db}} = \pi/2$. This ratio approaches unity, however, as the number of stages, either staggered or synchronously tuned, becomes large.

14-6 Low-noise-factor Circuit Design: The Cascode. In the quest for low-noise amplifiers for systems such as radar receivers, one rather remarkable circuit design has evolved.[1] It uses two triodes in the "cascode" arrangement shown in Fig. 14-6 (d-c connections are omitted). The general noise-factor relationships which have been presented in the earlier sections provide the means for demonstrating the effectiveness of the circuit.

The capacitances are usually those provided by the tubes plus wiring. Resistor G_1 represents only losses; the bandwidth is determined primarily by the generator resistance, tapped down on L_1 as needed. Resistor G_2 is provided entirely by the very low input resistance of the grounded-grid triode of stage 2.

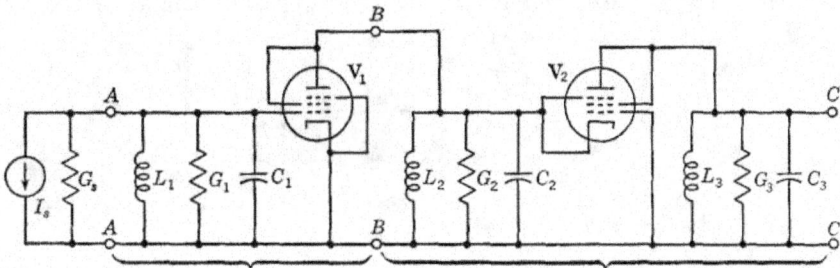

Fig. 14-6 The *cascode* amplifier (grounded cathode followed by grounded-grid stage).

The incentive for using triodes comes from their low noise compared with pentodes, owing to the absence of partition noise. Thus a 6AK5 pentode can have an R_{eq} in Eq. (13-4) of 1,880 ohms, but when the same tube is connected as a triode, the R_{eq} is only 385 ohms. The minimum noise figure when used as a pentode turns out to be 1.85 (2.7 db) at 30 Mc, whereas the same tube in the double triode circuit above will give a noise figure of 1.35 (1.3 db) at the same frequency. This might seem like a small improvement,[2] but it represents a substantial number of kilowatts if the high-powered transmitter is thereby effectively increased in output power by 37 per cent.

The trouble with triodes at radio frequencies is that the input conductance can become negative, thus leading to possible oscillation. This difficulty diminishes as the voltage amplification between plate and grid circuits is made small. Yet we should like the gain of the first tube to be high, so that noise contributions of subsequent tubes will not impair the noise figure. The apparent dilemma is neatly resolved by careful attention to the distinction between the voltage amplification, which determines the input conductance, and the available power gain, which determines the

---

[1] H. Wallman, A. B. Macnee, and C. P. Gadsden, A Low-noise Amplifier, *Proc. IRE*, vol. 36, pp. 700–708, June, 1948.

[2] If compared for equal *bandwidths*, e.g., 12 Mc, centered at 30 Mc, the noise figures become: pentode, 3.3 db; cascode, 1.35 db.

noise figure. It is perfectly possible for the latter to be high and the former to be low! Voltage amplification A–A to B–B is

$$A = \mu_1 \frac{R_2}{R_2 + r_{p1}} \tag{14-19}$$

where μ_1 = amplification factor of tube V_1 ($= 30$)
  r_{p1} = plate resistance of V_1 ($= 4,500$ ohms)
  $R_2 = 1/G_2 \cong 1/g_{m2}$ ($= 150$ ohms)
  g_{m2} = transconductance of V_2 ($= 6,670$ μmhos)
Thus $A = [(30)(150)]/(150 + 4,500) = 0.967$ for the example. Available power gain A–A to B–B (see Fig. 14-7) is

$$\mathcal{G}_1 = \frac{(\mu V_g)^2/4r_{p1}}{I_s{}^2/4G_s} = \frac{[\mu I_s/(G_s + G_1)]^2/4r_{p1}}{I_s{}^2/4G_s} \tag{14-20}$$

$$\cong \frac{\mu^2}{G_s r_{p1}} \qquad \text{for } G_s \gg G_1 \tag{14-20a}$$

For a triode-connected 6AK5 at 30 Mc,

$$G_1 = 10 \ \mu\text{mhos} \qquad \beta = 5$$

$$G_r = 12 \ \mu\text{mhos} \qquad R_{eq} = 385 \text{ ohms}$$

Therefore the optimum G_s computed as in Eq. (13-21) is 426 μmhos (2,350 ohms). Accordingly

$$\mathcal{G}_1 = \frac{(30)^2}{426 \times 10^{-6} \times 4,500} = 470$$

Thus the available power gain $\mathcal{G}_1$ is high, even though the voltage amplification A is so low that the first stage will be entirely stable.

The noise figure of the first stage can be computed as in Eq. (13-20), G_r being taken as 12 μmhos at 30 Mc. The result proves to be

$$F_1 = 1.35 \ (1.3 \text{ db})$$

From a formula which will not be derived here but which can be found in

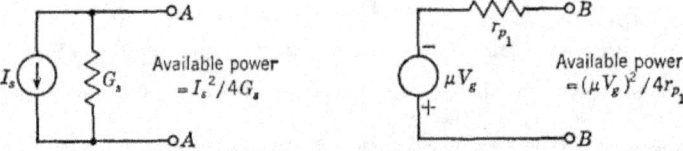

Fig. 14-7   Available powers in the circuit of Fig. 14-6.

the literature,[1] the noise figure for the second stage, where the triode has a grounded-grid connection, has the following calculated value:

$$F_2 = 3.74 \ (5.8 \ \text{db})$$

The over-all noise figure F_{12} can now be computed by using Eq. (14-11).

$$F_{12} = F_1 + \frac{F_2 - 1}{G_1}$$

$$= 1.35 + \frac{3.74 - 1}{470}$$

$$= 1.35 + 0.006$$

$$= 1.36$$

This over-all noise figure is contrasted with 1.85, which would be obtained if a pentode were used as an input stage.

14-7 Comparison of the Cascode with a Single-pentode Stage. It is pertinent to ask how the over-all available power gain would compare for the pentode as opposed to the double triode. The output terminals are those labeled C–C in Fig. 14-6, and included as part of the second stage is the conductance G_3. The value of G_3 will influence the value of the available gain; we shall use a G_3 of 500 μmhos (2,000 ohms), chosen by Wallman, Macnee, and Gadsden to meet the particular bandwidth requirement of the circuit L_3, C_3, G_3 in their application. For this G_3, the over-all available gain is

Pentode $G \cong 300$

Double triode $G \cong 200$

In either case, the available power gain is sufficiently high to make negligible the noise contributions of later stages.

Curiously enough, the available power gain of the two triodes is less than that of the first triode alone. The available power gain of a grounded-grid stage is always low, and especially so if there is a high load conductance (G_3 in this case). Hence the available power gain of the second stage is actually less than unity, but the over-all gain is still high enough.

Alternative double-triode configurations have been devised, but this one remains the most effective for operation at a fixed frequency. Other ar-

---

[1] Wallman, Macnee, and Gadsden, *op. cit.*, p. 704, eq. (23). See also G. E. Valley, Jr., and H. Wallman (eds.), "Vacuum Tube Amplifiers" (vol. 18, M.I.T. Radiation Laboratory Series), pp. 632–635, McGraw-Hill Book Company, Inc., New York, 1948.

rangements may prove more practical for variable-frequency operation, as in a television receiver or converter.[1]

Sometimes a neutralizing inductor is incorporated between plate and grid of V_1 in Fig. 14-6. Neutralizing is not needed to prevent oscillation, as usually required in tuned radio-frequency power amplifiers, but instead the purpose is to adjust admittances to secure minimum noise figure. As described by Wallman, Macnee, and Gadsden, an improvement of 0.2 db can be achieved at 30 Mc, and as much as 2.5 db at 180 Mc.

PROBLEMS

14-1. One way to connect a pair of triodes which would provide stability against oscillation would be to load the plate of the first triode with a resistance equal to the cathode input impedance of the grounded-grid stage of Fig. 14-6. Then the following stage could also be a grounded-cathode stage, as shown in Fig. P14-1. The gain of the two stages would then be identical with the gain of the two stages connected in cascode. To show that this connection is not as good as the cascode connection from a noise stand-

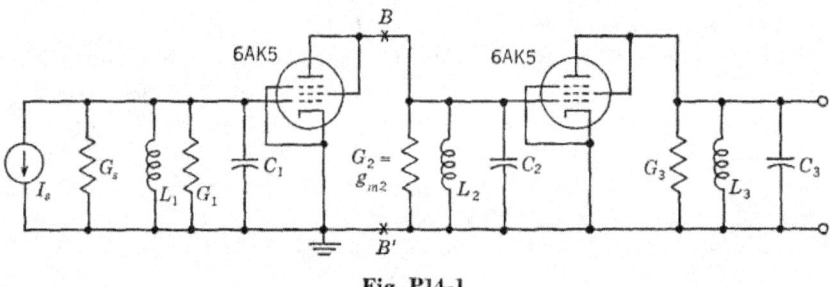

Fig. P14-1

point, calculate the following. Use the tube parameters and any other pertinent data given in the chapter.

  a. Using the value of G_s which gives a minimum noise factor for the input stage, calculate the available power gain of the first stage (up to the point B–B').

  b. Calculate the noise factor of the second stage.

  c. Calculate the over-all noise factor.

  d. Try to explain in physical terms why the noise factor you obtain in (*c*) is worse than that calculated for the cascode. (HINT: You might consider the noise generated by the second tube in the two circuits.)

14-2. An amplifier is fed from a 50-ohm source some distance away through a slightly lossy cable, as shown in Fig. P14-2. The amplifier is designed to be operated from a 50-ohm source and therefore is being fed from a proper impedance even though there is an evident mismatch at the receiving end of the line. You are to find the noise factor of the system by using the principles of Sec. 14-3. First you are to find the available power gain and noise factor of the transmission line itself. (SUGGESTION: Consider the Thévenin equivalent of the transmission line and matched source, and then use the basic defini-

[1] R. M. Cohen, Use of New Low-noise Twin Triode in Television Tuners, *RCA Rev.*, vol. 12, pp. 3–25, March, 1951.

tions of noise factor and available power gain.)   With these characteristics of the line, the over-all noise factor may be calculated directly.

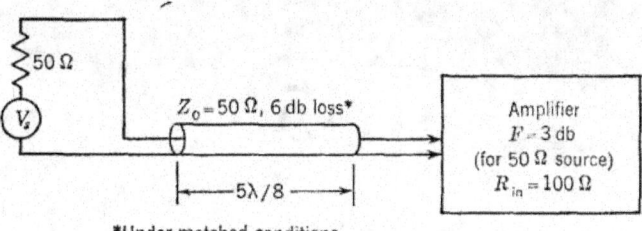

*Under matched conditions

Fig. P14-2

14-3. Show that the 3-db bandwidth and noise bandwidth become the same for maximally flat amplifiers if the number of staggered stages is allowed to increase without limit.

14-4a. Derive an integral equation which gives the average noise factor $\bar{F}$ if the noise factor F is known in the narrow band df and if $\mathcal{G}$ is known as a function of frequency. Check your result with Eq. (15-17).

b. In some practical situations $F(f)$ and $\mathcal{G}(f)$ are available only in graphical form from experimental data.   Explain how you would apply your equation from (*a*) to obtain $\bar{F}$ from a plot of $F(f)$ and $\mathcal{G}(f)$ versus f.

c. In the special case where $F(f)$ is a constant across the passband, as is true if the input circuit has a much wider passband than the rest of the amplifier, show that the output noise from the amplifier is simply $N_o = F(f_0)kTB_nA(f_0)$, where $A(f_0)$ is the transducer gain at the center frequency and B_n is the noise bandwidth of the entire amplifier.

15

Amplifier Measurements

A major purpose of the book has been a presentation of advanced design concepts and techniques which are especially powerful in the design of high-performance amplifiers, i.e., amplifiers which are intended to be operated at extremes of high gain, large bandwidth, high frequencies, fast transient response, low noise factor, etc. Such demands on design are accompanied by similar demands on measurement techniques in order to verify the design results and, earlier, upon measurements on the individual circuit components in order to provide the element values which go into the design formulation. This chapter therefore provides some selected measurement techniques which will help assure the successful achievement of the high performance made possible by the design techniques.

It is expected that the reader is already familiar with basic measurement procedures described in various reference texts.[1]

Measurements may be complicated by the usual problems of high-frequency measurements plus problems unique to high-gain amplifiers. For example, a measurement so seemingly simple as the measurement of the d-c bias on one stage of a 60-Mc i-f amplifier with a nominal gain of 100 db can lead to erroneous results. If the voltage is read when the amplifier shield cover is opened to insert the voltmeter leads, the reading will probably be wrong because of regeneration in the amplifier. Even bringing the voltmeter leads out with the shielding in place may not suffice unless the leads are carefully placed physically and properly decoupled. Consequently special care must be taken to ensure that the amplifier is truly operating normally when measurements are taken, or one may falsely condemn a design which in reality is good.

15-1 Measurement of R, L, and C. The standard measurements of passive elements used in an amplifier are covered by many texts, but their

---

[1] See, for instance, F. E. Terman and J. M. Pettit, "Electronic Measurements," 2d ed., chap. 8, Amplifier Measurements, McGraw-Hill Book Company, Inc., New York, 1952.

use in wideband amplifiers is greatly influenced by the "impurities" in
the actual realizations of an "ideal" R, L, or C. Consequently the measurement of the elements should take into account the imperfections. Figure 15-1 shows the simplest representation of a "resistor," "capacitor,"
and "inductor" valid over a reasonably wide frequency range.

The resistor (Fig. 15-1a) is contaminated by a shunt capacitance which
reduces its impedance at high frequencies. The capacitor typically is
about 0.6 pf for a ½-watt carbon
resistor. Special constructions make
possible resistors with $C \approx 0.02$ pf.[1]
The value of R is to some extent a
function of frequency and may be
measured with a radio-frequency
bridge appropriate to the value of
R being measured, e.g., resistance of
greater than 100 ohms can best be
measured on a parallel-substitution
bridge [2] or a transformer-ratio-arm
bridge.[3] Lower values of impedance
can be measured better on a series
substitution bridge.[4] The inductance

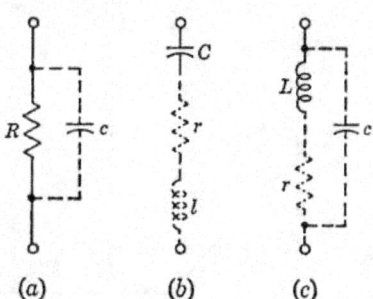

Fig. 15-1 Approximate equivalent circuits. (a) For a physical resistor. (b)
For a capacitor. (c) For an inductor.

of most high-frequency resistors is not usually troublesome unless low
resistance values are used at high frequencies—then the resistor lead
inductance must also be taken into account.

The choice and measurement of a capacitor depend upon its use. A
capacitor for use as a resonant element usually requires low loss and therefore utilizes an air, mica, or high-stability ceramic dielectric. Such a
capacitor may be measured with a Q meter [5] or a bridge in the frequency
range of interest. The leads used for measurement should be nearly the
same length as those to be used in the circuit. Such a measurement will
give the "effective capacitance," not the value shown as C in Fig. 15-1b.
If the capacitor is actually a capacitive reactance, the measured value of

[1] International Resistance Company type HFR, for example.

[2] The twin-T circuit is appropriate; see Terman and Pettit, *op. cit.*, pp. 84–87. A
commercial version is the General Radio Type 871-A. Another form is the Schering
circuit, Terman and Pettit, *op. cit.*, p. 74. A version commercially available is the
Boonton Type 250-A RX Meter. See *The Notebook* (Boonton Radio Corp., Boonton
N.J.), no. 2, pp. 1–4, Summer, 1954.

[3] See Terman and Pettit, *op. cit.*, pp. 116–117. Suitable commercial instruments
include the Wayne-Kerr Types B601 and 801.

[4] Terman and Pettit, *op. cit.*, pp. 79–81. A commercial instrument is the General
Radio Type 1606-A or its predecessor, Type 916-A.

[5] Terman and Pettit, *op. cit.*, pp. 90–91; also *The Notebook* (Boonton Radio Corp.,
Boonton, N.J.), no. 1, Spring, 1954, no. 4, Winter, 1955; no. 13, Spring, 1957. A commercial unit is the Boonton Type 260-A.

capacitance will always be greater than C. If the capacitor is to be used far below its resonant frequency, a low-frequency (e.g., 1-kc) measurement will suffice.

A capacitor to be used as a bypass capacitor (therefore as an approximate short circuit) poses a different sort of problem. Such a capacitor is often operated near or even above its series resonant frequency, and low-

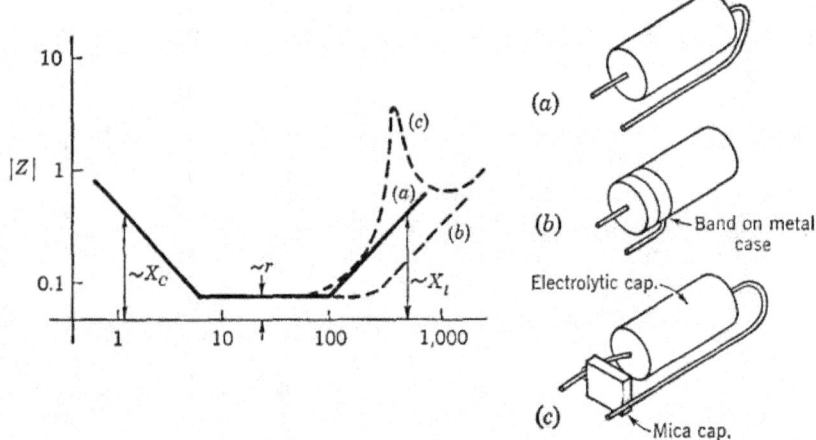

Fig. 15-2 Impedance of a typical capacitor over a wide frequency range. (*a*) Impedance measured between the lead ends. (*b*) Impedance measured between lead and (metallic) case. (*c*) Effect of paralleling capacitors.

frequency measurements are not often meaningful. Consider, for example, a cathode bypass capacitor in a video amplifier. For the capacitor to be effective, the condition to be fulfilled is $|g_m Z_k| \ll 1$ over the passband of the amplifier. Therefore $|Z_k| \ll 1/g_m$. The magnitude of the impedance of a typical electrolytic capacitor is shown in Fig. 15-2. The high-

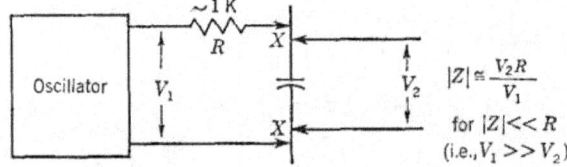

Fig. 15-3 Method of measuring impedance magnitude.

frequency impedance is largely due to the lead length of the capacitor and can be reduced by reducing the size of the loop through which the current passes. Note that paralleling a small and a large capacitor (Fig. 15-2*c*) usually is ineffective because the resulting resonance will produce a peak in $|Z|$ unless the Q is very low.

The simple measuring circuit for capacitors to be used in video ampli-
fiers (Fig. 15-3) allows direct measurement of $|Z|$. The impedance meas-
ured will be that effective between the two points X–X. A series substitution bridge could, of course, be used, but the measurements would be
more laborious.

For capacitors to be used at still higher frequencies, another useful measurement is the determination of the self-resonant frequency. The 0.01-μf
ceramic capacitor shown in Fig. 15-4 resonates at $\sim$10 Mc, as measured
by coupling an oscillator ("grid-dip meter") loosely to the resonant circuit

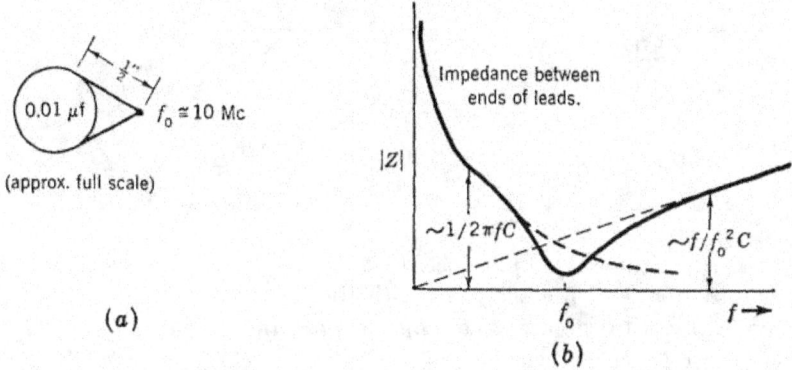

Fig. 15-4 (a) Resonant circuit formed of disk ceramic capacitor and its leads. (b) Impedance of capacitor and leads.

formed by short-circuiting the capacitor by its leads. Above this frequency the capacitor acts at its terminals (including $\frac{1}{2}$-in. lead wires) like
an inductor. (The inductance is mostly lead inductance.) Note that the
resonance is not necessarily bad as long as $|Z|$ stays sufficiently low in
the desired band; one might, in fact, intentionally choose the capacitor
and its leads to resonate within the band in order to achieve the lowest
possible bypass impedance. The impedance of the capacitor-inductor
combination may be approximated by taking the low-frequency value of
C and the self-resonant frequency f_0, as shown in the figure.

Knowledge of the interstage capacitance is fundamental to wideband
vacuum-tube amplifier design and should be known for transistor amplifier
design. The main difficulty of measurement is caused by the necessity
of measurement while the active device is operating. The capacitance is
small and is best measured by a device which uses a small enough signal
to prevent overdriving of a vacuum tube. Suitable instruments include
the Boonton RX Meter and the Tektronix LC Meter; these deliver less
than 0.1 volt across the unknown impedance. Larger voltages may over-
drive the grid circuit of the tube and cause nonlinear loading of the un-

known capacitance; erroneous measurements can result. A typical measurement with the LC Meter for a video interstage is shown in Fig. 15-5. The plate load resistor must be disconnected to keep from loading the LC Meter. The consequent removal of plate current from the tube does not significantly change the output capacitance. The tube in the next stage should operate under normal bias, however, to produce the proper input capacitance which is sensitive to the d-c operating conditions. Any

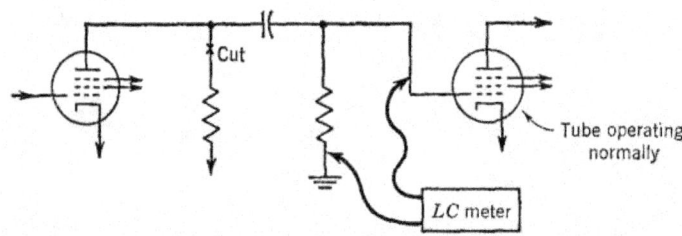

Fig. 15-5   Measurement of interstage capacitance with Tektronix LC Meter.

bypass capacitors should be effective at the frequency of measurement (in this case 135 kc). Measurement with the RX Meter is similar, but the plate load resistor need not be removed since the bridge can balance out the conductive component.

If the amplifier being measured has high gain, the capacitance measurement is complicated by the possibility of feedback from other stages affecting the measurement. One way to prevent this is to remove the tubes from other stages in the amplifier.

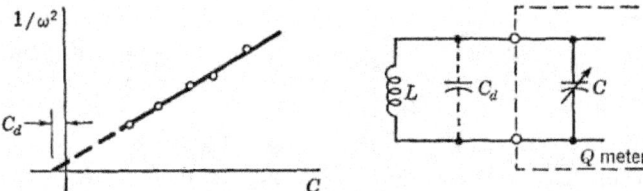

Fig. 15-6   A plot for determining the self-resonant frequency, true inductance, and distributed capacity of an inductor.

In some cases the total interstage capacitance may be deduced by merely measuring the cold capacitance of the interstage with the tubes removed. Since this is a purely passive circuit, the measurement is somewhat easier because no regeneration or overloading problems exist. The capacitance of the tubes may then be added in or taken from a table such as Table 10-7.

The measurement of inductors suitable for wideband amplifiers is readily accomplished with a Q meter or suitable bridge. The measurement should be made near the frequency of use, particularly if measured Q is important. The measured inductance is always somewhat higher than the "true" inductance L shown in Fig. 15-1c because of the distributed capacitance. Several methods of determining the value of L are given in texts;[1] one suitable method is to plot the capacitance required to resonate the inductor versus $1/\omega^2$, as shown in Fig. 15-6. Such a plot should result in a straight line which has a negative-capacitance intercept; i.e., the intercept is the negative capacitance required to cancel the distributed capacitance

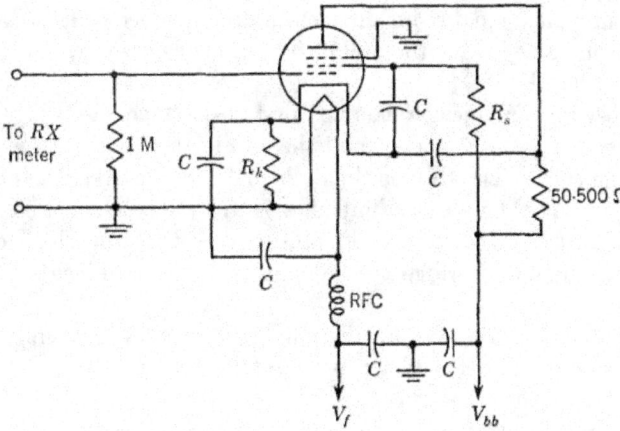

Fig. 15-7 Diagram of a jig for measuring the h-f input capacitance and conductance of a tube.

and give resonance at infinite frequency. The self-resonant frequency is also given by such a plot from the value of $1/\omega^2$ for zero added C.

15-2 Measurement of Vacuum-tube Parameters. The measurement of the low-frequency tube parameters necessary for amplifier design may be done on a tube bridge with considerable accuracy. The principle of operation is covered in several standard texts and will not be repeated here.[2] Measurement with the usual tube tester is at best only qualitative because the exact operating point is usually unknown.

On the other hand, measurement of the input and output admittance of a tube is not as easy or well documented. A test jig for measuring input admittance is shown in Fig. 15-7. For valid results all the leads carrying r-f currents should be as short as possible and the capacitances should all be chosen for minimum r-f impedance. Connections are shown for a tube with two cathode leads. The a-c component of plate current is returned to the cathode through one cathode lead. The grid circuit is completed

[1] See, for instance, Terman and Pettit, *op. cit.*, pp. 100–102.
[2] *Ibid.*, chap. 7.

through the other cathode lead, thereby reducing the input conductance due to cathode lead inductance. If the tube is not to be operated in this way in the circuit, the two cathode leads should be paralleled and the a-c plate current returned directly to ground. All leads to the jig should be decoupled so that no r-f potentials exist on them.

The most convenient instrument for the measurement is the RX Meter shown, but other parallel substitution bridges suitable for measuring high impedances, such as the Wayne-Kerr, may be used. The signal applied to the tube should not be greater than a few tenths of a volt to prevent overloading. Inadequate shielding or improper placement of bridge and jig leads is indicated when the null of the bridge changes when "grounded" wires are touched.

The procedure for measurement is first to balance the bridge with the jig removed. The jig is added, and the admittance of the jig is measured without the tube. The tube is then added to the jig and the total admittance measured. The input admittance is the difference between the total and jig admittance. The method presupposes that the electrical length between the tube and bridge is very small, so that line-length corrections need not be made.

The output admittance of a tube may be measured in a similar manner by putting the bridge terminals in the plate circuit with an r-f choke to carry plate current to the tube.

The measurement of the direct admittances as described as well as the measurement of the transfer admittances in the vhf and uhf region may be accomplished with a special transfer admittance meter. For wideband amplifier design even in the vhf region, knowledge of the direct admittances (particularly the input admittance) and the interelectrode capacitances usually suffices. Consequently, such elaborate instrumentation and measurement are not often necessary.

One additional tube parameter which is difficult to measure is the equivalent noise resistance R_{eq}. The best method of measuring this resistance is to measure the noise factor F of an amplifier input stage especially designed to make the effect of R_{eq} dominant in determining the noise factor. Reference to Eq. (13-20) shows that this may be accomplished by making the source conductance very large. Then F is very nearly

$$F \cong 1 + G_s R_{eq} \qquad G_s \gg G_r, G_1 \qquad (15\text{-}1)$$

A suitable preamplifier for the measurement is shown in Fig. 15-8; the noise diode, 5722, provides a calibrated source of noise current [1] across the 100-ohm resistor which also constitutes the necessary high value of G_s. The other equipment necessary is a high-gain bandpass amplifier and an

[1] Considerations in the use of a noise diode to measure amplifier noise figure are discussed in Sec. 15-7.

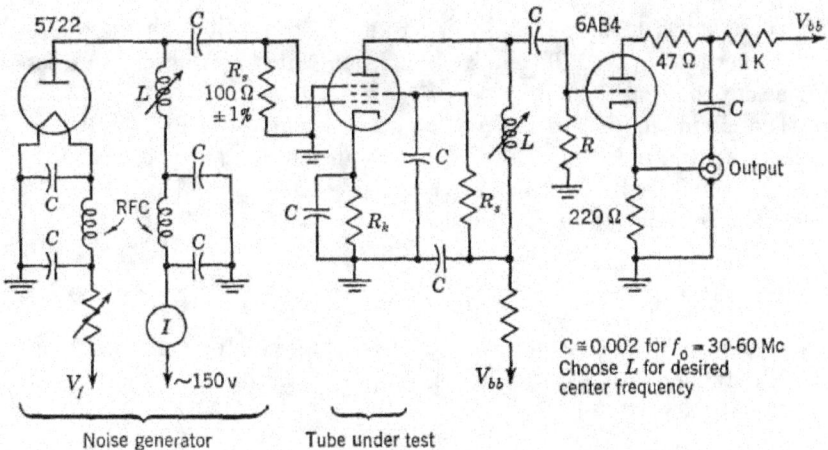

Fig. 15-8 A preamplifier for R_{eq} measurement.

output meter. The noise factor is measured as described in Sec. 15-7.
From the known values of F and G_s, R_{eq} may be calculated by using
Eq. (15-1).

15-3 Measurement of Low-frequency Transistor Parameters.
Several types of commercial equipment exist for the measurement of the
low-frequency transistor parameters, usually the h parameters. For
reasons outlined in Chap. 2 the h's are the easiest to measure, especially
in the common-base or common-emitter connections. Simple jigs may
also be used in conjunction with an oscillator and a-c millivoltmeter to
measure the h's. Figure 15-9 shows a jig for measuring h_{11e} and h_{21e}
with considerable precision if care is taken to ensure that the millivolt-
level signals are uncontaminated by spurious signals. The approximate
values of h_{11e} and h_{21e} for the values given in the figure are

$$h_{11e} = 10^6 V_1 \tag{15-2}$$

$$h_{21e} = 10^4 V_2 \tag{15-3}$$

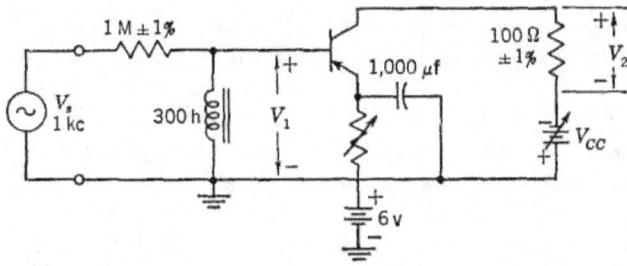

Fig. 15-9 A jig for measuring the low-frequency transistor h parameters h_{11e} and h_{21e}.
($V_s = 1$ volt.)

These values assume that $V_s \gg V_1$ and $V_s = 1$ volt. This is the same as saying that $h_{11e} \ll 10^6$ ohms, a condition which is true for most conditions of operation.

The circuit for measuring h_{12e} and h_{22e} is shown in Fig. 15-10 and is similar to the preceding circuit. The transformer shown is desirable to allow grounding the voltmeter. For the conditions of $V_s = 1$ volt, the parameters are given by

$$h_{22e} = 10^{-2}V_2 \tag{15-4}$$

$$h_{12e} = V_1 \tag{15-5}$$

Note that all the parameters are direct-reading; i.e., the voltmeter reading need only be multiplied by the appropriate power of 10 to get the desired

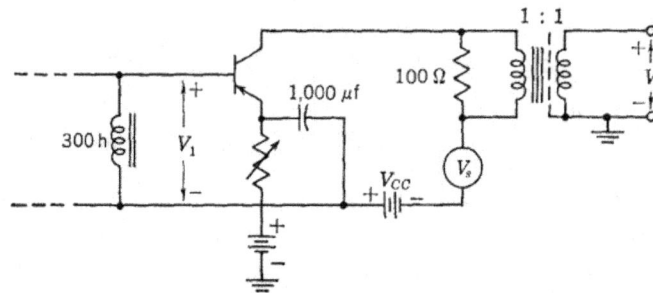

Fig. 15-10 A jig for measuring h_{12e} and h_{22e}. ($V_s = 1$ volt.)

parameter. In all cases the output of the voltmeter should be monitored with an oscilloscope, not only to determine whether spurious signals are present but also to determine whether the transistor is operating linearly. The measurements must be made at an extremely low signal level to ensure linear operation under all conditions of biasing. For example, the conditions shown for measuring h_{21e} apply a base a-c current of 1 μa rms, or 1.4 μa peak. If a transistor with $h_{21e} = 100$ were to be measured at a collector current of 100 μa, the drive current would be sufficiently high to completely cut off the collector current for a part of the cycle, resulting in very nonlinear operation. For the circuits shown, the input can be conveniently reduced by a factor of 10, which then increases the coefficients of Eqs. (15-2) and (15-5) tenfold. A good test to see whether or not the small-signal parameters are truly being measured on an active device is to take several measurements at different signal levels. If the signal level is too large, the measured parameter values will change with level.

If the T-circuit parameters (Fig. 2-9) are desired, the most convenient way of obtaining them is through the h parameters. Table 15-1 gives conversion formulas for finding the T parameters from the common-

Table 15-1

$$\beta = \frac{\alpha}{1 - \alpha} = h_{21e}$$

$$r_c = \frac{h_{21e} + 1}{h_{22e}}$$

$$r_e = \frac{h_{12e}}{h_{22e}}$$

$$r_b = h_{11e} - \frac{h_{12e}(h_{21e} + 1)}{h_{22e}}$$

$$\alpha = \frac{h_{21e}}{h_{21e} + 1}$$

emitter h parameters. The formulas contain the approximation that $r_c(1 - \alpha) \gg r_e$, an inequality which is usually very good.

15-4 Measurement of Transistor High-frequency Parameters.
Because many different high-frequency transistor equivalent circuits are possible, there are a multitude of different ways to measure the h-f characteristics of a transistor. The discussion herein will be centered about measuring the elements of the hybrid-pi circuit, which has been used throughout this text and is repeated in Fig. 15-11. Relatively simple measurements will be described which are nevertheless capable of adequately describing the operation of the transistor for engineering purposes. For more precise measurements the reader may refer to literature describing the specific measuring equipment, e.g., the Rhode and Schwartz Diagraph, the General Radio transfer-function meter, and the Wayne-Kerr r-f bridges.

One of the most useful h-f parameters which can be readily measured with simple equipment is the frequency f_t defined in Chap. 2, i.e., the frequency at which the common-emitter current gain $|h_{21e}|$ becomes unity. A suitable jig for measuring $|h_{21e}|$ is shown in Fig. 15-12. The r-f oscillator and 20-kilohm resistor provide a constant-current source for the base of the transistor. The collector current is measured as the drop across a 50-ohm line termination by means of the crystal detector. Because of the high frequency of operation of the jig, careful attention must be paid to lead dress, lead length, and unwanted capacitances. The method of operation of the jig is to connect a short circuit from the base to the emitter contacts of the socket (a

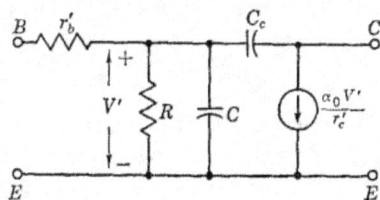

Fig. 15-11 The high-frequency pi equivalent circuit.

defunct transistor with a wire between collector and base is convenient).
The r-f oscillator output is then increased until some reference level is
established on the output meter.   This reference level must represent
only a few millivolts of r-f signal to ensure linear operation when the
transistor is measured.   The setting of the attenuator is noted for this reference reading (say N db).   The short circuit is now removed from the
jig and a transistor inserted and properly biased for the condition of operation desired.   The attenuation of the attenuator is now increased (say to

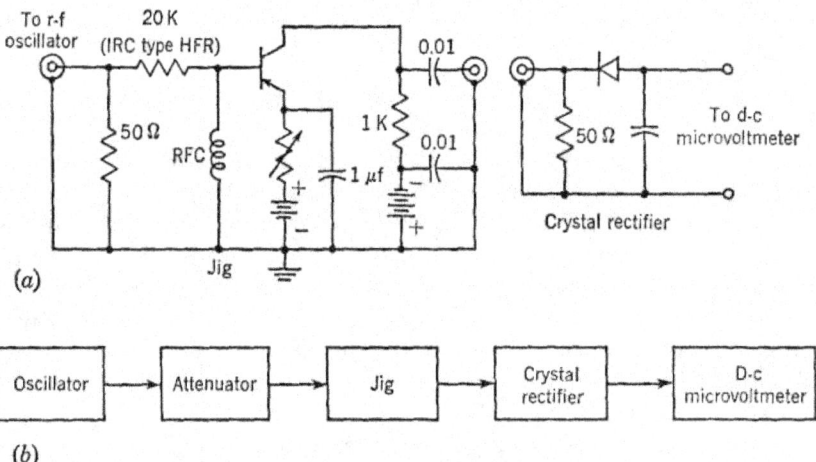

(a)

(b)

Fig. 15-12   A jig and measurement setup for measuring f_T.

M db) until the reference reading is again obtained on the output meter.
The current gain of the transistor is then $M - N$ db.   The accuracy of the
measurement depends upon $h_{11e} \ll 20$ kilohms (so that the transistor is
effectively driven by a current source) and the accuracy of the attenuator
but does not depend upon the linearity of the output meter.   A small error
is incurred by the finite impedance in the collector circuit.

   To find f_t, several such measurements are made of $|h_{21e}|$ to find a region
in which $|h_{21e}|$ is decreasing inversely with frequency.   For a normal
transistor this is a broad frequency range.   The value of f_t then is given
by the equation

$$f_t = |h_{21e}(f_1)|f_1 \qquad\qquad (15\text{-}6)$$

where $|h_{21e}(f_1)|$ is the value of $|h_{21e}|$ at the frequency f_1.   Once the
proper frequency range is determined for a given transistor type, only one
measurement is required to determine f_t.   This is an advantage of this
measurement as compared with the measurement of f_α, for example, where
several measurements are required to find a -3-db point.   The measurement of f_t should usually be conducted at such a frequency that the cur-

rent gain is of the order of 5. At frequencies much higher than this the equivalent circuit may not be too satisfactory because of anomalous effects.

The measurement of collector capacitance C_c may be done with any of several bridges, a Q meter, or the Tektronix LC Meter. The latter permits direct reading of C_c on a meter used with the simple jig shown in Fig. 15-13. The guard terminal on the LC Meter is used to keep the signal current through the upper 10-mh choke to a minimum so that the meter may be zeroed with the transistor removed. Upon insertion of the transistor the meter reads the direct capacitance added between the input terminal of the meter and ground. The added capacitance is in part the

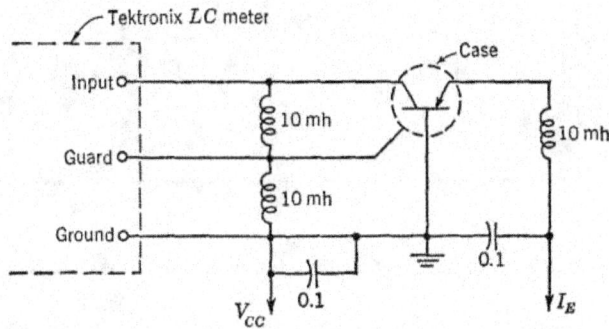

Fig. 15-13 A jig for measuring C_c with the Tektronix LC Meter.

capacitance from the collector lead to the case, a capacitance which is not properly part of C_c. This parasitic capacitance may be removed from the meter indication by connecting the case of the transistor to the guard terminal of the meter (on the assumption, of course, that the case is not connected internally to the transistor).

The extrinsic base resistance r_b' is probably the most difficult element in the equivalent circuit to measure accurately. A method which gives a good approximate result for transistors in which C_c is not extremely small is shown in Fig. 15-14. In this jig an r-f voltage applied to the collector of the transistor causes current to flow primarily through C_c and then through r_b' and the 100-ohm external resistor, as in Fig. 15-14b. The magnitude of the current is measured in switch position b as the drop across the 100 ohms. The drop across r_b' is measured in position a. On the assumption that the same current flows through both resistances, r_b' is

$$ r_b' = \frac{100 V_a}{V_b} \tag{15-7} $$

where V_a and V_b are the voltage readings in the respective switch positions. (Note that if V_b is set to 1 mv, for example, the voltmeter reads

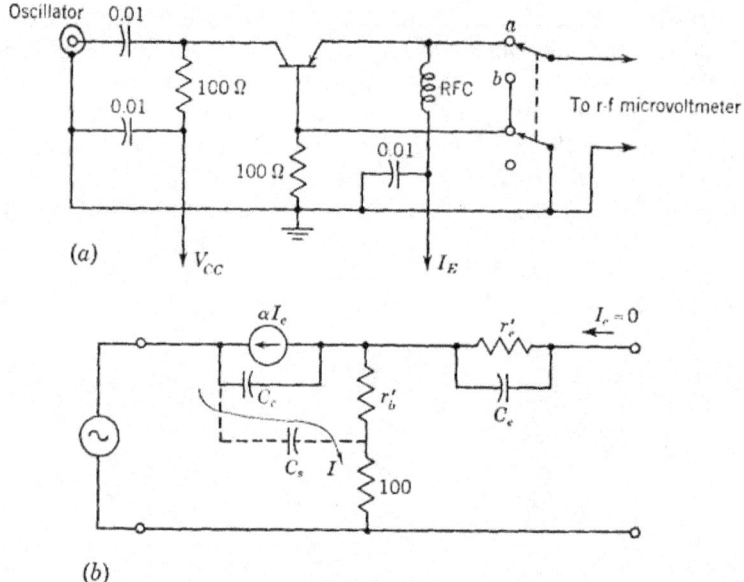

(a)

(b)

Fig. 15-14 A jig and its equivalent circuit for measuring r_b'.

r_b' directly, with 100 ohms being full scale.)   The accuracy of the measurement depends upon the capacitance shown as C_s being small compared with C_e so that the current through the two resistors is truly equal. Grounding the case of the transistor, if possible, helps reduce C_s.

In principle, these measurements give all the necessary parameters for the equivalent circuit of Fig. 15-11 since R may be calculated if the d-c emitter current and the low-frequency current gain are known.   For more accurate results a direct measurement of R would be desirable, and is actually possible.   A plot of the input impedance h_{11e} of the transistor for all frequencies is shown in Fig. 15-15.   The values of both r_b' and R may be obtained from such a plot, which is a semicircle of radius $R/2$ centered at $r_b' + R/2$.   In practice the data for h_{11e} are measured by using an r-f bridge and are plotted as shown in Fig. 15-15.   A circle is then fitted to

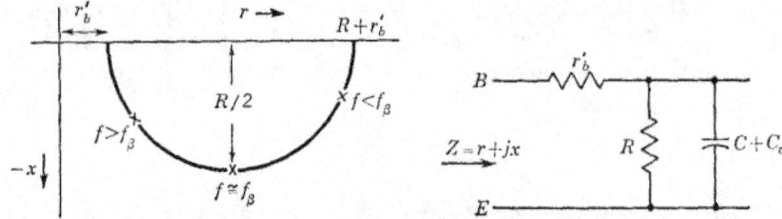

Fig. 15-15 The circle diagram for finding r_b' and R.

the measured data by trial and error in such a way as best to approximate the data. The intersection of the circle and the real axis gives the values of r'_b and $r'_b + R$. This method of obtaining r'_b, although more laborious than the preceding method, probably leads to a more nearly correct result because the value of resistance obtained is less influenced by direct capacitance from collector to base. Since this capacitance is not negligible compared with C_c in vhf transistors, particularly mesa units, the circular plot is particularly applicable to these units.

If an accurate value of r'_b is obtained by the preceding means, then a different method of obtaining C_c may be useful, particularly in the case

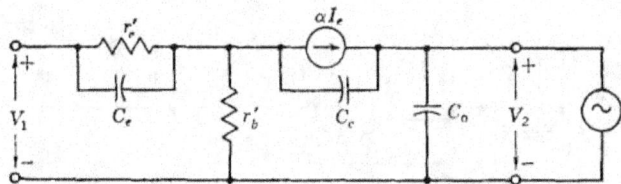

Fig. 15-16 The measurement of high-frequency h_{12b}.

of mesa transistors that have a collector capacitance which may be approximated by splitting it into two parts (Fig. 15-16). One is the ordinary collector capacitance C_c; the other, C_o, is also caused by the collector depletion region but does not cause the current due to V_2 to flow through r'_b. Therefore the capacitance designated C_o acts like a direct capacitance between collector and base. In a mesa unit the two capacitances might be of the same order of magnitude. The previous measurement will measure the two capacitances in parallel, thus resulting in too large a value for C_c. This difficulty may be obviated by measuring $|h_{12b}| = |V_1/V_2|$. For the equivalent circuit in Fig. 15-16

$$|h_{12b}| = \frac{\omega r'_b C_c}{|j\omega r'_b C_c + 1|} \cong \omega r'_b C_c \qquad \text{assuming } h_{12b} \ll 1 \qquad (15\text{-}8)$$

If the values of ω, r'_b, and h_{12b} are known, the value of C_c may be calculated. The measurement must be at a frequency sufficiently high for the current flowing through r'_c (see Fig. 2-20) to be negligible in comparison with the current through C_c. The practical difficulty in the measurement is obtaining a high impedance at the emitter in measuring V_1.

15-5 Amplifier Gain Measurement. Two basic methods of amplifier gain measurement are shown in Fig. 15-17. The first method is a very straightforward method of measuring voltage gain but suffers from several faults which should be understood. The method is to connect the amplifier under test to a generator of suitable frequency range and then to measure directly the input and output signal voltages. If the amplifier has much gain, the signal level at the input may be too low to measure with an ordinary a-c millivoltmeter. Of course the signal level may be increased

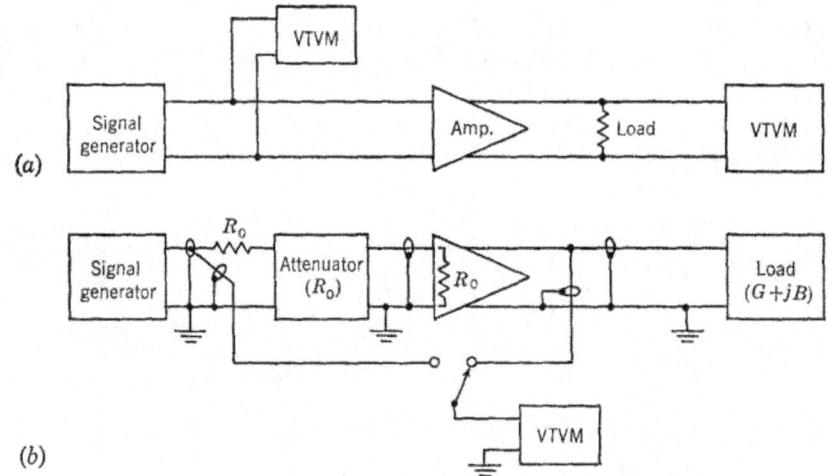

Fig. 15-17   Two methods of measuring amplifier gain.

but this may cause overloading of the amplifier under test. A convenient way of finding the proper level at which to take the gain measurement is to plot the output of the amplifier as a function of the input, as shown in Fig. 15-18. From this curve it is easy to see that too low a level will con-

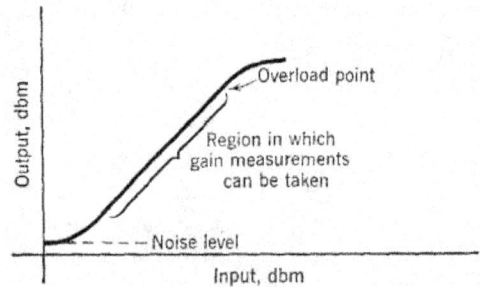

Fig. 15-18   A typical amplifier input-output characteristic.

taminate the measured output signal with noise and too high a level will cause nonlinearity of the amplifier.

The direct measurement is also unsatisfactory if the frequency response of the amplifier is desired, because the measured gain-vs.-frequency curve will also contain the difference in the gain-vs.-frequency characteristics of the two voltmeters. Even using the same voltmeter at input and output will not solve the problem, because the frequency characteristics often are not exactly the same on different scale ranges.

The second measuring method shown in Fig. 15-17*b* is much more satisfactory because the accuracy of measurement is almost completely dependent upon the accuracy of the passive attenuator used. Since such attenuators may be obtained with high accuracy, wide frequency range

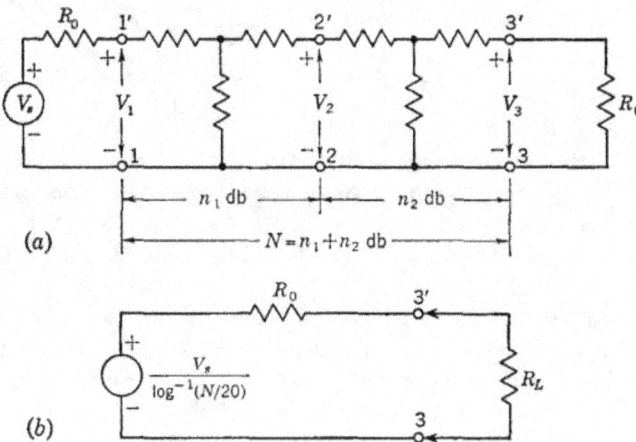

Fig. 15-19 The typical form of a wideband attenuator and its Thévenin equivalent.

(0 to 500 Mc, typically), and almost unlimited attenuation, reliance upon the attenuator is practical if the operation of the attenuator is well understood. An attenuator suitable for testing wideband amplifiers is usually of the form shown in Fig. 15-19*a*, where the number of attenuating sections is varied to change the over-all attenuation. Typical values of R_0 are 50 to 100 ohms to match common coaxial cables. The attenuation marked upon the unit is the voltage (or current) ratio at input and output when the attenuator is terminated in R_0, that is,

$$n_1 = 20 \log \frac{V_1}{V_2} \quad \text{db}$$

Looking into terminals 1–1′ to the right or into terminals 3–3′ to the left, one sees R_0 as long as the opposite end of the attenuator is terminated in R_0. Because of this property, the entire attenuator of Fig. 15-19*a* may be replaced by a Thévenin equivalent, good for any frequency, as shown in Fig. 15-19*b*. Note that the voltage V appearing across R_L and the available power P_{av} from the attenuator are

$$V_3 = \frac{V_s R_L}{R_0 + R_L} \frac{1}{\log^{-1}(N/20)} \tag{15-9}$$

$$P_{\mathrm{av}} = \frac{V_s^2}{4[\log^{-1}(N/20)]R_0} \tag{15-10}$$

Or if the power is expressed in terms of dbm (decibels with reference to a level of 1 mw), the available power is simply

$$P_{\mathrm{av}} = P_{\mathrm{av},s} - N \qquad \text{all powers in dbm} \tag{15-11}$$

where $P_{\mathrm{av},s}$ is the power available from the source V_s and R_0.

If the attenuator does not provide the correct source resistance for the amplifier, the source resistance may be raised or lowered by adding series or shunt resistance, respectively, at the output of the attenuator. The new output voltage or power is easily computed by using the Thévenin equivalent of Fig. 15-19b.

By the method of Fig. 15-17 the voltage gain of the amplifier may be found if the input resistance of the amplifier is known. In the usual case the input impedance of the amplifier is considerably greater than the 50 to 100 ohms of the attenuator so that the attenuator is adequately terminated by merely connecting R_0 in shunt with the amplifier terminals. The attenuator is adjusted to give equal readings in positions 1 and 2 of the voltmeter switch. The voltage gain is then equal to the attenuator setting *plus* 6 *db*. Note that neither the linearity nor the calibration of the voltmeter enters into the reading. Consequently, if the gain vs. frequency is measured, the voltmeter frequency response does not cause error. (Note also that omitting the resistor at the input of the attenuator removes the need for adding 6 db to the attenuator reading.)

Measurement of the various kinds of power gain may also be most easily done with the aid of an attenuator. For example, the transducer gain may be found by calculating the available power input to the amplifier, P_1, from Eq. (15-10). The attenuator should be connected to the amplifier with no terminating resistor added. The transducer gain in Fig. 15-17 is then

$$A_T = \frac{V_2^2 G}{P_1} \tag{15-12}$$

The available power gain is measured in the same way, but the load must be adjusted for maximum power output from the amplifier. In either case the source impedance presented by the attenuator and any series or shunt impedance must be correct for the amplifier. If the amplifier were to be operated from a 600-ohm line, then a 550-ohm resistor should be added in series with a 50-ohm attenuator to simulate the actual source.

The measurement of gain or frequency response of very high-gain amplifiers is complicated by the problem of regeneration. All connections made to the amplifier should be coaxial, with tight-fitting connectors. It is often useful to employ cables with a double shield braid to reduce leakage and input-output coupling caused by shield porosity. In extreme cases, at least the cable between the attenuator and amplifier input may have to be cable with a copper-tubing outside conductor. The input and output connections should be kept physically separated to reduce the chance of coupling. The number of ground loops should be kept to a minimum; this may mean in some cases that the power-line ground provided with much test equipment must be removed from all but one line connection.

Two tests may be used to discover whether or not the setup is adequately shielded. The easiest is to reduce the output of the signal source to as near zero as possible by adding attenuation at the input. Any signal except random noise remaining in the output may indicate either the presence of a spurious signal, e.g., a strong nearby transmitter, or leakage from the generator or cables. In an extreme case of leakage, the amplifier may oscillate. Terminating the input in a shielded, coaxial resistor should stop the oscillation if cable leakage is at fault.

A more sensitive test for regeneration is to plot several gain-vs.-frequency curves for different values of amplifier gain, i.e., different gain-control settings. If no regeneration is present, a well-designed amplifier should have a frequency response which is substantially independent of gain. In the presence of regeneration the band shape and frequency of maximum gain usually shift markedly, the bandwidth usually becoming narrower as the gain is increased. Some small amount of band-shape change should be expected, however, because of the change of the active-element parameters with gain and because of a small, irreducible amount of feedback, due primarily to the active elements.

A very convenient way of conducting this latter test is to apply a swept-frequency oscillator to the input of the amplifier and to observe the output with an oscilloscope. By this means, a display of relative gain vs. frequency may be obtained, and the changes in band shape as gain is varied may be seen directly.

The importance of eliminating regeneration cannot be too strongly stressed, because the presence of even a small amount will completely obscure the careful design of an amplifier.

Sometimes the gain of a single stage or group of stages in an amplifier is desired. One cannot connect a voltmeter or other instrument across an interstage without more or less upsetting that interstage. Even a diode probe has several picofarads of capacitance and would seriously detune most wideband stages. One method of avoiding the detuning problem is to convert the following stage into a broadband voltmeter-amplifier stage, as shown in Fig. 15-20. In the figure the stage to be measured is tube (a) and the network between (a) and (b). Since the grid circuit of (b) is undisturbed by the added conductance in the plate circuit, the tuning of the interstage is undisturbed. The frequency response of (b) may be measured

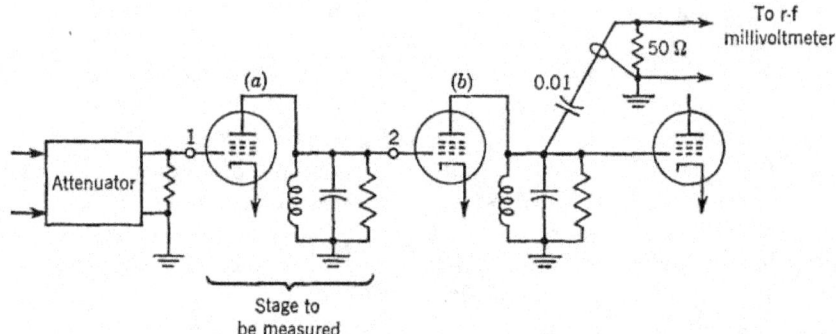

Fig. 15-20 Measurement of the gain of a single stage by converting the following stage into a very wideband amplifier.

by attaching the input to point 2—for many purposes the gain of (b) may be regarded as constant with frequency because of the large amount of loading in the plate circuit. The gain of the combination is found by subtracting the gain of (b) from the over-all gain measured at point 1 (gains measured in decibels).

Sometimes an indication of the signal being applied to a tube in a band-

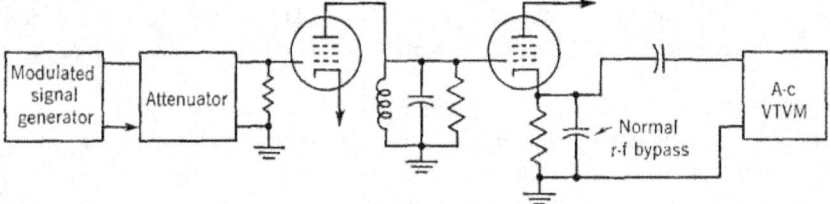

Fig. 15-21 Measurement of a modulated r-f signal by the detected component at a cathode.

pass amplifier can be obtained by measuring the detected signal in the cathode circuit of the tube. The signal is caused by the slight nonlinearity of the tube operating even with small signals. The equipment is set up as shown in Fig. 15-21. A modulated r-f signal is used for the input. The signal is passed through the stage in question, and a sensitive voltmeter is connected to the cathode of the following stage. (A standing-wave-detector amplifier is useful for the a-c millivoltmeter, because it may be tuned to the modulation frequency.) The detected signal is usually proportional to the square of the input voltage; hence the measuring stage either must be calibrated by connecting the attenuator output directly to the last grid or must be used only to obtain a reference reading; e.g., the frequency response of the interstage is given by recording the attenuator readings required to produce a constant output as frequency is varied.

15-6 Amplifier Transient-response Measurement. The basic equipment necessary for a measurement of the transient response is shown in Fig. 15-22. Here the input is the test waveform, which is usually either a step (simulated by a square wave) or an impulse (short pulse). The output of the generator is reduced by an attenuator which must have constant

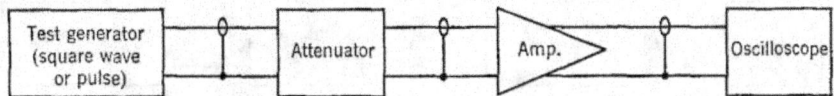

Fig. 15-22 Measurement setup for determining transient response.

attenuation and a constant time delay throughout the passband of the amplifier. The test signal is amplified by the amplifier being measured and displayed on a wideband oscilloscope. The measurement in principle is very simple; the usual troubles are associated with undesired output-input coupling (see the previous section) and imperfections in the test equipment.

The square-wave test signal is the most widely used because it is easiest to interpret and because the attainment of a sufficiently short pulse for impulse testing is often difficult. For measurement of rise time the square-wave frequency is unimportant as long as the frequency is not so high that the amplifier output never reaches its 100 per cent value. The rise time and flat top of the input square wave are very important. The rise time should be short and free of overshoot compared with the amplifier. To preserve these characteristics, the output cables from the generator must be correctly terminated to eliminate reflections.

The oscilloscope must have a short rise time and freedom from overshoot, also. A convenient check is to connect the oscilloscope directly to the output of the attenuator and measure the over-all rise time T_1 and overshoot. (The attenuation is, of course, reduced to give sufficient signal.) The overshoot should be negligible for accurate measurements.

The rise time measured at the output of the amplifier, T_o, is related approximately to the amplifier rise time T_A by (see Sec. 4-10)

$$\simeq \sqrt{T_o{}^2 - T_1{}^2} \tag{15-13}$$

where T_1 is the rise time of the generator plus oscilloscope. It is very important to test the amplifier for overloading in making transient measurements because the overloaded amplifier still has a square-wave output. A simple test is to see whether or not a 6-db change in attenuation produces a 2:1 change in output. Look also for changes in the overshoot, if any.

Sometimes measurement on one or two stages of an amplifier is desired. A so-called "low-capacitance probe" is not usually effective because its capacitance is an appreciable fraction of the total interstage capacitance. The method shown in Fig. 15-20 may be adapted to video amplifiers to make internal measurements.

Many square-wave and pulse generators are constructed with inadequate shielding, which may give rise to spurious outputs. One sign of trouble is a

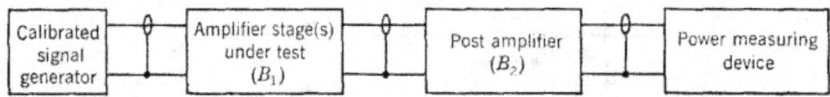

Fig. 15-23 Setup for measuring noise factor.

waveform synchronized to the square wave even when the attenuation is increased to remove the square wave. Such interference may be reduced by changing system grounds. Occasionally, one may have to install line filtering in the offending generator.

The sag in an amplifier may be measured by decreasing the square-wave frequency until enough sag is present to be readily measured on the oscilloscope. Again, it is important that the amplifier be operating linearly.

15-7 Measurement of Noise Factor.[1] The measurement of amplifier noise factor is complicated by the need for accurately calibrated signals at very low power levels. For example, the thermal noise generated in a resistor at room temperature and in a 1-Mc bandwidth is about 4×10^{-15} watt of available power. To measure the noise factor, an accurately calibrated source of about this magnitude must be available. Also, needless to say, all interfering and extraneous signals must be reduced to a level considerably below this at the input of the amplifier. This means that unusual care must be taken in shielding the input leads (and output leads if the output is at the same frequency as the input).

The arrangement for making the noise-factor measurement is shown in Fig. 15-23. Here a calibrated signal generator which provides the proper source impedance for the amplifier is connected to the input of the amplifier. For the measurement shown, the output of the amplifier is connected to a postamplifier which has sufficient gain to bring the noise signal up to a

[1] For further background information, see IRE Standards on Methods of Measuring Noise in Linear Twoports, 1959, *Proc. IRE*, vol. 48, pp. 60–68, January, 1960; also Terman and Pettit, *op. cit.*, pp. 353–379.

level measurable by the power-measuring device. The latter must truly measure h-f *power* and is usually a bolometer or thermocouple. For a first example, assume that the bandwidth of the postamplifier is much smaller than the bandwidth of the amplifier being measured. The power output of the system is then measured with the signal generator connected but the signal output turned to zero. The power read is

$$P_1 = FAkT_0B_2 \qquad (15\text{-}14)$$

where F is the average noise factor of the amplifier under test in the frequency band B_2, A is the transducer gain of the two amplifiers, and kT_0B_2 is the available noise power from the signal generator in the band B_2. [Compare with Eq. (14-3), which can be rewritten to give the *actual* rather than the *available* power output.]

The signal from the signal generator is now increased until a new and higher reading P_2 is obtained at the output. Since the signal and noise are uncorrelated, the new power is given by

$$P_2 = FAkT_0B_2 + AP_s \qquad (15\text{-}15)$$

where P_s is the *available* signal power from the signal generator. The gain, which is relatively difficult to measure accurately and maintain constant, may be eliminated from Eqs. (15-14) and (15-15), giving an equation for F containing only externally measurable quantities.

$$F = \frac{1}{P_2/P_1 - 1} \frac{P_s}{kT_0B_2} \qquad (15\text{-}16a)$$

The temperature of the source must be the standard temperature, 290°K according to the standards for measuring noise factor. The ratio of P_2/P_1 should be reasonably large for accuracy, but, because of amplifier overload problems, the ratio is commonly made 2, in which case F becomes

$$F = \frac{P_s}{kT_0B_2} \qquad \text{for } \frac{P_2}{P_1} = 2 \qquad (15\text{-}16b)$$

If B_2 is much less than B_1, as assumed, so that F is constant in the band B_2, the noise factor measured is the single frequency or spot noise factor at the center frequency of B_2. If this corresponds to the center frequency of B_1, then the measured value is essentially the F calculated in Chap. 13. The integrated noise factor is given by the equation

$$\bar{F} = \frac{\displaystyle\int_0^\infty F(f)A(f)\,df}{\displaystyle\int_0^\infty A(f)\,df} \qquad (15\text{-}17)$$

where $F(f)$ and $A(f)$ are the spot noise factor and transducer gain at the frequency f. The integrated noise factor can be measured by the same technique as before except that B_2 must be much larger than B_1. If no postamplifier is used, the noise factor measured will also be the integrated noise factor because the detector responds to noise generated throughout the bandwidth of the amplifier. The integrated noise factor may also be found by measuring $F(f)$ and $A(f)$ and performing the indicated integration graphically.

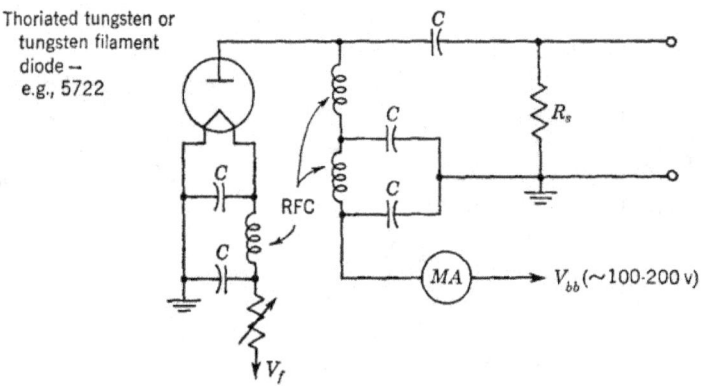

Fig. 15-24 Schematic diagram of a typical diode noise generator.

One difficulty with the preceding method is the necessity of measuring the noise bandwidth, although the 3-db bandwidth is a good approximation for a multistage amplifier (see Sec. 14-5). The necessity for measuring the bandwidth may be removed by using a signal source with an available power distributed uniformly over the bandwidth of the amplifier. Such a signal source may be obtained by connecting a suitable diode across a resistance to provide the proper value of source resistance for the amplifier. A typical diode unit is shown in Fig. 15-24. From Eq. (13-3), the mean-square value of the noise current in the plate of a temperature-limited diode is

$$\overline{i_n{}^2} = 2qI_{\text{dc}}B \qquad (13\text{-}3)$$

Therefore the available power from the generator of Fig. 15-24 is (on the assumption that the r_p of the diode is much larger than R_s)

$$P_s = \frac{qI_{\text{dc}}BR_s}{2} \qquad (15\text{-}18)$$

If the diode noise source is used in place of the calibrated signal generator, the noise factor becomes

$$F = \frac{1}{P_2/P_1 - 1} \frac{qI_{dc}R_s}{2kT_0}$$

$$= 20I_{dc}R_s \frac{1}{P_2/P_1 - 1} \qquad \text{for } T_0 = 290°K \qquad (15\text{-}19)$$

The powers P_1 and P_2 are the output power without the diode being turned on and the output power with the diode on. Although the bandwidth of the system does not appear in the equation, the bandwidth of the second

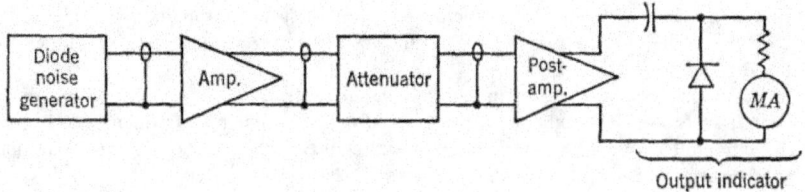

Fig. 15-25 Setup for measuring noise factor with a random-noise source and without a power meter.

amplifier enters into determining whether the spot or integrated noise factor is measured, as discussed before.

The accuracy of the diode noise generator is excellent if care is taken to ensure that no spurious signals enter through power-supply leads and if the diode is operated below the frequency at which self-resonances in the leads and transit time become important. In the type 5722 diode shown in Fig. 15-24, lead resonances begin to become important in the 100- to 200-Mc region if operated at the 50-ohm impedance level. Special coaxial diodes are available which permit operation up to the 1,000-Mc region when transit-time corrections are made. Other random-noise sources which may be considered for particular applications are heated resistors and gaseous-discharge tubes. The latter are particularly useful in the microwave region.

The random-noise generator not only has the advantage of eliminating the necessity of a bandwidth measurement but can also eliminate the need for a true power-measuring device. This is an advantage because the stability and convenience of most power-measuring devices leave something to be desired. The procedure in using the setup shown in Fig. 15-25 is to obtain a reference reading with the noise diode current off on the d-c meter which reads the average value of the detected output signal. The diode generator is then turned on and the attenuator setting increased by 3 db. The d-c plate current of the noise diode is adjusted (by adjusting the heater current) so that the reference reading is again obtained. This procedure makes $P_2 = 2P_1$, so that the noise factor is

$$F = 20I_{dc}R_s$$

The fact that the signal applied to the detector has the same character for both power measurements allows use of an output indicator which does not respond proportionally to power. The linearity of the output meter is unimportant also, since only repeatability of readings is required.

PROBLEMS

15-1. Assume that a given 0.01-μf ceramic capacitor comprises an ideal inductanceless capacitor in series with the lead inductance. Find the series resonant frequency of the capacitor and its leads if they are of $\frac{1}{2}$-in. No. 20 copper wire. (Use the inductance calculated for a straight wire.[1])

15-2. A single-layer solenoid inductor is calculated and carefully wound to give a certain inductance. The inductor is then measured at a single high frequency and found to have too much inductance. Explain why this could occur even though the calculations and measurement are both correct.

15-3. The resistance of a deposited carbon resistor does not change rapidly with frequency, but if such a resistor is measured on a series substitution bridge (i.e., measured in terms of series R and X), the measured resistance is found to vary considerably with frequency. Assume that the resistor may be represented by a parallel combination of R_p and a small capacitance C_p, and derive an equation giving the R_s and C_s, which will be measured by a series substitution bridge at a given frequency. If the measured resistor is to be used in a single-tuned interstage, show how the measured values of R_s and C_s relate to the interstage-element values.

15-4. In an effort to prevent r-f energy from reaching the input stages of a 300-Mc amplifier, Mr. Jones installs the lowpass filter of Fig. P15-4 in the supply leads to the input stages of his amplifier. He calculates that it will supply more than adequate filtering at 300 Mc, but when he installs it, he discovers that the attenuation is much less than he hoped.

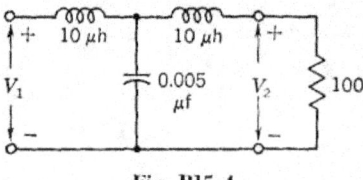

Fig. P15-4

Calculate the attenuation (V_1/V_2 expressed in decibels) for the filter as shown. Then calculate the approximate attenuation if measurement shows that the self-resonant frequencies of the inductors and capacitor are 100 Mc and 30 Mc, respectively. (HINT: Replace each inductor by a capacitor which has the equivalent reactance at 300 Mc. The capacitor can be similarly replaced by an inductor.) Advise Mr. Jones of changes to improve his filter. (These must be physically realizable changes; i.e., don't tell him to buy a capacitor with no inductance!)

[1] For example, see F. E. Terman, "Radio Engineers' Handbook," McGraw-Hill Book Company, Inc., New York, 1943.

15-5. The value of g_m is needed to compare some amplifier calculations with the measured performance. Design a test jig for measuring the low-frequency value of g_m if a tube bridge is not obtainable. You have available an audio oscillator, an a-c millivoltmeter, and any necessary precision resistors.

15-6. The parameter h_{21} is defined as I_2/I_1 for $V_2 = 0$. In the test setup of Fig. 15-9 V_2 is not quite zero because of the necessary presence of the 100-ohm resistor required to measure I_2. Compute the error in the measured h_{21e} caused by the 100 ohms if the transistor actually has the parameters of the typical unit in Table 2-1.

15-7. Devise a pair of jigs similar to those of Figs. 15-9 and 15-10 which would be suitable for measuring the y parameters of a transistor in the common-emitter connection. Include suitable values for the measuring resistors, and provide means of biasing. Use the typical transistor parameters of Table 2-1 to find typical values of the voltages which must be measured in each parameter measurement.

15-8. Devise a measurement jig similar to that of Fig. 15-12 to measure f_α. Give suitable values for all the elements of the jig. Referring to Fig. 2-20, find a correction factor to apply to the cutoff frequency measured in the preceding jig. The correction factor is necessary because the collector current consists of two components: (1) the desired component due to the current generator αI_e, and (2) an undesired component due to the collector capacitance. (The current through r_c' is usually small enough to be neglected.)

15-9. The gain of two amplifiers is to be measured with the setup of Fig. 15-17 except that the resistor R_0 at the input of the attenuator is omitted. Calculate the amplifier voltage gain for the following four cases. Assume in each case that the voltmeter reads the same at the input of the attenuator and the output of the amplifier.

a. The amplifiers terminate the attenuator in R_0. The attenuator reads 8 db in one case and 63 db in the other.

b. The amplifiers terminate the attenuator in an open circuit. The two attenuator readings are as in (*a*).

(You will probably have to refer to a handbook for attenuator data.)

15-10. The transducer gain of an amplifier of unknown input impedance is to be measured as shown in Fig. 15-17. The attenuator has $R_0 = 50$ ohms. The amplifier is designed to be run from a 500-ohm source and into a resistance of 1,500 ohms. Show by a sketch how you would arrange to make the measurement, and compute the amplifier gain if the attenuator reads 88 db for equal input and output voltages.

15-11. The noise factor of an amplifier to be used from a 250-ohm source is to be measured with the setup of Fig. 15-23. The signal generator is calibrated in its open-circuit terminal voltage, and it has an internal impedance of 10 ohms. To make the amplifier have the correct source impedace, a 240-ohm resistor is connected between the generator and amplifier input. The following data are available; find the noise factor.

Amplifier under test is made up of two staggered single-tuned triples with an over-all 3-db bandwidth of 8 Mc.

Postamplifier is made up of two synchronously tuned single-tuned stages with an over-all 3-db bandwidth of 0.5 Mc.

Power output with signal generator disconnected = 3 mw.

Power output with signal generator connected but turned off = 1 mw.

Power output is 5 mw with the signal generator on and set to 4.5 μv.

Index